Oxidative Stress in Applied Basic Research and Clinical Practice

For further volumes:
http://www.springer.com/series/8145

Ashok Agarwal · Nabil Aziz
Botros Rizk
Editors

Studies on Women's Health

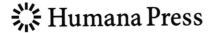 Humana Press

Editors
Ashok Agarwal
Center for Reproductive Medicine
Cleveland Clinic
Cleveland, OH
USA

Botros Rizk
College of Medicine, OB/Gyn Department
University of South Alabama
Mobile, AL
USA

Nabil Aziz
Gynaecology and Reproductive Medicine
Liverpool Women's Hospital
Liverpool
UK

ISBN 978-1-62703-040-3   ISBN 978-1-62703-041-0 (eBook)
DOI 10.1007/978-1-62703-041-0
Springer New York Heidelberg Dordrecht London

Library of Congress Control Number: 2012942255

© Springer Science+Business Media New York 2013
This work is subject to copyright. All rights are reserved by the Publisher, whether the whole or part of the material is concerned, specifically the rights of translation, reprinting, reuse of illustrations, recitation, broadcasting, reproduction on microfilms or in any other physical way, and transmission or information storage and retrieval, electronic adaptation, computer software, or by similar or dissimilar methodology now known or hereafter developed. Exempted from this legal reservation are brief excerpts in connection with reviews or scholarly analysis or material supplied specifically for the purpose of being entered and executed on a computer system, for exclusive use by the purchaser of the work. Duplication of this publication or parts thereof is permitted only under the provisions of the Copyright Law of the Publisher's location, in its current version, and permission for use must always be obtained from Springer. Permissions for use may be obtained through RightsLink at the Copyright Clearance Center. Violations are liable to prosecution under the respective Copyright Law.
The use of general descriptive names, registered names, trademarks, service marks, etc. in this publication does not imply, even in the absence of a specific statement, that such names are exempt from the relevant protective laws and regulations and therefore free for general use.
While the advice and information in this book are believed to be true and accurate at the date of publication, neither the authors nor the editors nor the publisher can accept any legal responsibility for any errors or omissions that may be made. The publisher makes no warranty, express or implied, with respect to the material contained herein.

Printed on acid-free paper

Humana Press is a brand of Springer

Springer is part of Springer Science+Business Media (www.springer.com)

To my father Professor RC Aggarwal for instilling the virtues of honesty, dedication, and hard work. To Professor CJ Dominic (late) (Banaras Hindu University), Professor Kevin Loughlin (Harvard Medical School), and Dr. Anthony Thomas (Cleveland Clinic) for making an indelible positive impression on my life

Ashok Agarwal

I dedicate our book to my dear father, Mitry Botros Rizk—a philosopher and thinker—for his inspiration, my mother, Dr. Isis Mahrous Rofail—a marvelous clinician and surgeon—for her unlimited love and sacrifice, my dear father-in-law, Dr. George Nawar Moussa—an accomplished scientist—for his spiritual guidance, and to my mother-in-law, Aida Erian Attalla—a talented artist—for her genuine care and sincerity

Botros Rizk

I would like to dedicate this book to my father Professor Fahim Aziz and my mother who inspired me to peruse excellence remembering humility at all times. I would like also to dedicate it to my wife Howayda and my daughters Sally and Amy for their continued support

Nabil Aziz

Foreword

Books can be separated into two categories: those that are exploratory and forward looking, that propose to define new pathways and to "change the way we think about a topic" and those that are exploratory and forward looking. This book is a good blend of both types. Its subject matter is oxidative stress in women health. Oxidative stress and how to prevent its impact on human diseases has become a hot scientific and clinical topic over the last decade. There is mounting evidence that oxidative stress or an imbalance in the oxidant/antioxidant activity in female organs and tissues plays a pivotal role in the development of infertility, cancer, and placental-related diseases of pregnancy.

Evolution for mammals living on dry land has been closely linked to adaptation to changes in O_2 concentration in the environment and thus it is not surprising that the oxygen metabolism by human tissues has such an impact on human reproduction. Basic science studies have shown that most of the O_2 used during the oxidation of dietary organic molecules is converted into water via the combined action of the enzymes of the respiratory chain. Around 1–2 % of the O_2 consumed escapes this process and is diverted into highly reactive oxygen free radicals (OFR) and other reactive oxygen species (ROS) at a rate dependent on the prevailing oxygen tension. When the production of OFR exceeds the cellular natural protection, indiscriminate damage can occur to proteins, lipids, and DNA. The consequences may range from the activation of stress-response proteins through to apoptosis or necrosis.

Human reproduction and early fetal development in utero is influenced and modulated by a constant adaptation to ambient oxygen concentrations. Oxidative stress influences the entire reproductive life span of a woman and beyond during the menopause phase of her life. In human reproduction it influences the oocyte and embryo quality and thus the fertilization rates. ROS appears to play a significant role in the modulation of gamete interaction and also for successful fertilization to take place. Oxidative stress exacerbates the development of endometriosis by inducing chemoattractants and endometrial cell growth-promoting activity and the oxidative proinflammatory state of the peritoneal fluid is an important mediator of endometriosis. Thus oxidative stress affects multiple physiological processes from oocyte maturation to fertilization, embryo development and pregnancy but there is also new evidence that free radicals in small concentrations are essential for normal cellular function and modulate the activity of many transcription factors.

There is increasing evidence indicating that failure of placentation leading to a specific human pregnancy disease such as preeclampsia is associated with an imbalance of free radicals, which will further affect placental development and function and may subsequently have an influence on both the fetus and its mother. Maternal metabolic disorders, for example diabetes, and lifestyle factors such as alcohol consumption and cigarette smoking, which are associated with an increased production of free radicals species, are also known to be associated with a higher incidence of miscarriage and fetal structural defects. Furthermore, the teratogenicity of drugs such as thalidomide has recently been shown to involve free radical-mediated oxidative damage, indicating that the human fetus can be irreversibly damaged by oxidative stress.

The chapters in this book cover all the important aspects of the relationship between oxidative stress and women health and the editors have to be congratulated on their vision and the way in which this book provides a comprehensive review of this important topic.

Eric Jauniaux

Preface

God created man in excellent health. While there is extensive documentation on life and death in the ancient civilizations, there is scarce literature on when man started suffering from disease. (1) The ancient Egyptians were the first to extensively document medical interventions. The ancient Egyptians left for us impressive illustrations of neurosurgery, childbirth, and circumcision. Some of these date back to 2600 BC. (2) We also have historical documentation of infertility in the Old Testament related to events between 2000 and 1700 BC that are very well known in the history of the Patriarchs, Abraham, Isaac, and Jacob.

Most medical books focus on a medical problem or symptom and then attempt to discuss the different medical evaluations and their medical and surgical treatments. Our book on oxidative stress and women's health is an exception. This book focuses on the involvement of a medical issue, oxidative stress, in many pathological situations and diseases. At the beginning, the editors worked diligently to find the concept that ties all these very different pathologies. The book is the first comprehensive book that explores the involvement of oxidative stress in oogenesis, pregnancy, placental functions, as well as endometriosis and cardiovascular disease. While some of these issues such as endometriosis are easier to understand, others, we must admit, await further investigations. Our book is the beginning of the exploration of a new field and in a few years, many of the discussed topics will be totally redefined but we take great pride in presenting the first idea. To our dear readers, we put in your hands a novel book that we hope that will stimulate your scientific curiosity and interest and prove to be an enjoyable reading experience.

References

1. Jauniaux E, Rizk B (eds) (2012) Pregnancy after assisted reproductive technology. Cambridge University Press, Cambridge
2. Rizk B, Silber SJ, Serour GI, Nagib OZ, Abou Abdallah M (2012) Chapter 42: Judaism, Christianity and Islam and in vitro fertilization and cloning. In: Rizk B, Sallam H (eds) Clinical infertility and in-vitro fertilization. Jaypee Brothers Medical Publishers, New Delhi, pp. 372–396

Contents

1   **Maternal Nutrition, Oxidative Stress and Prenatal Devlopmental Outcomes** 1
Kaïs Hussain Al-Gubory

2   **Methods for Detection of ROS in the Female Reproductive System** 33
Rakesh K. Sharma, Nathan Reynolds, Mitali Rakhit and Ashok Agarwal

3   **Oxidative Stress and The Endometrium** 61
Botros Rizk, Marwa Badr and Christina Talerico

4   **Oxidative Stress, Oogenesis and Folliculogenesis** 75
Malcolm A. Paine, Elizabeth H. Ruder, Terryl J. Hartman, Jeffrey Blumberg and Marlene B. Goldman

5   **Placental Vascular Morphogenesis and Oxidative Stress** 95
Amani Shaman, Beena J. Premkumar and Ashok Agarwal

6   **The Use of Antioxidants in Pre-eclampsia** 115
Jean-François Bilodeau

7   **Recurrent Pregnancy Loss and Oxidative Stress** 131
Nabil Aziz

8   **Premature Rupture of Membranes and Oxidative Stress** 143
Anamar Aponte and Ashok Agarwal

9   **Endometriosis and Oxidative Stress** 149
Lucky H. Sekhon and Ashok Agarwal

10	Oxidative Stress Impact on the Fertility of Women with Polycystic Ovary Syndrome. 169
	Anamar Aponte and Ashok Agarwal

11	The Menopause and Oxidative Stress 181
	Lucky H. Sekhon and Ashok Agarwal

12	Oxidative Stress in Assisted Reproductive Technologies. 205
	Catherine M. H. Combelles and Margo L. Hennet

13	Antioxidant Strategies to Overcome OS in IVF-Embryo Transfer 237
	Mitali Rakhit, Sheila R. Gokul, Ashok Agarwal and Stefan S. du Plessis

14	Oxidative Insult After Ischemia/Reperfusion in Older Adults. . . . 263
	Tinna Traustadóttir and Sean S. Davies

15	Relationship of Oxidative Stress with Cardiovascular Disease . . . 285
	Richard E. White, Scott A. Barman, Shu Zhu and Guichun Han

16	Female Infertility and Free Radicals: Potential Role in Endometriosis and Adhesions 315
	Zeynep Alpay Savasan

17	Impact of Life Style Factors on Oxidative Stress 335
	Peter T. Campbell

About the Editors 359

Index 361

Contributors

Ashok Agarwal Center for Reproductive Medicine, Cleveland Clinic, Cleveland, OH, USA

Kaïs Hussain Al-Gubory Department of Animal Physiology and Livestock Systems, INRA, Joint Research Unit of Developmental Biology and Reproduction, Jouy-en-Josas, France

Anamar Aponte Center for Reproductive Medicine, Cleveland Clinic, Cleveland, OH, USA

Nabil Aziz Gynaecology and Reproductive Medicine, Liverpool Women's Hospital, Crown Street, Liverpool L87 SS, UK

Marwa Badr Division of Reproductive Endocrinology and Infertility, Department of Obstetrics and Gynecology, University of South Alabama, Mobile, AL, USA; Ain Shams University, Cairo, Egypt

Scott A. Barman Department of Pharmacology and Toxicology, Medical College of Georgia, Georgia Health Sciences University, Augusta, GA, USA

Jean-François Bilodeau Centre de Recherche du CHUQ, CHUL, Obstetrics and Gynecology, Laval University, Québec, Canada

Jeffrey B. Blumberg Antioxidants Research Laboratory, Jean Mayer USDA Human Nutrition Research Center on Aging, Tufts University, Boston, MA, USA

Peter T. Campbell Tumor Repository, American Cancer Society, Epidemiology Research Program, NW Atlanta, GA, USA

Catherine M. H. Combelles Department of Biology, Middlebury College, Middlebury, VT, USA

Sean S. Davies Department of Pharmacology, Vanderbilt University, Nashville, TN, USA

Stefan S. du Plessis Medical Physiology, Faculty of Health Sciences, University of Stellenbosch, Tygerberg, Western Cape, South Africa

Sheila R. Gokul Center for Reproductive Medicine, Cleveland Clinic, Cleveland, OH, USA

Marlene B. Goldman Obstetrics & Gynecology and Community & Family Medicine, Dartmouth-Hitchcock Medical Center, Lebanon, NH, USA

Guichun Han Department of Veterinary Physiology and Pharmacology, Texas A&M University Hospital, Michael E. DeBakey Institute, College of Veterinary Medicine, College Station, TX, USA

Terryl J. Hartman Nutritional Sciences, Diet Assessment Center, Penn State University, University Park, PA, USA

Margo L. Hennet Department of Biology, Middlebury College, Middlebury, VT, USA

Malcolm Paine Department of Obstetrics and Gynecology, Dartmouth Hitchcock Medical Center, Lebanon, NH, USA

Beena J. Premkumar Cleveland Clinic Foundation, Center for Reproductive Medicine, Cleveland, OH, USA

Mitali Rakhit Center for Reproductive Medicine, Cleveland Clinic, Cleveland, OH, USA

Nathan Reynolds Department of Biological Sciences, Western Michigan University, Kalamazoo, MI, USA

Botros R. M. B. Rizk College of Medicine, OB/Gyn Department, University of South Alabama, Mobile, AL, USA

Elizabeth H. Ruder Cancer Prevention, Division of Cancer Epidemiology and Genetics, National Cancer Institute, NIH, Rockville, MD, USA

Zeynep Alpay Savasan Department of Obstetrics and Gynecology, Hutzel Women's Hospital, Wayne State University, Detroit, MI, USA

Lucky H. Sekhon Mount Sinai School of Medicine, OB/GYN, New York, NY, USA

Amani Shaman Cleveland Clinic Foundation, Center for Reproductive Medicine, Cleveland, OH, USA

Rakesh Sharma Center for Reproductive Medicine, Cleveland Clinic, Cleveland, OH, USA

Christina Talerico University of South Alabama College of Medicine, Mobile, AL, USA

Tinna Traustadóttir Department of Biological Sciences, Northern Arizona University, Flagstaff, AZ, USA

Richard E. White Pharmacology and Toxicology, Medical College of Georgia, Augusta, GA, USA

Shu Zhu Department of Pharmacology and Toxicology, Georgia Health Sciences University, Augusta, GA, USA

Chapter 1
Maternal Nutrition, Oxidative Stress and Prenatal Devlopmental Outcomes

Kaïs Hussain Al-Gubory

Abstract Aerobic organisms have adapted themselves to a coexistence with reactive oxygen species (ROS) by developing various and interdependent antioxidant systems that includes enzymatic and non-enzymatic antioxidants. Dietary antioxidants also play important roles in protecting the developing organisms from ROS damage, and both dietary and enzymatic antioxidants are components of interrelated systems that interact with each other to control ROS production. Oxidative stress can arise from an imbalance between generation and elimination of ROS leading to excessive ROS levels that damage all biomolecules. Tightly controlled ROS generation is one of the central elements in the mechanisms of cellular signaling and maintenance of signal transduction pathways involved in cell function, growth, and differentiation. Oxidative stress is considered to be a promoter of several prenatal developmental disorders and complications, importantly defective embryogenesis, embryopathies, embryonic mortality, spontaneous abortion, recurrent pregnancy loss, fetal growth restriction, intrauterine fetal death, low birth weight, preeclampsia, and preterm delivery. Environmental chemicals in food, water, and beverage may contribute to such adverse prenatal developmental outcomes and increase the susceptibility of offspring to disease via impairment of the antioxidant defense systems and enhancement of ROS generation. This chapter deals with the state of knowledge on the association between ROS, oxidative stress, antioxidants, and prenatal developmental outcomes. The importance of maternal antioxidant-rich foods in eliciting favorable effects on women health and prenatal development outcomes is highlighted.

K. H. Al-Gubory (✉)
Department of Animal Physiology and Livestock Systems,
INRA, Joint Research Unit of Developmental Biology
and Reproduction, Jouy-en-Josas, France
e-mail: kais.algubory@jouy.inra.fr

Keywords Reactive oxygen species · Oxidative stress · Antioxidant enzymes · Dietary antioxidants · Environmental chemicals · Maternal foods · Developmental defects · Prenatal developmental outcomes · Pregnancy disorders and complications

1.1 Introduction

Reactive oxygen species (ROS) are molecules and free radicals produced during metabolic pathway that uses energy released by the oxidation of nutrients to produce adenosine triphosphate (ATP). The concept of the paradox of aerobic life [1] or the paradox of ROS [2] has been emerged for our acceptance that life of aerobic organisms is a paradox. Tightly controlled ROS generation is an important constitutive process and is one of the central elements in cell signaling and maintenance of redox homeostasis [3] and signal transduction pathways involved in cell function, growth, differentiation, or death [4–6]. To maintain redox homeostasis and to cope with injury from ROS-induced oxidative damage, aerobic organisms have adapted to a coexistence with ROS through the development of various and interdependent antioxidant systems that includes enzymatic and non-enzymatic antioxidants [7]. Various dietary antioxidants also play important roles in protecting cells from ROS damage, and both dietary and enzymatic antioxidants are components of interrelated systems that interact synergically with each other to control ROS production. The balance between ROS and antioxidant systems determines the degree of oxidative damage to biological macromolecules. Abnormally high ROS generation leads to irreversible alteration of biomacromolecules, mainly lipids, proteins, and nucleic acids, ultimately causes mitochondrial dysfunction, mitochondrial ROS-induced ROS release (RIRR), and cell death by apoptosis [8].

ROS and antioxidants cross-talk is important component of the mammalian reproductive functions, such as ovarian follicular development, luteal steroidogenesis, endometrium receptivity, embryonic development, implantation, placental development, and growth [9, 10]. Maintaining equilibrium between antioxidants and ROS is crucial for the survival of the developing organisms. High levels of ROS during embryonic, fetal, and placental development are a feature of pregnancy [11, 12]. ROS-induced oxidative stress has emerged as a likely promoter of several prenatal developmental disorders and complications, such as defective embryogenesis, embryopathies, embryonic mortality, abortion, idiopathic recurrent pregnancy loss, hydatidiform mole, fetal growth restriction, intrauterine fetal death, low birth weight, preeclampsia, and preterm delivery [13–15]. Maternal malnutrition and/or maternal exposure to environmental chemicals may contribute to such adverse prenatal developmental outcomes and increase the susceptibility of offspring to disease. This occurs, at least in part, via impairment of the antioxidant systems and enhancement of ROS generation which alters cellular signaling and/or damage biomacromolecules. The links among ROS, oxidative stress, antioxidants, the female reproductive system, and adverse prenatal developmental outcomes, constitute therefore important issues in human and animal reproductive medicine.

Maternal nutrition is one of the most important lifestyle factors determining embryonic/fetal development. Malnutrition plays major roles in programing the offspring susceptibility to oxidative stress and disease [16]. Man-made chemicals are of increasing public health concern owing to an increasing number of such environmental pollutants arising from industrial and agricultural activities. Prenatal exposure to various environmental chemicals occurs through the consumption of contaminated food, water, and beverage. Maternal exposure to environmental chemicals during the critical periods of pregnancy [17], pass across the placental barrier into the fetal blood, transferred to the fetus, and may affect developmental outcomes [18]. This chapter deals with the state of knowledge on the association between oxidative stress, antioxidants, and prenatal developmental outcomes. In recent years, antioxidant nutritional interventions will provide sensible strategies against maternal malnutrition and/or environmental chemical insults. The importance of antioxidant-rich foods in eliciting favorable effects on women health and prenatal development outcomes is highlighted.

1.2 Reactive Oxygen Species (ROS)

Reactive oxygen species (ROS) includes oxygen free radicals, such as the superoxide anion radical ($\cdot O_2^-$), hydroxyl radical ($\cdot OH$), and nitric oxide ($NO\cdot$), and also non-radical reactive oxygen derivatives, such as hydrogen peroxide (H_2O_2) and peroxynitrite ($ONOO^-$). A free radical is a chemical species containing one or more unpaired electrons in its outer orbital. The ROS are ubiquitous, highly reactive, and diffusible molecules. ROS may promote oxidative damage to proteins, lipids, and DNA when they are overabundant. The consequences of these attacks are respectively, loss of enzymes activity, cell membrane alterations, DNA lesions, and oxidative mutagenesis. The generation of $\cdot O_2^-$ by a single electron donation to dioxygen molecule (O_2) is the first step in the formation and propagation of ROS within and out of the cell. The toxicity $\cdot O_2^-$ is based on generation of very reactive ROS, so-called downstream products of $\cdot O_2^-$. The $\cdot O_2^-$ radical is the precursor of most ROS and could be a mediator in oxidative chain reactions. Dismutation of $\cdot O_2^-$ produces H_2O_2 and O_2 in accordance with McCord and Fridovich [19] reaction:

$$\cdot O_2^- + \cdot O_2^- + 2H+ \rightarrow H_2O_2 + O_2$$

In the presence of free unbound iron and/or copper ions, H_2O_2 interact with $\cdot O_2^-$ in a Haber–Weiss reaction [20] which is considered as a common source of the highly reactive and oxidative $\cdot OH$. The Haber–Weiss cycle consists of the following reactions:

$$Fe^{3+} + \cdot O_2^- \rightarrow Fe^{2+} + O_2$$
$$Fe^{2+} + H_2O_2 \rightarrow Fe^{3+} + OH^- + \cdot OH \text{ (Fenton reaction)}$$
$$Cu^{2+} + \cdot O_2^- \rightarrow Cu^+ + O_2$$
$$Cu^+ + H_2O_2 \rightarrow Cu^{2+} + OH^- + \cdot OH$$

The net reaction:

$$\cdot O_2^- + H_2O_2 \rightarrow \cdot OH + OH^- + O_2$$

The resulting Fe^{2+} could propagate the oxidative degradation to new amino acid residues by reacting with H_2O_2 to form further $\cdot OH$. As a consequence of such reactions, uncontrolled generation of highly reactive ROS can lead to oxidative damage to cellular biomolecules and cell death by apoptosis [21, 22]. Since there is no known pathway to remove $\cdot OH$, the generation of $\cdot OH$ from $\cdot O_2^-$ and H_2O_2 is considered the central mechanism by which H_2O_2 induces damage to DNA, proteins, and lipids in biological systems [7].

Other important cellular ROS messenger is $NO\cdot$. It is produced from L-arginine in a reaction catalyzed by NO synthase (NOS) and acts as a regulator of many physiologic events [23]. NO is also an important prooxidant and toxic factor. Prooxidant action of $NO\cdot$ is often attributed to NO-derived species such as $ONOO^-$ which possess strong oxidative properties. Indeed, $NO\cdot$ may react with $\cdot O_2^-$ in a reaction controlled by the rate of diffusion of both radicals to form $ONOO^-$ [24] as follows:

$$NO\cdot + \cdot O_2^- \rightarrow ONOO^-$$

The molecule $ONOO^-$ can diffuse freely within and out of the cell, and react with lipids, proteins, and DNA, leading to cell membrane lipid peroxidation [25], DNA damage, and apoptosis [26]. Importantly, protein S-nitrosylation inhibits the activity of enzymes known to be S-nitrosylated by $NO\cdot$ [27], such as caspase-3 [28]. Decrease in caspase-3 S-nitrosylation is associated with an increase in caspase activity [29]. It is important to highlight that the balance between $NO\cdot$ and $\cdot O_2^-$ and consequent $ONOO^-$ production determine the degree of cellular oxidative damage [30].

1.3 Cellular Sources of Reactive Oxygen Species (ROS)

The main site of ROS production, mainly $\cdot O_2^-$, H_2O_2, and $\cdot OH$, is the mitochondrial electron transfer chain. The machinery of mitochondrial oxidative metabolism and ATP synthesis is associated with the inevitable ROS production and propagation within and out of the mitochondria. The inner membrane mitochondrial respiratory chain consists of four multimeric complexes (complexes I-IV), coenzyme Q (CoQ), and cytochrome C (Cyt C). The production of $\cdot O_2^-$ occurs during the passage of electrons through the mitochondrial electron transport system, so-called respiratory chain oxidative phosphorylation (OXPHOS). This process is the major ATP synthetic pathway in eukaryotes [31]. It comprises the electron transport chain that establishes a proton gradient across the mitochondrial inner membrane by oxidizing the reduced nicotinamide adenine dinucleotide (NADH) produced from the Krebs cycle. The main substrate for NADH is supplied by cellular glucose, fatty acids, and amino acids via three interconnected pathways [31]. The reduced flavin adenine dinucleotide (FADH2) adds its electrons to the electron transport system at a lower level than NADH. During OXPHOS, electrons are transferred from the reducing equivalent NADH-FADH2 to O_2 in redox reactions via a chain of respiratory H^+ pumps. These pumps (complexes I-IV) establish an H^+ gradient across the inner mitochondrial membrane, and the energy of this gradient is then used to drive ATP synthesis by complex V [32]. During the electron transfer, $\cdot O_2^-$ radicals are mainly generated at complexes I and III [33]. Mitochondria are endowed with a NOS (mtNOS) [34, 35] and thus $\cdot NO$ and its derivatives are also produced by mitochondria where they have multiple effects that impact on cell physiology and death [36–38]. Mitochondrial ROS released into cytosol function as a second messenger to activate RIRR in neighboring mitochondria and potentially leading to apoptosis [8]. ROS are produced by various enzymatic pathways, mainly membrane-bound NADH and NADPH oxidases, xanthine oxidase, lipoxygenases, cyclooxygenases, and cytochromes P450 [39, 40]. ROS are also generated outside mitochondria during biotransformation of xenobiotics and drugs, inflammation, UV exposition, ionic irradiation, and lipid peroxidation of plasma membrane and other membrane–lipid structures [41].

1.4 Antioxidant Systems

Aerobic organisms are protected against ROS-induced oxidative damage by a network of highly complex and integrated endogenous enzymatic and non-enzymatic antioxidant systems. The key enzymes synthesized in the cell and directly involved in the control of ROS production are superoxide dismutases (SODs), catalase (CAT), glutathione peroxidases (GPXs), glutathione reductase (GSR), glucose-6-phosphate dehydrogenase (G6PD), and isocitrate dehydrogenases (IDHs). Glutathione, nicotinamide adenine dinucleotide phosphate ($NADP^+$), the reduced form of $NADP^+$ (NADPH), and sulfur-containing amino acids are the major non-enzymatic antioxidant systems that play key roles in protecting cells from oxidative stress. Aerobes are also protected from oxidative damage by various dietary antioxidants which are concentrated in foods of plant origin, mainly fruits and vegetables.

1.4.1 Endogenous Enzymatic Antioxidant Systems

The dismutation of $\cdot O_2^-$, generated in different cellular compartments, into H_2O_2 and O_2 is the first step that plays a vital role in the control of cellular $\cdot O_2^-$ production. Three distinct isoforms of SOD have been identified and characterized in mammals. Copper-zinc containing SOD (Cu, Zn-SOD or SOD1), a dimeric protein, was the first SOD discovered and characterized in eukaryotes cytoplasm [19]. Manganese containing SOD (Mn-SOD or SOD2) is located in the mitochondrial matrix [42]. The extracellular superoxide dismutase (EC-SOD) is a Cu- and Zn-containing tetrameric glycoprotein which is the major SOD in extracellular fluids such as plasma and lymph [43].

The control of H_2O_2 production is the second step that plays a vital role against ROS propagation. To prevent production and propagation of $\cdot OH$ and $ONOO^-$, H_2O_2 generated after SOD catalyzing reaction is quickly converted into H_2O and O_2 by GPXs which have selenocysteine within its active site, and therefore they are selenium dependent for antioxidant activity. GPXs detoxify H_2O_2 to H_2O through the oxidation of Glutathione [44]. In addition, GPXs catalyze the degradation of lipid peroxides (LPO) and can metabolize lipid hydroperoxides to less reactive hydroxy fatty acids [44]. GPXs are therefore the primary antioxidant enzymes that protect biomembranes and cellular components against oxidative damage [45]. Glutathione exists in the reduced (GSH) and disulfide-oxidized (GSSG) states. GSR catalyzes the reduction of GSSG to GSH with NADPH as the reducing agent [44]; and it is therefore essential for the glutathione redox cycle that maintains adequate levels of GSH for GPX catalytic activity. GPXs present in the cytoplasm and the mitochondria, and CAT which found primarily within peroxisomes, both catalyze the conversion of H_2O_2 to H_2O.

NADPH required for the regeneration of GSH [46] is crucial for scavenging mitochondrial ROS through GSR and peroxidase systems. G6PD-produced NADPH pathway [47] is therefore crucial in the defense mechanism against oxidative stress. As a source of NADPH, mitochondrial and cytosolic NADP-IDHs play an important role in cellular defense against oxidative damage [48, 49]. The protective role of mitochondrial $NADP+$ dependent IDH against ROS-induced oxidative damage may be attributed to increased levels of a NADPH, needed for regeneration of GSH in the mitochondria [48]. The cytoplasmic NADP-linked IDH generates NADPH via oxidative decarboxylation of isocitrate, and thus potentially involved in the maintenance of the cellular redox state and plays an important role in the protection of cytoplasm components against ROS-induced oxidative stress [49].

The antioxidant enzymes and the small molecular weight non-enzymatic antioxidants represent coordinately operating cellular systems controlling ROS production [10] and prevent their propagation within and out of the cell (see Fig. 1.1).

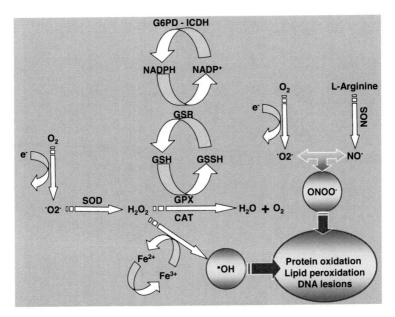

Fig. 1.1 Schematic representation of cellular reactive oxygen species (ROS) production and their control by enzymatic antioxidant systems. The production of superoxide anion ($\cdot O_2^-$) by a single electron (e) donation to molecular oxygen (O_2) is the initial step in the formation and propagation of other ROS within and out of the cell. The major enzymes, namely copper-zinc containing superoxide dismutase (Cu, Zn-SOD or SOD_1), manganese containing SOD (MnSOD or SOD_2), catalase (CAT), glutathione peroxidase (GPX), glutathione reductase (GSR), glucose-6-phosphate dehydrogenase (G6PD), and isocitrate dehydrogenases (ICDH) represent coordinately operating network of defenses against ROS-induced oxidative stress and cell damage. Hydrogen Peroxide (H_2O_2); water (H_2O); hydroxyl radical ($\cdot OH$); peroxynitrite ($ONOO^-$); nitric oxide synthase (NOS); nitric oxide ($NO\cdot$); reduced/oxidized glutathione (GSH/GSSG); reduced/oxidized nicotinamide adenine dinucleotide phosphate (NADPH/NADP)

1.4.2 Endogenous Non-Enzymatic Antioxidant Systems

The non-enzymatic antioxidants, NADH, NADPH, and GSH (as indicated above), and sulfur-containing amino acids are the key elements in the system of enzymatic antioxidant defenses against ROS-induced oxidative stress [50–54]. GSH has important roles as an independent free radical scavenger and antioxidant enzyme cofactor, as well as a cysteine carrier–storage form. Beside their role in protein synthesis, the sulfur-containing amino acids methionine and cysteine are also precursors of the key small molecular weight antioxidants, GSH and taurine. GSH is a tripeptide comprises glutamate, cysteine, and glycine. Cysteine is the rate-limiting substrate of GSH synthesis. Cysteine is a thiol-containing amino acid, which is also called a semi-essential amino acid because humans can synthesize it from an essential amino acid, methionine [55, 56]. Cysteine can also be obtained from the diet, usually as cystine which is a dimer of cysteine. Taurine is

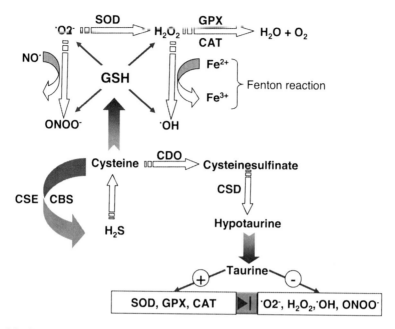

Fig. 1.2 Cytoprotective role of hydrogen sulfide (H_2S) against reactive oxygen species (ROS)-induced oxidative damage. H_2S is formed from cysteine by cystathionine β-synthase (CBS) and cystathionine γ-lyase (CSE). Then H_2S protects cells by two mechanisms: it enhances the production of the reduced form of glutathione (GSH) by enhancing cystine/cysteine transport and redistributes them to the mitochondria. Molecular oxygen (O_2); superoxide anion ($^•O_2^-$); hydrogen peroxide (H_2O_2); water (H_2O); hydroxyl radical ($^•OH$); peroxynitrite ($ONOO^-$); nitric oxide ($NO^•$); superoxide dismutase (SOD); catalase (CAT); glutathione peroxidase (GPX), cysteine dioxygenase (CDO), cysteinesulnate decarboxylase (CSD)

synthesized from cysteine via the actions of cysteine dioxygenase (CDO), which gives rise to cysteinesulnate, and cysteinesulnate decarboxylase (CSD), which decarboxylates cysteinesulnate to hypotaurine [57, 58]. Hypotaurine is further oxidized to taurine (see Fig. 1.2). Taurine prevents oxidant-induced cell damage by direct upregulation of key antioxidant enzymes and attenuation of ROS generation [52, 53], and plays an important cytoprotective role [59].

Hydrogen sulfide (H_2S), an endogenous antioxidant gas [60], is recognized as an important signaling molecule [61, 62] with therapeutic potential [63, 64]. H_2S has been shown to be produced by various mammalian organs [65], including the reproductive tissues [66]. It is formed from cysteine by pyridoxal-5′-phosphate (PLP)-dependent enzymes, namely cystathionine β-synthase (CBS) and cystathionine γ-lyase (CSE) [67]. Importantly, H_2S protects cells against oxidative stress [68] by two mechanisms: it increases the production of GSH by enhancing cystine/cysteine transport and redistributes them to the mitochondria [69]. The cytoprotective role of H_2S [70–72] is therefore crucial for preventing ROS accumulation and oxidative damage (see Fig. 1.2).

1.4.3 Exogenous Antioxidant Systems

Nutrient antioxidants are diverse compounds characterized by their ability to quench ROS. Common dietary antioxidants include vitamins, trace elements, and polyphenols. Dietary antioxidants are classified into two types, depending on whether they are soluble in water (hydrophilic) or in lipids (hydrophobic). Water-soluble antioxidants are present in the cytoplasm and the blood plasma, while lipid-soluble antioxidants are mainly present in cell membrane and cellular organelles where they protect cells from lipid peroxidation.

Carotenoids (vitamin A), Ascorbate (vitamin C), and tocopherols (vitamin E) are among the major dietary antioxidants that scavenge directly extracellular ROS and provide a major source of protection against their damaging effects [73–78]. Carotenoids, such as ß-carotene, lutein, α-carotene, zeaxanthin, cryptoxanthin, and lycopene, are fat-soluble antioxidants [73, 74]. ß-carotene is the main source of provitamin A. Green leafy vegetables, carrots, and other yellow root vegetables, and yellow and orange fruits contain ß-carotene. ß-carotene, and lycopene are important biological compounds that can inactivate electronically excited molecules, a process termed quenching, and may also participate in free radical reactions [79]. Vitamin C is a water-soluble antioxidant which is present in various fruits and vegetables. Ascorbic acid has been shown to interact with the tocopheroxyl radical and to regenerate the reduced tocopherol [80]. α, β, γ, and δ-tocopherol and the corresponding four tocotrienols are naturally occurring fat-soluble dietary vitamins with antioxidant properties [81]. The natural sources of vitamins E are vegetable oils. The most biologically active form of the vitamin E homologs is α-tocopherol. It contributes in the protection of cell membrane polyunsaturated fatty acids (PUFA) against ROS-induced peroxidative damage [82], and thus plays important role in membrane stability and functions [83–85]. It is important to highlight that vitamins C and E interact synergistically in association with GSH-related enzymes to control the production of lipid peroxidation products (see Fig. 1.3).

Trace elements are present in minute quantities in the body and are required in low concentrations in the diet [86]. Some elements form part of the active site necessary for the antioxidant enzyme function, act as cofactors in the regulation of antioxidant enzymes, or shaping the enzyme configuration necessary for its activity. Among trace elements, Cu, Zn, Mn, and Se are essential micronutrients with wide range of functions in the body including the synthesis and activity of antioxidant enzymes, namely Cu, Zn-SOD (cytosolic SOD1), Mn-SOD (mitochondrial SOD2), and four Se-GPX isoforms (cytosolic GPX1, gastrointestinal GPX or GPX2, plasma GPX or GPX3, phospholipid hydroperoxide GPX or GPX4).

Polyphenols are the most abundant antioxidants in the human diet. They are a wide variety of organic molecules that naturally occurring in vegetables, fruits, and plant-derived beverages such as tea, red wine, and olive oil [87]. Polyphenols are characterized by the presence of several groups involved in phenolic structures

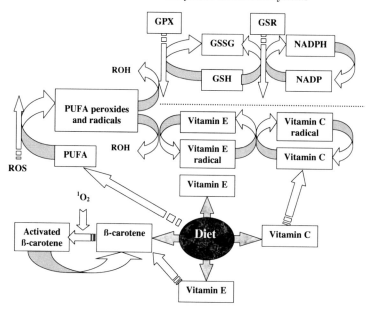

Fig. 1.3 The reaction sequences between dietary antioxidant vitamins and enzymatic antioxidants ensuring cellular defense against reactive oxygen species (ROS) and lipid peroxide. Glutathione peroxidase (GPX); glutathione reductase (GSR); reduced/oxidized glutathione (GSH/GSSG); reduced/oxidized nicotinamide adenine dinucleotide phosphate (NADPH/NADP), polyunsaturated fatty acids (PUFA); singlet oxygen ($^1O_2$); alcohol (ROH). Adapted from Machlin and Bendich [80]

and include the flavonoids and phenolic acids such as catechins, resveratrol, quercetin, anthocyanins, hesperitin derivatives, phytic, caffeic, and chlorogenic acids. Pigmented fruits, such as pomegranates, grapes, apples, berries, pears, cantaloupe melon, and watermelon, and vegetables, such as broccoli, cabbage, celery, onion, and parsley, are rich in polyphenol antioxidants. Phytic acid, by virtue of chelating free iron, is a potent inhibitor of ˙OH formation by the Fenton reaction and accelerates O_2-mediated Fe^{2+} depletion and suppress iron-mediated oxidative processes. Various polyphenolic compounds with antioxidant activity are potential contributors to explain the human health benefits of diets rich in fruits and vegetables. Dietary polyphenols, in addition to their direct ROS scavenging action, can protect cells against ROS oxidative damage by increasing the expression of mRNA encoding the key antioxidant enzymes [88].

1.5 ROS and Antioxidants Cross-Talk in Prenatal Developmental Outcomes

The establishment of pregnancy in mammals can be divided into four developmental periods: (1) fertilization of the oocyte and early embryonic development, (2) implantation of the developing blastocyst, (3) post-implantation embryonic development, and (4) Fetal and placental development. Mitochondrial OXPHOS is a vital metabolic pathway that uses energy released by the oxidation of nutrients to produce ATP to meet high energy demand of the conceptus (embryo/fetus and associated placental membranes) during its metabolic and developmental processes [89, 90]. During early embryonic development, mitochondria also provide ATP for the regeneration of NADPH and GSH [91].

Before implantation, embryonic development and growth occurs under uterine hypoxic environment and embryonic tissues cannot produce large amount of ATP. The switch from preimplantation embryonic anaerobic metabolism to post-implantation aerobic metabolism induced by conceptus vascularization onset and uteroplacental blood flow exposed embryonic and extraembryonic tissues to ROS produced as normal by-products of OXPHOS. These changes take place during the period of organogenesis, when the conceptus is potentially vulnerable to environmental factors. Post-implantation conceptus may be therefore particularly vulnerable to oxidative stress early in pregnancy. The control of ROS production is important in development through cellular signaling pathways involved in proliferation, differentiation or apoptosis, whereas high ROS levels induces oxidative stress and can alter embryonic development [92]. Mitochondrial dysfunction due to pathologic and/or environmental insults triggers apoptosis in the embryo and compromises early developmental processes and birth outcome [91, 93]. Abnormal or reduced mitochondrial activity alters ROS production and reduces implantation rates in women [94]. Therefore, antioxidant status of the conceptus and its surrounding play vital roles in protecting embryonic and extraembryonic tissues from the deleterious effects of endogenous ROS and those generated after in utero exposure to environmental factors during the critical period of early prenatal development.

Embryonic implantation is promoted by a network of signaling molecules that mediate cell-to-cell communications between the receptive endometrium and embryonic trophectoderm. Animal studies, mainly in rodents, indicate that ROS production from both mitochondrial and non-mitochondrial sources, functions as a physiologic component of the early embryonic development. GSH-dependent antioxidant mechanisms are developmentally regulated in the inner blastocyst cell mass and H_2O_2 is a potential mediator of apoptosis in the blastocyst [95]. Pre- and post-implantation mouse embryos generate and release ROS [96, 97]. An abrupt drop in SOD activity and a concomitant rise in $\cdot O_2^-$ levels occur in mouse blastocyst at the perihatching stage [98]. The uterine $\cdot O_2^-$ burst at proestrus in rat suggested its involvement in regulating uterine edema and cell proliferation [99]. A peak of $\cdot O_2^-$ in the uterus at day five of mice pregnancy suggested a contribution of this radical in vascular permeability at the initiation of implantation [100]. NADPH-dependent $\cdot O_2^-$ production pathway associated with the uterus of pregnant mice increases across the

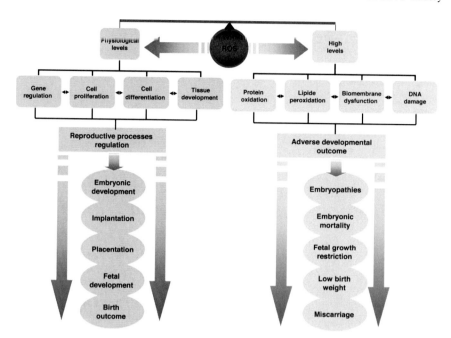

Fig. 1.4 Schematic overview of physiologic roles of reactive oxygen species (ROS) in normal reproductive processes or induced adverse developmental outcomes. Controlled physiologic level of ROS plays important roles in developmental processes and birth outcome through the regulation of gene expression, cell proliferation and differentiation, and tissue development. High levels of ROS damage cellular macromolecules and membranes which lead to adverse developmental outcomes, disorders, and miscarriage

preimplantation stages [101]. Antioxidant enzyme systems have been shown as components of the developing embryo and its receptive uterine endometrium [102–107] and may have vital role in regulating fetal development and survival through the control of placental ROS production and propagation [108–112].

Physiologic levels of ROS play important role in the regulation of reproductive processes, including embryonic development, uterine receptivity, embryonic implantation, placental development and endocrine functions, and fetal development, whereas high ROS levels have adverse developmental outcome, including embryopathies, embryonic mortality, fetal growth restriction, low birth weight, and miscarriage (see Fig. 1.4).

1.6 Maternal Dietary Antioxidants and Prenatal Developmental Outcomes

Maternal dietary antioxidants are essential nutrients for embryonic and fetal development and health outcomes. Adequate maternal antioxidant status before and during pregnancy could prevent oxidative mechanisms induced by poor dietary nutritional factors, importantly micronutrient antioxidants. Maternal under- or poor nutrition are associated with multiple micronutrient deficiencies, a major cause of IUGR and miscarriage [113]. Maternal nutritional factors play a major role in programing the susceptibility of offspring to oxidative stress and disease [16]. Maternal diet composition is important determinant of the balance between antioxidants and ROS that affect prenatal growth and development, [114, 115]. Maternal antioxidant status has a positive influence on birth weight [116].

Under conditions of low dietary intake, maternal nutritional deficiencies, or suboptimal maternal nutritional status, trace elements and/or antioxidant vitamins during pregnancy can be compromised [117–119]. Deficiencies and or alterations in trace elements and antioxidant vitamins, associated with oxidant/antioxidant imbalance, impair fetoplacental development, and have persistent effects on fetal development and neonatal tissues [120–124]. Trace elements deficiency influences prenatal development through several mechanisms including ROS-induced oxidative damage that is secondary to compromised antioxidant defense systems. The activities of Cu, Zn-SOD (SOD1), Mn-SOD (SOD2) and Se-GPX are sensitive to dietary deficiency in Cu, Zn, Mn, or Se. Low SOD activity in Cu-deficient rat embryos exposes the embryo to excessive ROS production, resulting in subsequent cellular oxidative stress, a shift in the redox state to a more oxidized environment, embryonic damage, and developmental defects [125]. Maternal Cu deficiency is associated with multiple fetal developmental defects that can affect the central nervous system, cardiovascular, and skeletal systems, and result in poor immunocompetence and behavioral abnormalities of the offspring [126]. Cu deficiency induce decreased SOD activity and increased $\cdot O_2^-$ and $ONOO^-$ concentrations that can lead to nitration of proteins and alterations in protein function and activity [127]. A decrease in NO bioavailability can inhibit NO-mediated intracellular signaling [128] and adversely affect embryonic and fetal development [129]. Deficiencies of Zn, Cu, and Mn have been implicated in human reproductive disorders like infertility, pregnancy wastage, pregnancy induced hypertension, premature rupture of extraembryonic membranes, and low birth weight [130]. Malnutrition is associated with oxidative stress in small for gestational age neonates born at term to malnourished mothers [131]. Poor maternal dietary antioxidants have adverse prenatal developmental outcomes and play a role in programing the susceptibility of offspring to disease (see Fig. 1.5).

Antioxidant vitamins and minerals interventions during prenatal development are important for preventing the damage caused by ROS in utero [132]. Animal studies have shown that one selected antioxidant or combined with other supplements decreases embryonic mortality and improves birth outcomes [133–135]. Blood α-tocopherol concentrations are lower in abnormal pregnancies than those in normal ones, suggesting that vitamin E requirements increase throughout pregnancy [136]. Nevertheless, the association between dietary antioxidants and birth outcomes is not conclusive in human or animal studies [137–139]. Prophylactic use of some micronutrients may be useful in preventing adverse pregnancy outcomes [140]. Nevertheless, early human pregnancy supplementation with pharmacologic doses of antioxidant vitamin E is associated with a decrease in birth

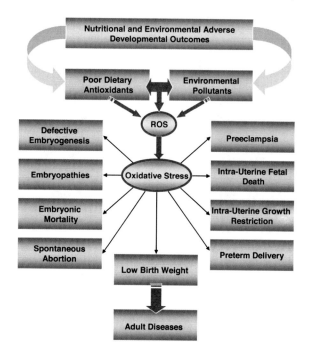

Fig. 1.5 Reactive oxygen species (ROS)-induced oxidative stress caused by poor dietary antioxidant intakes and/or environmental pollutants in foods, and the subsequent adverse developmental outcomes and pregnancy-associated disorders and complications

weight [141]. Maternal supplementation with vitamins C and/or E does not prevent preeclampsia in high-risk pregnant women [142, 143]. Furthermore, high maternal consumption of selected dietary antioxidants does not protect the child from development of advanced β cell autoimmunity in early childhood [144]. The reason for the failure of such supplementations to improve the prenatal developmental outcomes, and maternal and child health could be imbalanced administration of antioxidant vitamins and/or trace elements. The benefits of using dietary supplement that contains one or a limited number of antioxidants on prenatal developmental outcomes remain far from efficient and inconclusive. Consumption of various antioxidants in foods is likely more effective than limited number of antioxidants or large doses of a single antioxidant given for a finite period.

The diet is an important source of diverse antioxidants. Antioxidant-rich foods, specifically high consumption of fruits, vegetables, cereals, and plant-derived oils can prevent and delay the development of chronic age-related diseases [145]. An interesting example worth mentioning is the traditional Mediterranean diet which is rich in antioxidants components, mainly vitamins, essential trace elements, and polyphenols [146]. Mediterranean diet is widely reported as a nutritional model that keeps overall maintenance of human health and protection from disease [147–149] through the lowering of prooxidant-induced oxidative stress [150–152]. This diet is essentially composed of high intake of legumes, fruits, vegetables, cereals, and nuts, high to moderate consumption of fish, moderate consumption of dairy products and wine, and low consumption of meat and meat products.

Mediterranean diet seems to play an important role in preventing chronic diseases [153, 154]. Mediterranean-type diet can be proposed as a mean to improve fertility [155] and decrease risk of premature labor and gestational diabetes [156]. Despite the fact that plant roots, stems, leaves, flowers, fruits, and seeds consumption can improves dietary quality in women before conception and during the first trimester of pregnancy, little is known about how these wide variety of antioxidant-rich edible dietary sources affects favorably prenatal developmental outcomes.

1.7 Environmental Contaminants and Prenatal Developmental Outcomes

Industrial and agricultural activities contribute to the release of large quantities of various chemical in the environment and have resulted in widespread air, soil, and water contamination. Exposure to such environmental contaminants is inevitable as it occurs through the consumption of contaminated food, water, beverages. There is substantial evidence that many human chronic diseases results from exposure to environmental factors early in development [157]. Environmental chemicals disrupt endocrine function and contribute to alterations in growth and development [158, 159]. In utero exposure to environmental chemicals can mediate adverse prenatal developmental and birth defect [160]. Various environmental chemicals, to which the mother is exposed during the critical periods of pregnancy [17], pass across the placental barrier into the fetal blood stream and can be transferred to the fetus [18]. The fetus is particularly vulnerable to the adverse effects of environmental chemicals [17]. Embryonic and fetal period are vulnerable to oxidative stress and many environmental pollutants may contribute to adverse prenatal developmental outcomes and can lead to chronic health problems later in life (see Fig. 1.5), at least in part, via ROS generation which damage cellular macromolecules and/or alter signal transduction pathways [18, 161]. Environmental chemical exposure during development induces oxidative stress and fetal toxicity that adversely affects fetal ovarian development [162], contribute to birth defects [163] and may ultimately lead to cancer later on in life [164].

The developing organism is susceptible to oxidative stress induced by chemical contaminants due to its inadequate antioxidant defense systems [165]. The toxicity, dosage, and timing of exposure to environmental chemicals are important determinants of adverse in utero effects on the developing organs and tissues. ROS-induced oxidative stress has emerged as a promoter of prenatal developmental disorders, such as embryopathies, embryonic mortality, abortion, fetal growth restriction, and low birth weight [9, 10]. Chemicals in food may contribute to such adverse prenatal outcomes and increase the susceptibility of offspring to disease [164] via impairment of the antioxidant defense systems and enhancement of ROS generation [16]. However, the relationship between oxidative stress induced by environmental chemicals and the adverse prenatal developmental outcomes is not clear and cannot be investigated in human pregnancies for evident ethical reasons. Animals, mainly rodents, have been often used to examine the adverse developmental outcomes of chemical contaminants.

The man-made chemicals, importantly the estrogen-mimic bisphenol-A (BPA) to which the mother is exposed during the critical periods of pregnancy, pass across the placenta and can be transferred to the fetus [166–168]. BPA is the main monomeric chemical used in the production of polycarbonate plastic, the manufacture of food-storage containers and the epoxy resin that form the lining of food and beverage cans and dental sealants. This endocrine disruptor [169] can be leached from plastic, food cans, and containers and widely spread in the environment and food chain. Of particular concern is the contamination by BPA during pregnancy that is evidenced by its presence in maternal and umbilical cord blood, amniotic fluid, fetal and placental tissues [170, 171]. The mouse brain, kidney, liver, and testes display high levels of H_2O_2 following BPA exposure [172]. In these organs, BPA-induced oxidative stress occurs, at least in part, through distribution of the redox control systems, mainly the levels of GSH/GSSG [172]. BPA exposure during embryonic/fetal development in rodents induced oxidative stress and ultimately leads to underdevelopment of fetal brain, kidney, and testis [173] and disturbs postnatal reproductive functions [174, 175]. Prenatal BPA exposure of ewes, at levels similar to that seen in human maternal circulation resulted in low birth weight [176]. All these perturbations raise the question of the impact of maternal exposure to BPA on the development process, the risk of fetal growth restriction and consequently the risk of developing endocrine and reproductive disorders throughout adult life.

The actions of estrogen and progesterone via their uterine receptors orchestrate the changes in the mammalian endometrium that make it receptive to the blastocyst implantation and the establishment of pregnancy. Exposure of mice to BPA early in pregnancy acts at the uterus to disrupt intrauterine implantation, consistent with an estrogenic BPA action [177]. Estrogen biosynthesis is catalyzed by aromatase cytochrome P450, the product of the CYP19 gene. BPA is an estrogen receptor antagonist and inhibit estrogen synthesis by downregulation of CYP19. The BPA-induced changes in the expression of sex steroid receptors have consequences for the hormonal responsiveness of the mice uterus to both endogenous and exogenous hormones, and the potential for predisposition of the organ to disease later in life [175]. The activities of antioxidant enzymes in the sheep endometrium are regulated by estradiol [107]. BPA has been shown to downregulate CYP19 expression in placental cells [178]. The early luteo-placental shift in progesterone production plays crucial role in maintenance of pregnancy beyond the corpus luteum life span in humans [179] and sheeps [180], allowing continued prenatal development. Increased placentome progesterone content early in pregnancy [111] indicates that the steroidogenic capacity of placentome cells increased as pregnancy advances. ROS inhibit steroidogenesis by blocking cholesterol transport into mitochondria [181]. The increases in antioxidant enzyme activities and progesterone production in ovine placentomes during early pregnancy [111] indicates that antioxidant systems act as a protective mechanism against ROS-induced oxidative damage during placentation. By interfering with steroidogenesis signaling pathways, BPA could therefore deregulate the developmental expression of enzymatic antioxidant defense systems.

Environmental metals affect reproduction and development at every stage of the reproductive process [182]. The most common heavy metals implicated in human toxicity include mercury, lead, and cadmium. Prenatal exposure to such heavy metals induces oxidative stress through impairment of the antioxidant defense systems in the brain, liver, and kidney of the developing fetuses [183–187]. The source of exposure to mercury is through concentration of methylmercury (MeHg) in the fish and marine food products. Inorganic Hg, present in the environment is converted to MeHg within microorganisms in aquatic sediments, allowing it to enter the food chain. The developing central nervous system is highly sensitive to metal-induced ROS production and prenatal exposure to MeHg disrupts the postnatal development of the glutathione antioxidant system in the mouse brain [187]. Omega-3 polyunsaturated fatty acids (ω3PUFAs) are crucial for the biogenesis of membrane phospholipids and are sources of energy during normal prenatal development and function of organs and tissues [188], especially the brain [189–191]. PUFAs are also precursor of prostaglandin synthesis and thus have indirect roles in the control of uterine function and female fertility [192, 193]. Maternal exposure to MeHg during pregnancy, mainly from fish consumption, may diminish the efficacy of the beneficial action of nutrients, importantly ω3PUFAs [194, 195].

Pesticides, a broad class of chemicals, including herbicides, fungicides, insecticides, are widely employed in intensive agriculture to increase crop yields. The excessive use of pesticides has direct impacts on environmental and water quality and can affect public health. Exposure to pesticide has been associated with adverse developmental health and birth defects [196]. Contamination of soil and ground water from agricultural runoff [197, 198] results in the exposure of human to various pesticides and other agricultural chemicals directly and/or indirectly. Fruits and vegetables can become contaminated by pesticides, particularly insecticides, as a result of treatments in many crops [199]. The aquatic environment receives great amounts of man-made chemicals pollutants [200], including agricultural pesticides, that cause ROS production and oxidative stress in fish organism [201]. Animal and human exposure to pesticides has been shown to enhance oxidative stress, lipid peroxidation, and DNA damage [202–206]. The greatest risk of exposure to pesticides is during early pregnancy. During prenatal development, pesticide residues appear in maternal and cord blood, pass through the placental barrier [207, 208] and induce oxidative stress [209].

The link between food contamination by environmental chemicals from various industrial and agricultural activities, oxidative stress and the associated adverse prenatal developmental outcomes is therefore one of the major current concerns in reproductive medicine and health endpoints in human (see Fig. 1.6).

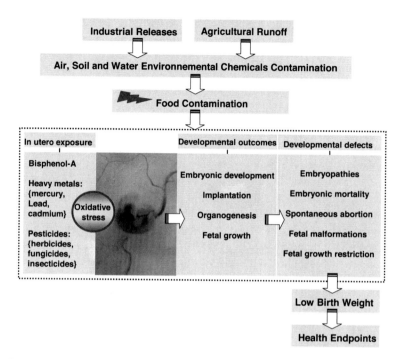

Fig. 1.6 The links between food contamination by environmental chemicals from various industrial and agricultural activities, oxidative stress early in development, adverse prenatal developmental outcomes, and pregnancy-associated disorders and complications

1.8 Proteomics in Nutrition, Oxidative Stress, and Prenatal Developmental Outcomes

The genomic approach has emerged to provide a better comprehension of the pathophysiologic conditions leading to pregnancy associated disorders, thereby providing a perspective for improving prenatal development outcomes. Nevertheless, it is insufficient to accurately predict protein expression patterns and function from quantitative messenger RNA (mRNA) due to post-transcriptional regulation mechanisms (mRNA export, surveillance, silencing and turnover) and post-translational modifications which can determine protein activity, localization, turnover, and interactions with other proteins [210]. Since protein expression is the functional outcome of gene transcription and translation, proteomics offers a unique mean for analyzing the expressed genome under pathophysiologic conditions. Of particular interest to investigators in the field of nutrition and reproductive medicine is proteomic approach to measure the secretome, i.e., those proteins that are produced by cells and secreted in biological fluids under a variety of cellular conditions [211]. Various biological fluids, including serum, plasma, urine, and amniotic fluid, have been employed in human pregnancy research [212], to predict human preterm labor and offspring outcomes [213, 214] and pregnancy-associated disorders [215–218]. It is also important to highlight the role that might play proteomic in defining the human embryonic secretome, i.e., proteins produced by the embryo and secreted into the surrounding medium, followed by the identification of specific proteins critical for implantation [219] in ongoing research focused on maternal nutrition, oxidative stress, and prenatal developmental outcome. Secretome may also become a tool for identification of embryo quality in clinical practice. Proteomic approach can be used to obtain new information about the antioxidant pathways that are critical to normal endometrial, luteal, and placental functions [10]. For example, examination of proteome of the cytotrophoblast of early pregnancy [220], the placenta derived from assisted reproductive technology [221], the endometrium and [222] and the CL [223] obtained during embryo implantation may be used to identify oxidative stress biomarkers.

Proteomics is expected to have an impact in solving malnutrition-associated disorders and pathologies, such as intrauterine fetal restriction, obesity, diabetes, and cardiovascular disease [224]. It could be also expected that proteomics may provide new insights into antioxidant alterations/adaptations depending on maternal nutritional status and in response to exposure to various environmental insults during prenatal development. Proteomic technology will not only provide physicians and biologists with an improved understanding of the underlying biological processes involved in the establishment and outcomes of pregnancy, but also to identify proteins responsible for prenatal developmental outcomes and health.

1.9 Conclusions

Growing body of the literature indicates that oxidative stress plays a role in prenatal defects and pregnancy-related disorders [9, 10, 14, 15]. It is nevertheless true that the relationship between ROS-induced oxidative stress and the prenatal developmental dysfunctions is not clear and cannot be adequately investigated in human pregnancies for evident ethical reasons. Furthermore, there is a lack of fundamental insights regarding cellular, biochemical, and molecular adaptive responses of the developing organism to the in utero exposition to various environmental chemicals. Therefore, animal models of normal and disturbed prenatal development are essential to fill in these important gaps in our knowledge. Understanding the developmental changes in antioxidant expression, as well as the cellular and molecular mechanisms of antioxidant regulation, in the female reproductive tract is needed. Such studies will provide insights about ROS-mediated antioxidant adaptive responses in normal and pathologic pregnancies and will facilitate treatment of prenatal developmental complications.

Maternal nutritional status is one of the most important lifestyle factors determining embryonic/fetal development and pregnancy outcomes. Conception, implantation, placentation, and fetal organogenesis are the critical stages potentially affected by nutrition during the periconception period [140]. Furthermore, preconception and early pregnancy malnutrition play major roles in programing the offspring susceptibility to disease, which may be mediated by deficiencies of antioxidant vitamins and essential trace elements [225]. In recent years, nutritional interventions will be used as primary prevention strategies against environmental chemical insults and associated diseases [226]. Early nutritional interventions to improve maternal antioxidant status are likely to have lifelong health benefits for the offspring. Improving women's health through diet, rich in antioxidants before conception and during the critical periods of early pregnancy would be a practical strategy for prevention of prenatal developmental dysfunctions in populations at high risk of reproductive disorders, as well as in populations with inadequate or poor baseline dietary consumption of natural antioxidants from common fruits [227] and vegetables [228]. A wide variety of plant-derived antioxidant compounds, such as vitamins, carotenoids, polyphénols, and trace elements are provided by fruits and vegetables. It is therefore wise and crucial to show that maternal diet varied and balanced in fruit and vegetables is able to provide adequate antioxidant defense necessary for prenatal developmental outcomes.

Diet and environmental conditions are widely different between countries and between geographic areas within a given country, or even between individuals within a given area. Maternal malnutrition before and during pregnancy is more common in poorer areas of the developing and underdeveloping countries, particularly among poorer peoples. Epidemiological and clinical studies have linked environmental factors, such as diet and lifestyle, to diseases and aging in association with ROS-induced oxidative stress due to insufficient ROS detoxification [229]. Environment quality and human and livestock health are closely interrelated. Industrial and intensive farming and agricultural activities in both developed and developing countries, and in the near future, in the emerging countries, contribute to the release of large quantities of chemical pollutants in the environment and have already resulted in widespread soil and water contamination. Exposure to these pollutants is inevitable as it occurs through the consumption of contaminated food, water and beverage, and by air inhalation. In the future, the greatest concerns of human beings are the ability to feed properly and to have healthy lifestyle. These concerns are dependent on our ability to generate safe nutrients and high quality products while minimizing adverse environmental impacts on prenatal development and reproductive processes arising from exposure to man-made chemicals.

The association among oxidative stress, the female reproductive system, and adverse prenatal developmental outcomes is an important issue in reproductive medicine. It is increasingly accepted that research focused on the impact of one family of antioxidants in prenatal development and pregnancy outcomes is no longer sufficient for developing practical and effective prevention strategies against oxidative stress and associated developmental dysfunctions and complications. Antioxidant and prooxidant status before and during early pregnancy reflect the balance between the endogenous antioxidant systems and prooxidants, and could be an index for estimating the risk of oxidative developmental damage. The special attention given to proteins as biomarkers of oxidative stress (redox proteomics) and diseases [230] and to the role of oxidative stress in human diseases [231], pediatric [232] and veterinary [233] medicine has increased the need to develop reliable methods to quantify oxidative stress markers. In the near future, there will be a growing need to standardize the analytical techniques of measurement of oxidative stress biomarkers. Standardization of oxidative stress bioanalytical techniques may offer accurate, reliable, and practical tools for clinicians and researchers working in the field of free radical biology and medicine.

References

1. Davies KJ (1995) Oxidative stress: the paradox of aerobic life. Biochem Soc Symp 61:1–31
2. Thannickal VJ (2003) The paradox of reactive oxygen species: injury, signaling, or both? Am J Physiol Lung Cell Mol Physiol 284:L24–L25
3. Dröge W (2002) Free radicals in the physiological control of cell function. Physiol Rev 82:47–95
4. Haddad JJ (2004) Redox and oxidant-mediated regulation of apoptosis signaling pathways: immuno-pharmaco-redox conception of oxidative siege versus cell death commitment. Int Immunopharmacol 4:475–493
5. Janssen-Heininger YM, Mossman BT, Heintz NH et al (2008) Redox-based regulation of signal transduction: principles, pitfalls, and promises. Free Radic Biol Med. 45:1–17
6. Valko M, Leiter D, Moncol J et al (2007) Free radicals and antioxidants in normal physiological functions and human disease. Int J Biochem Cell Biol 39:44–84
7. Halliwell B, Gutteridge JMC (2007) Free radicals in biology and medicine, 4th edn. Clarendon Press, Oxford
8. Zorov DB, Juhaszova M, Sollott SJ (2006) Mitochondrial ROS-induced ROS release: an update and review. Biochim Biophys Acta 1757:509–517
9. Agarwal A, Gupta S, Sekhon L, Shah R (2008) Redox considerations in female reproductive function and assisted reproduction: from molecular mechanisms to health implications. Antioxid Redox Signal 10:1375–1403
10. Al-Gubory KH, Fowler PA, Garrel C (2010) The roles of cellular reactive oxygen species, oxidative stress and antioxidants in pregnancy outcomes. Int J Biochem Cell Biol 42:1634–1650
11. Myatt L, Cui X (2004) Oxidative stress in the placenta. Histochem Cell Biol 122:369–382
12. Poston L, Raijmakers MTM (2004) Trophoblast oxidative stress, antioxidants and pregnancy outcome-A Review. Placenta 25 (Suppl A):S72–S78
13. Agarwal A, Allamaneni SSR (2004) Role of free radicals in female reproductive diseases and assisted reproduction. Reprod Biomed Online. 9:338–347
14. Agarwal A, Gupta S, Sharma R (2005) Oxidative stress and its implications in female infertility—a clinician's perspective. Reprod Biomed Online. 11:641–650
15. Gupta S, Agarwal A, Banerjee J, Alvarez J (2007) The role of oxidative stress in spontaneous abortion and recurrent pregnancy loss: A systematic review. Obstet Gynecol Survey. 62:335–347
16. Luo ZC, Fraser WD, Julien P et al (2006) Tracing the origins of "fetal origins" of adult diseases: programming by oxidative stress? Med Hypotheses 66:38–44
17. Luo ZC, Liu JM, Fraser WD (2010) Large prospective birth cohort studies on environmental contaminants and child health—Goals, challenges, limitations and needs. Med Hypotheses 74:318–324

18. Barr DB, Needham A, Bishop LL (2007) Concentrations of xenobiotic chemicals in the maternal-fetal unit. Reprod Toxicol 23:260–266
19. McCord I, Fridovich JM (1969) Superoxide dismutase. an enzymatic function for erythrocuprein. J Biol Chem 244:6049–6055
20. Kehrer JP (2000) The Haber-Weiss reaction and mechanisms of toxicity. Toxicology 149:43–50
21. Simon HU, Haj-Yehia A, Levi-Schaffer F (2000) Role of reactive oxygen species (ROS) in apoptosis induction. Apoptosis 5:415–418
22. Higuchi Y (2003) Chromosomal DNA fragmentation in apoptosis and necrosis induced by oxidative stress. Biochem Pharmacol 66:1527–1535
23. Moncada S, Palmer RMJ, Higgs EA (1991) Nitric oxide: physiology, pathophysiology and pharmacology. Pharmacol Rev 43:109–142
24. Koppenol WH, Moreno JJ, Pryor WA, Ischiropoulos H, Beckman JS (1992) Peroxynitrite, a cloaked oxidant formed by nitric oxide and superoxide. Chem Res Toxicol 5:834–842
25. Radi R, Beckman JS, Bush KM, Freeman BA (1991) Peroxynitrite induced membrane lipid peroxidation: the cytotoxic potential of superoxide and nitric oxide. Arch Biochem Biophys 288:481–487
26. Marla SS, Lee J, Groves JT (1997) Peroxynitrite rapidly permeates phospholipids membranes. Proc Natl Acad Sci. 94:14243–14248
27. Stamler JS (1994) Redox signaling: nitrosylation and related target interactions of nitric oxide. Cell 78:931–936
28. Li J, Billiar TR, Talanian RV, Kim YM (1997) Nitric oxide reversibly inhibits seven members of the caspase family via S-nitrosylation. Biochem Biophys Res Commun 240:419–424
29. Mannick JB, Hausladen A, Liu L et al (1999) Fas-induced caspase denitrosylation. Science 284:651–654
30. Kirkinezos IG, Moraes CT (2001) Reactive oxygen species and mitochondrial diseases. Semin Cell Dev Biol 12:449–457
31. Brown GC (1992) Control of respiration and ATP synthesis in mammalian mitochondria and cells. Biochem J. 284:1–13
32. Adam-Vizi V, Chinopoulos C (2006) Bioenergetics and the formation of mitochondrial reactive oxygen species. Trends Pharmacol Sci 27:639–645
33. Turrens JF (2003) Mitochondrial formation of reactive oxygen species. J Physiol 552: 335–344
34. Lacza Z, Puskar M, Figueroa JP et al (2001) Mitochondrial nitric oxide synthase is constitutively active and is functionally upregulated in hypoxia. Free Radic Biol Med. 31:1609–1615
35. Lacza Z, Snipes JA, Zhang J et al (2003) Mitochondrial nitric oxide synthase is not eNOS, nNOS or iNOS. Free Radic Biol Med. 35:1217–1228
36. Yoon SJ, Choi KH, Lee KA (2002) Nitric oxide-mediated inhibition of follicular apoptosis is associated with HSP70 induction and Bax suppression. Mol Reprod Dev 61:504–510
37. Jee BC, Kim SH, Moon SY (2003) The role of nitric oxide on apoptosis in human luteinized granulosa cells. Immunocytochemical evidence. Gynecol Obstet Invest. 56:143–147
38. Brown GC (2007) Nitric oxide and mitochondria. Front Biosci. 12:1024–1033
39. Dröge W (2002) Free radicals in the physiological control of cell function. Physiol Rev 82:47–95
40. Zangar RC, Davydov DR, Verma S (2004) Mechanisms that regulate production of reactive oxygen species by cytochrome P450. Toxicol Appl Pharmacol 199:316–331
41. Jezek P, Hlavatá L (2005) Mitochondria in homeostasis of reactive oxygen species in cell, tissues, and organism. Int J Biochem Cell Biol 37:2478–2503
42. Weisiger RA, Fridovich I (1973) Mitochondrial superoxide dismutase. Site of synthesis and intramitochondrial localisation. J Biol Chem 248:4791–4793
43. Marklund SL (1982) Human copper-containing superoxide dismutase of high molecular weight. Proc Natl Acad Sci. 79:7634–7638

44. Chance B, Sies H, Boveris A (1979) Hydroperoxide metabolism in mammalian organs. Physiol Rev 59:527–605
45. Brigelius-Flohé R (1999) Tissue-specific functions of individual glutathione peroxidases. Free Radic Biol Med. 27:951–965
46. Kirsch M, De Groot H (2001) NAD(P)H, a directly operating antioxidant? FASEP. J 15:1569–1574
47. Pandolfi PP, Sonati F, Rivi R et al (1995) Targeted disruption of the house keeping gene encoding glucose 6-phosphate dehydrogenase (G6PD): G6PD is dispensable for pentose synthesis but essential for defense against oxidative stress. EMBO J 14:5209–5215
48. Jo SH, Son MK, Koh HJ et al (2001) Control of mitochondrial redox balance and cellular defense against oxidative damage by mitochondrial $NADP^+$-dependent isocitrate dehydrogenase. J Biol Chem 276:16168–16176
49. Lee SM, Koh HJ, Park DC et al (2002) Cytosolic NADP(+)-dependent isocitrate dehydrogenase status modulates oxidative damage to cells. Free Radic Biol Med. 32:1185–1196
50. Schafer FQ, Buettner GR (2001) Redox environment of the cell as viewed through the redox state of the glutathione disulfide/glutathione couple. Free Radic Biol Med. 30:1191–1212
51. Dickinson DA, Forman HJ (2002) Glutathione in defense and signaling: lessons from a small thiol. Ann N Y Acad Sci 973:488–504
52. Yildirim Z, Kiliç N, Ozer C et al (2007) Effects of taurine in cellular responses to oxidative stress in young and middle-aged rat liver. Ann N Y Acad Sci 1100:553–561
53. Parvez S, Tabassum H, Banerjee BD, Raisuddin S (2008) Taurine prevents tamoxifen-induced mitochondrial oxidative damage in mice. Basic Clin Pharmacol Toxicol 102: 382–387
54. Métayer S, Seiliez I, Collin A et al (2008) Mechanisms through which sulfur amino acids control protein metabolism and oxidative status. J Nutr Biochem 19:207–215
55. Dröge W (2005) Oxidative stress and ageing: is ageing a cysteine deficiency syndrome? Philos Trans R Soc. 360:2355–2372
56. Elshorbagy AK, Nurk E, Gjesdal CG et al (2008) Homocysteine, cysteine, and body composition in the Hordaland Homocysteine Study: does cysteine link amino acid and lipid metabolism? Am J Clin Nutr 88:738–746
57. Stipanuk MH (2004) Role of the liver in regulation of body cysteine and taurine levels: a brief review. Neurochem Res 29:105–110
58. Ubuka T, Okada A, Nakamura H (2008) Production of hypotaurine from L-cysteinesulfinate by rat liver mitochondria. Amino Acids 35:53–58
59. Nishimura T, Sai Y, Fujii J et al (2010) Roles of TauT and system A in cytoprotection of rat syncytiotrophoblast cell line exposed to hypertonic stress. Placenta 31:1003–10009
60. Whiteman M, Armstrong JS, Chu SH et al (2004) The novel neuromodulator hydrogen sulfide: an endogenous peroxynitrite scavenger? J Neurochem 90:765–768
61. Ali MY, Ping CY, Mok YY et al (2006) Regulation of vascular nitric oxide in vitro and in vivo; a new role for endogenous hydrogen sulphide? Br J Pharmacol 149:625–634
62. Li L, Moore PK (2007) An overview of the biological significance of endogenous gases: new roles for old molecules. Biochem Soc Trans 35:1138–1141
63. Elrod JW, Calvert JW, Morrison J et al (2007) Hydrogen sulfide attenuates myocardial ischemia-reperfusion injury by preservation of mitochondrial function. Proc Natl Acad Sci U S A. 104:15560–15565
64. Szabo C (2007) Hydrogen sulphide and its therapeutic potential. Nat Rev 6:917–935
65. Zhao W, Ndisang JF, Wang R (2003) Modulation of endogenous production of H_2S in rat tissues. Can J Physiol Pharmacol 81:848–853
66. Patel P, Vatish M, Heptinstall J et al (2009) The endogenous production of hydrogen sulphide in intrauterine tissues. Reprod Biol Endocrinol 7:10
67. Szabó C (2007) Hydrogen sulphide and its therapeutic potential. Nat Rev Drug Discov 6:917–935

68. Kimura H. Hydrogen sulfide: its production, release and functions. Amino Acids. 2010 Feb 27 [Epub ahead of print]
69. Kimura Y, Goto Y, Kimura H (2010) Hydrogen sulfide increases glutathione production and suppresses oxidative stress in mitochondria. Antioxid Redox Signal 12:1–13
70. Tyagi N, Moshal KS, Sen U et al (2009) H2S protects against methionine-induced oxidative stress in brain endothelial cells. Antioxid Redox Signal 11:25–33
71. Yin WL, He JQ, Hu B et al (2009) Hydrogen sulfide inhibits MPP(+)-induced apoptosis in PC12 cells. Life Sci 85:269–275
72. Mishra PK, Tyagi N, Sen U et al (2010) H2S ameliorates oxidative and proteolytic stresses and protects the heart against adverse remodeling in chronic heart failure. Am J Physiol Heart Circ Physiol 298:H451–H456
73. Young AJ, Phillip D, Lowe GL. In: Krinsky NI, Mayne ST, Sies H, eds. (2004) Carotenoid antioxidant activity. Carotenoids in Health and Disease. Marcel Dekker Inc, New York, 105–126 (Chap. 5)
74. Young AJ, Lowe GM (2001) Antioxidant and prooxidant properties of carotenoids. Arch Biochem Biophys 385:20–27
75. Bramley PM, Elmadfa I, Kafatos A et al (2000) Vitamin E. J Sci Food Agric 80:913–938
76. Johnson LJ, Meacham SL, Kruskall LJ (2003) The antioxidants—vitamin C, vitamin E, selenium, and carotenoids. J Agromedicine. 9:65–82
77. Debier C, Larondelle Y (2005) Vitamins A and E: metabolism, roles and transfer to offspring. Br J Nutr 93:153–174
78. Seifried HE, Anderson DE, Fisher EI, Milner JA (2007) A review of the interaction among dietary antioxidants and reactive oxygen species. J Nutr Biochem 18:567–579
79. Stahl W, Nicolai S, Briviba K et al (1997) Biological activities of natural and synthetic carotenoids: induction of gap junctional communication and singlet oxygen quenching. Carcinogenesis 18:89–92
80. Machlin LJ, Bendich A (1987) Free radical tissue damage: protective role of antioxidant nutrients. FASEB J 1:441–445
81. Herrera E, Barbas C (2001) Vitamin E: action, metabolism and perspectives. J Physiol Biochem 57:43–56
82. Tappel A, Zalkin H (1960) Inhibition of lipid peroxidation in microsomes by vitamin E. Nature 185:35
83. Valk EE, Hornstra G (2000) Relationship between vitamin E requirement and polyunsaturated fatty acid intake in man: a review. Int J Vitam Nutr Res 70:31–42
84. Gavazza MB, Catalá A (2006) The effect of alpha-tocopherol on lipid peroxidation of microsomes and mitochondria from rat testis. Prostaglandins Leukot Essent Fatty Acids 74:247–254
85. Traber MG, Atkinson J (2007) Vitamin E, antioxidant and nothing more. Free Radic Biol Med 43:4–15
86. Mertz W (1981) The essential trace elements. Science 213:1332–1338
87. Scalbert A, Johnson IT, Saltmarsh M (2005) Polyphenols: antioxidants and beyond. Am J Clin Nutr 81(1 Suppl):215S–217S
88. Rahman I, Biswas SK, Kirkham PA (2006) Regulation of inflammation and redox signaling by dietary polyphenols. Biochem Pharmacol 72:1439–1452
89. Wilding M, Dale B, Marino M et al (2011) Mitochondrial aggregation patterns and activity in human oocytes and preimplantation embryos. Hum Reprod 16:909–917
90. Smith LC, Thundathil J, Filion F (2005) Role of mitochondrial genome in preimplantation development and assisted reproductive techniques. Reprod Fert Develop 17:15–22
91. Dumollard R, Carroll J, Duchen MR, Campbell K, Swann K (2009) Mitochondrial function and redox state in mammalian embryos. Semin Cell Dev Biol 20:346–353
92. Dennery PA (2007) Effects of oxidative stress on embryonic development. Birth Defects Res Part C Embryo Today 81:155–162
93. Dumollard R, Duchen M, Carroll J (2007) The role of mitochondrial function in the oocyte and embryo. Curr Top Dev Biol 77:21–49

94. Bartmann AK, Romão GS (2004) Ramos Eda S, Ferriani RA. Why do older women have poor implantation rates? A possible role of the mitochondria. J Assist Reprod Genet 21: 79–83
95. Pierce GB, Parchment RE, Lewellyn AL (1991) Hydrogen peroxide as a mediator of programmed cell death in the blastocyst. Differentiation. 46:181–186
96. Nasr-Esfahani MH, Aitken JR, Johnson MH (1990) Hydrogen peroxide levels in mouse oocytes and early cleavage stage embryos developed in vitro or in vivo. Development 109:501–507
97. Gagioti S, Colepicolo P, Bevilacqua E (1995) Post-implantation mouse embryos have the capability to generate and release reactive oxygen species. Reprod Fertil Dev 7:1111–1116
98. Thomas M, Jain S, Kumar GP, Laloraya MA (1997) Programmed oxyradical burst causes hatching of mouse blastocysts. J Cell Sci 110:1597–1602
99. Laloraya M, Kumar GP, Laloraya MM (1991) Changes in the superoxide radical and superoxide dismutase levels in the uterus of Rattus norvegicus during the estrous cycle and a possible role for superoxide radical in uterine oedema and cell proliferation at proestrus. Biochem Cell Biol 69:313–316
100. Laloraya M, Kumar GP, Laloraya MM (1989) A possible role of superoxide anion radical in the process of blastocyst implantation in Mus musculus. Biochem Biophys Res Commun 161:762–770
101. Jain S, Saxena D, Kumar GP, Laloraya M (2000) NADPH dependent superoxide generation in the ovary and uterus of mice during estrous cycle and early pregnancy. Life Sci 66:1139–1146
102. Harvey MB, Arcellana-Panlilio MY, Zhang X, Schultz GA, Watson AJ (1995) Expression of genes encoding antioxidant enzymes in preimplantation mouse and cow embryos and primary bovine oviduct cultures employed for embryo coculture. Biol Reprod 53:532–540
103. El Mouatassim S, Guérin P, Ménézo Y (1999) Expression of genes encoding antioxidant enzymes in human and mouse oocytes during the final stages of maturation. Mol Hum Reprod 5:720–725
104. Guérin P, El Mouatassim S, Ménézo Y (2001) Oxidative stress and protection against reactive oxygen species in the pre-implantation embryo and its surroundings. Hum Reprod 7:175–189
105. Orsi NM, Leese HJ (2001) Protection against reactive oxygen species during mouse pre-implantation embryo development: role of EDTA, oxygen tension, catalase, superoxide dismutase and pyruvate. Mol Reprod Dev 59:44–53
106. Blomberg LA, Long EL, Sonstegard TS et al (2005) Serial analysis of gene expression during elongation of the peri-implantation porcine trophectoderm (conceptus). Physiol Genomics 20:188–194
107. Al-Gubory KH, Bolifraud P, Garrel C (2008) Regulation of key antioxidant enzymatic systems in the sheep endometrium by ovarian steroids. Endocrinology 149:4428–4434
108. Takehara Y, Yoshioka T, Sasaki J (1990) Changes in the levels of lipoperoxide and antioxidant factors in human placenta during gestation. Acta Med Okayama 44:103–111
109. Qanungo S, Sen A, Mukherjea M (1999) Antioxidant status and lipid peroxidation in human feto-placental unit. Clin Chim Acta 285:1–12
110. Qanungo S, Mukherjea M (2000) Ontogenic profile of some antioxidants and lipid peroxidation in human placental and fetal tissues. Mol Cell Biochem 215:11–19
111. Garrel C, Fowler PA, Al-Gubory KH (2010) Developmental changes in antioxidant enzymatic defences against oxidative stress in sheep placentomes. J Endocrinol 205:107–116
112. Al-Gubory KH, Garrel C, Delatouche L et al (2010) Antioxidant adaptive responses of extraembryonic tissues from cloned and non-cloned bovine conceptuses to oxidative stress during early pregnancy. Reproduction 140:175–181
113. Fall CH, Yajnik CS, Rao S et al (2003) Micronutrients and fetal growth. J Nutr 133(Suppl 2):1747S–1756S

114. Ashworth CJ, Antipatis C (2001) Micronutrient programming of development throughout pregnancy. Reproduction 122:527–535
115. Moore VM, Davies MJ (2005) Diet during pregnancy, neonatal outcomes and later health. Reprod Fertil Dev 17:341–348
116. Osorio JC, Cruz E, Milanés M et al (2011) Influence of maternal redox status on birth weight. Reprod Toxicol 31:35–40
117. Keen CL, Hanna LA, Lanoue L et al (2003) Developmental consequences of trace mineral deficiencies in rodents: acute and long-term effects. J Nutr 133:1477S–1480S
118. Gamsbling L, McArdle HJ (2004) Iron, copper and fetal development. Proc Nutr Soc 63:553–562
119. Uriu-Adams JY, Keen CL (2010) Zinc and reproduction: effects of zinc deficiency on prenatal and early postnatal development. Birth Defects Res B Dev Reprod Toxicol 89:313–325
120. Kharb S (2000) Vitamin E and C in preeclampsia. Eur J Obstet Gynecol Reprod Biol 93:37–39
121. Lee BE, Hong YC, Lee KH et al (2004) Influence of maternal serum levels of vitamins C and E during the second trimester on birth weight and length. Eur J Clin Nutr 58:1365–1371
122. Atamer Y, Koçyigit Y, Yokus B et al (2005) Lipid peroxidation, antioxidant defense, status of trace metals and leptin levels in preeclampsia. Eur J Obstet Gynecol Reprod Biol 119:60–66
123. Ahn YM, Kim YJ, Park H et al (2007) Prenatal vitamin C status is associated with placental apoptosis in normal-term human pregnancies. Placenta 28:31–38
124. Jansson T, Powell TL (2007) Role of the placenta in fetal programming: underlying mechanisms and potential interventional approaches. Clinica Science 113:1–13
125. Hawk SN, Lanoue L, Keen CL et al (2003) Copper-deficient rat embryos are characterized by low superoxide dismutase activity and elevated superoxide anions. Biol Reprod 68:896–903
126. Keen CL, Clegg MS, Hanna LA et al (2003) The plausibility of micronutrient deficiencies being a significant contributing factor to the occurrence of pregnancy complications. J Nutr 133:1597S–1605S
127. Gow AJ, Farkouh CR, Munson DA et al (2004) Biological significance of nitric oxide-mediated protein modifications. Am J Physiol Lung Cell Mol Physiol 287:L262–L268
128. Bagi Z, Toth E, Koller A, Kaley G (2004) Microvascular dysfunction after transient high glucose is caused by superoxide-dependent reduction in the bioavailability of NO and BH(4). Am J Physiol Heart Circ Physiol 287:H626–H633
129. Tiboni GM, Giampietro F, Di Giulio C (2003) The nitric oxide synthesis inhibitor N{omega}-nitro-L-arginine methyl ester (l-NAME) causes limb defects in mouse fetuses: protective effect of acute hyperoxia. Pediatr Res 54:69–76
130. Pathak P, Kapil U (2004) Role of trace elements zinc, copper and magnesium during pregnancy and its outcome. Indian J Pediatr 71:1003–1005
131. Gupta P, Narang M, Banerjee BD, Basu S (2004) Oxidative stress in term small for gestational age neonates born to undernourished mothers: a case control study. BMC Pediatr 4:14–23
132. Jenkins C, Wilson R, Roberts J et al (2000) Antioxidants: their role in pregnancy and miscarriage. Antioxidant Redox Signal 2:623–628
133. Cederberg J, Simán CM, Eriksson UJ (2001) Combined treatment with vitamin E and vitamin C decreases oxidative stress and improves fetal outcome in experimental diabetic pregnancy. Pediatr Res 49:755–762
134. Cederberg J, Eriksson UJ (2005) Antioxidative treatment of pregnant diabetic rats diminishes embryonic dysmorphogenesis. Birth Defects Res A Clin Mol Teratol 73:498–505
135. Jishage K, Arita M, Igarashi K et al (2001) α-Tocopherol transfer protein is important for the normal development of placental labyrinth trophoblasts in mice. J Biol Chem 276:1669–1672

136. Brigelius-Flohé R, Kelly FJ, Salonen JT et al (2002) The European perspective on vitamin E: current knowledge and future research. Am J Clin Nutr 76:703–716
137. Tarin JJ, Perez-Albala S, Pertusa JF, Cano A (2002) Oral administration of pharmacological doses of vitamins C and E reduces reproductive fitness and impairs the ovarian and uterine functions of female mice. Theriogenology 57:1539–1550
138. Rumbold A, Crowther CA (2005) Vitamin C supplementation in pregnancy Cochrane database Syst Rev Issue 1. Art. No.: CD004072. doi: 10.1002/14651858.CD004072.pub2
139. Rumbold A, Crowther CA (2005) Vitamin E supplementation in pregnancy. Cochrane Database Syst Rev Issue 2. Art. No.: CD004069. doi: 10.1002/14651858.CD004069.pub2
140. Cetin I, Berti C, Calabrese S (2010) Role of micronutrients in the periconceptional period. Hum Reprod Update 16:80–95
141. Boskovic R, Gargaun L, Oren D et al (2005) Pregnancy outcome following high Vitamin E supplementation. Reprod Toxicol 20:85–88
142. Beazley D, Ahokas R, Livingston J et al (2005) Vitamin C and E supplementation in women at high risk for preeclampsia: a double-blind, placebo-controlled trial. Am J Obstet Gynaecol 192:520–521
143. Debier C (2007) Vitamin E during pre- and postnatal periods. Vitm Horm 76:357–373
144. Uusitalo L, Kenward MG, Virtanen SM et al (2008) Intake of antioxidant vitamins and trace elements during pregnancy and risk of advanced beta cell autoimmunity in the child. Am J Clin Nutr 88:458–464
145. Astley SB, Lindsay DG (2002) European Research on the Functional Effects of Dietary Antioxidants (EUROFEDA). Conclusions Mol Aspects Med 23:287–291
146. Visioli F (2000) Antioxidants in Mediterranean diets. World Rev Nutr Diet 87:43–55
147. Sköldstam L, Hagfors L, Johansson G (2003) An experimental study of a Mediterranean diet intervention for patients with rheumatoid arthritis. Ann Rheum Dis 62:208–214
148. Trichopoulou A, Costacou T, Bamia C, Trichopoulos D (2003) Adherence to a Mediterranean diet and survival in a Greek population. N Engl J Med 348:2599–2608
149. Mitrou PN, Kipnis V, Thiebaut AC et al (2007) Mediterranean dietary pattern and prediction of all-cause mortality in a US population: results from the NIH-AARP Diet and Health Study. Arch Intern Med 167:2461–2468
150. Stachowska E, Wesołowska T, Olszewska M et al (2005) Elements of Mediterranean diet improve oxidative status in blood of kidney graft recipients. Br J Nutr 93:345–352
151. Fitó M, Guxens M, Corella D et al (2007) Effect of a traditional Mediterranean diet on lipoprotein oxidation: a randomized controlled trial. Arch Intern Med 167:1195–1203
152. Dai J, Jones DP, Goldberg J et al (2008) Association between adherence to the Mediterranean diet and oxidative stress. Am J Clin Nutr 88:1364–1370
153. Sofi F, Abbate R, Gensini G et al (2008) Adherence to Mediterranean diet and health status: meta-analysis. BMJ 337–344
154. Martinez-Gonzalez MA, Bes-Rastrollo M, Serra-Majem L et al (2009) Mediterranean food pattern and the primary prevention of chronic disease: recent developments. Nutr Rev 67(Suppl 1):S111–S116
155. Vujkovic M, de Vries JH, Lindemans J et al (2010) The preconception Mediterranean dietary pattern in couples undergoing in vitro fertilization/intracytoplasmic sperm injection treatment increases the chance of pregnancy. Fertil Steril 94:2096–2101
156. Barger MK (2010) Maternal nutrition and perinatal outcomes. J Midwifery Womens Health 55:502–511
157. Bateson P, Barker D, Clutton-Brock T et al (2004) Developmental plasticity and human health. Nature 430:419–421
158. Miller KP, Borgeest C, Greenfeld C et al (2004) In utero effects of chemicals on reproductive tissues in females. Toxicol Appl Pharmacol 198:111–131
159. Sanderson JT (2006) The steroid hormone biosynthesis pathway as a target for endocrine-disrupting chemicals. Toxicol Sci 94:3–21
160. Buczyńska A, Tarkowski S (2005) Environmental exposure and birth outcomes. Int J Occup Med Environ Health 18:225–232

161. Nicol CJ, Zielenski J, Tsui LC, Wells PG (2000) An embryoprotective role for glucose-6-phosphate dehydrogenase in developmental oxidative stress and chemical teratogenesis. FASEB J 14:111–127
162. Fowler PA, Dorà NJ, McFerran H et al (2008) In utero exposure to low doses of environmental pollutants disrupts fetal ovarian development in sheep. Mol Hum Reprod 14:269–280
163. Wells PG, Bhuller Y, Chen CS et al (2005) Molecular and biochemical mechanisms in teratogenesis involving reactive oxygen species. Toxicol Appl Pharmacol 207(2 Suppl):354–366
164. Wan J, Winn LM (2006) In utero-initiated cancer: the role of reactive oxygen species. Birth Defects Res C Embryo Today. 78:326–332
165. Davis JM, Auten RL (2010) Maturation of the antioxidant system and the effects on preterm birth. Semin Fetal Neonatal Med 15:191–195
166. Takahashi O, Oishi S (2000) Disposition of orally administered 2,2-Bis(4-hydroxyphenyl) propane (Bisphenol A) in pregnant rats and the placental transfer to fetuses. Environ Health Perspect 108:931–935
167. Shin BS, Yoo SD, Cho CY et al (2002) Maternal-fetal disposition of bisphenol a in pregnant Sprague-Dawley rats. J Toxicol Environ Health A 65:395–406
168. Zalko D, Soto AM, Dolo L et al (2003) Biotransformations of bisphenol A in a mammalian model: answers and new questions raised by low-dose metabolic fate studies in pregnant CD1 mice. Environ Health Perspect 111:309–319
169. Maffini MV, Rubin BS, Sonnenschein C, Soto AM (2006) Endocrine disruptors and reproductive health: the case of bisphenol-A. Mol Cell Endocrinol 254–255:179–186
170. Vandenberg LN, Hauser R, Marcus M et al (2007) Human exposure to bisphenol A (BPA). Reprod Toxicol 24:139–177
171. Le HH, Carlson EM, Chua JP, Belcher SM (2008) Bisphenol A is released from polycarbonate drinking bottles and mimics the neurotoxic actions of estrogen in developing cerebellar neurons. Toxicol Lett 176:149–156
172. Kabuto H, Hasuike S, Minagawa N, Shishibori T (2003) Effects of bisphenol A on the metabolisms of active oxygen species in mouse tissues. Environ Res 93:31–35
173. Kabuto H, Amakawa M, Shishibori T (2004) Exposure to bisphenol A during embryonic/fetal life and infancy increases oxidative injury and causes underdevelopment of the brain and testis in mice. Life Sci 74:2931–2940
174. Rubin BS, Murray MK, Damassa DA et al (2001) Perinatal exposure to low doses of bisphenol A affects body weight, patterns of estrous cyclicity, and plasma LH levels. Environ Health Perspect 109:675–680
175. Markey CM, Wadia PR, Rubin BS et al (2005) Long-term effects of fetal exposure to low doses of the xenoestrogen bisphenol-A in the female mouse genital tract. Biol Reprod 72:1344–1351
176. Savabieasfahani M, Kannan K, Astapova O et al (2006) Developmental programming: differential effects of prenatal exposure to bisphenol-A or methoxychlor on reproductive function. Endocrinology 147:5956–5966
177. Berger RG, Foster WG, de Catanzaro D (2010) Bisphenol-A exposure during the period of blastocyst implantation alters uterine morphology and perturbs measures of estrogen and progesterone receptor expression in mice. Reprod Toxicol 30:393–400
178. Huang H, Leung LK (2009) Bisphenol A downregulates CYP19 transcription in JEG-3 cells. Toxicol Lett 189:248–252
179. Csapo AI, Pulkkinen MO, Ruttner B, Sauvage JP, Wiest WG (1972) The significance of the human corpus luteum in pregnancy maintenance. I. Preliminary studies. Am J Obstet Gynecol 112:1061–1067
180. Al-Gubory KH, Solari A, Mirman B (1999) Effects of luteectomy on the maintenance of pregnancy, circulating progesterone concentrations and lambing performance in sheep. Reprod Fertil Dev 11:317–322

181. Behrman HR, Aten RF (1991) Evidence that hydrogen peroxide blocks hormone-sensitive cholesterol transport into mitochondria of rat luteal cells. Endocrinology 128:2958–2966
182. Thompson J, Bannigan J (2008) Cadmium: toxic effects on the reproductive system and the embryo. Reprod Toxicol 25:304–315
183. Dreiem A, Gertz CC, Seegal RF (2005) The effects of methylmercury on mitochondrial function and reactive oxygen species formation in rat striatal synaptosomes are age-dependent. Toxicol Sci 87:156–162
184. Uzbekov MG, Bubnova NI, Kulikova GV (2007) Effect of prenatal lead exposure on superoxide dismutase activity in the brain and liver of rat fetuses. Bull Exp Biol Med 144:783–785
185. Chater S, Douki T, Favier A, Garrel C et al (2008) Influence of static magnetic field on cadmium toxicity: study of oxidative stress and DNA damage in pregnant rat tissues. Electromagn Biol Med 27:393–401
186. Chater S, Douki T, Garrel C et al (2008) Cadmium-induced oxidative stress and DNA damage in kidney of pregnant female rats. C R Biol 331:426–432
187. Stringari J, Nunes AK, Franco JL et al (2008) Prenatal methylmercury exposure hampers glutathione antioxidant system ontogenesis and causes long-lasting oxidative stress in the mouse brain. Toxicol Appl Pharmacol 227:147–154
188. Herrera E (2002) Implications of dietary fatty acids during pregnancy on placental, fetal and postnatal development—a review. Placenta 23(Suppl A):S9–S19
189. Alessandri JM, Guesnet P, Vancassel S et al (2004) Polyunsaturated fatty acids in the central nervous system: evolution of concepts and nutritional implications throughout life. Reprod Nutr Dev 44:509–538
190. Mccann JC, Ames BN (2005) Is docosahexaenoic acid, an n-3 long-chain polyunsaturated fatty acid, required for development of normal brain function? An overview of evidence from cognitive and behavioral tests in humans and animals. Am J Clin Nutr 82:281–295
191. Innis SM (2007) Dietary (n3) fatty acids and brain development. J Nutr 137:855–859
192. Wathes DC, Abayasekara DR, Aitken RJ (2007) Polyunsaturated fatty acids in male and female reproduction. Biol Reprod 77:190–201
193. Coyne GS, Kenny DA, Childs S, Sreenan JM, Waters SM (2008) Dietary n-3 polyunsaturated fatty acids alter the expression of genes involved in prostaglandin biosynthesis in the bovine uterus. Theriogenology 70:772–782
194. Al-Ardhi FM, Al-Ani MR (2008) Maternal fish consumption and prenatal methylmercury exposure: a review. Nutr Health 19:289–297
195. Davidson PW, Strain JJ, Myers GJ et al (2008) Neurodevelopmental effects of maternal nutritional status and exposure to methylmercury from eating fish during pregnancy. Neurotoxicology 29:767–775
196. Sharp DS, Eskenazi B (1986) Delayed health hazards of pesticide exposure. Ann Rev Public Health. 7:441–471
197. Bengtsson S, Berglöf T, Kylin H (2007) Near infrared reflectance spectroscopy as a tool to predict pesticide sorption in soil. Bull Environ Contam Toxicol 78:295–298
198. Anderson B, Phillips B, Hunt J et al (2011) Pesticide and toxicity reduction using an integrated vegetated treatment system. Environ Toxicol Chem. doi:10.1002/etc.471
199. Keikotlhaile BM, Spanoghe P, Steurbaut W (2010) Effects of food processing on pesticide residues in fruits and vegetables: a meta-analysis approach. Food Chem Toxicol 48:1–6
200. Valavanidis A, Vlahogianni T, Dassenakis M, Scoullos M (2006) Molecular biomarkers of oxidative stress in aquatic organisms in relation to toxic environmental pollutants. Ecotoxicol Environ Saf 64:178–189
201. Slaninova A, Smutna M, Modra H, Svobodova Z (2009) A review: oxidative stress in fish induced by pesticides. Neuro Endocrinol Lett. 30:2–12
202. Koner BC, Banerjee BD, Ray A (1998) Organochlorine pesticide-induced oxidative stress and immune suppression in rats. Indian J Exp Biol 36:395–398
203. Sahoo A, Samanta L, Chainy GB (2000) Mediation of oxidative stress in HCH-induced neurotoxicity in rat. Arch Environ Contam Toxicol 39:7–12

204. Ranjbar A, Pasalar P, Abdollahi M (2002) Induction of oxidative stress and acetylcholinesterase inhibition in organophosphorous pesticide manufacturing workers. Hum Exp Toxicol 21:179–182
205. Shadnia S, Azizi E, Hosseini R et al (2005) Evaluation of oxidative stress and genotoxicity in organophosphorus insecticide formulators. Hum Exp Toxicol 24:439–445
206. Rastogi SK, Satyanarayan PV, Ravishankar D, Tripathi S (2009) A study on oxidative stress and antioxidant status of agricultural workers exposed to organophosphorus insecticides during spraying. Indian J Occup Environ Med. 13:131–134
207. Falcon M, Olive J, Osuna E et al (2004) HCH and DDT residues in human placentas in Murcia (Spain). Toxicology 195:203–208
208. Pathak R, Suke SG, Ahmed RS et al (2008) Endosulfan and other organochlorine pesticide residues in maternal and cord blood in North Indian population. Bull Environ Contam Toxicol 81:216–219
209. Pathak R, Suke SG, Ahmed T et al (2010) Organochlorine pesticide residue levels and oxidative stress in preterm delivery cases. Hum Exp Toxicol 29:351–358
210. Mann M, Jensen ON (2003) Proteomic analysis of post-translational modifications. Nat Biotechnol 21:255–261
211. Hathout Y (2007) Approaches to the study of the cell secretome. Expert Rev Proteomics 4:239–248
212. Shankar R, Gude N, Cullinane F et al (2005) An emerging role for comprehensive proteome analysis in human pregnancy research. Reproduction 129:685–696
213. Buhimschi IA, Buhimschi CS (2008) Proteomics of the amniotic fluid in assessment of the placenta. Relevance for preterm birth. Placenta 29 (Suppl A):S95–S101
214. Di Quinzio MK, Georgiou HM, Holdsworth-Carson SJ et al (2008) Proteomic analysis of human cervico-vaginal fluid displays differential protein expression in association with labor onset at term. J Proteome Res 7:1916–1921
215. Ferrero S, Gillott DJ, Remorgida V et al (2007) Proteomic analysis of peritoneal fluid in women with endometriosis. J Proteome Res 6:3402–3411
216. Fowler PA, Tattum J, Bhattacharya S et al (2007) An investigation of the effects of endometriosis on the proteome of human eutopic endometrium: a heterogeneous tissue with a complex disease. Proteomics 7:130–142
217. Buhimschi IA, Zhao G, Funai EF, et al. (2008) Proteomic profiling of urine identifies specific fragments of SERPINA1 and albumin as biomarkers of preeclampsia. Am J Obstet Gynecol 199:551.e1–551.e16
218. Romero R, Espinoza J, Rogers WT et al (2008) Proteomic analysis of amniotic fluid to identify women with preterm labor and intra-amniotic inflammation/infection: the use of a novel computational method to analyze mass spectrometric profiling. J Matern Fetal Neonatal Med 21:367–388
219. Katz-Jaffe MG, McReynolds S, Gardner DK, Schoolcraft WB (2009) The role of proteomics in defining the human embryonic secretome. Mol Hum Reprod 15:271–277
220. Hoang VM, Foulk R, Clauser K, Burlingame A, Gibson BW, Fisher SJ (2001) Functional proteomics: examining the effects of hypoxia on the cytotrophoblast protein repertoire. Biochemistry 40:4077–4086
221. Zhang Y, Zhang YL, Feng C et al (2008) Comparative proteomic analysis of human placenta derived from assisted reproductive technology. Proteomics 8:4344–4356
222. Domínguez F, Garrido-Gómez T, López JA et al (2009) Proteomic analysis of the human receptive versus non-receptive endometrium using differential in-gel electrophoresis and MALDI-MS unveils stathmin 1 and annexin A2 as differentially regulated. Hum Reprod 24:2607–2617
223. Arianmanesh M, McIntosh R, Lea RG, Fowler PA, Al-Gubory KH (2011) Ovine corpus luteum proteins, with functions including oxidative stress and lipid metabolism, show complex alterations during implantation. J Endocrinol (2011 in press)
224. Wang J, Li D, Dangott LJ, Wu G (2006) Proteomics and its role in nutrition research. J Nutr 136:1759–1762

225. Franco Mdo C, Ponzio BF, Gomes GN et al (2009) Micronutrient prenatal supplementation prevents the development of hypertension and vascular endothelial damage induced by intrauterine malnutrition. Life Sci 85:327–333
226. Hennig B, Ettinger AS, Jandacek RJ et al (2007) Using nutrition for intervention and prevention against environmental chemical toxicity and associated diseases. Environ Health Perspect 115:493–495
227. Wolfe KL, Kang X, He X et al (2008) Cellular antioxidant activity of common fruits. J Agric Food Chem 56:8418–8426
228. Song W, Derito CM, Liu MK et al (2010) Cellular antioxidant activity of common vegetables. J Agric Food Chem 58:6621–6629
229. Limón-Pacheco J, Gonsebatt ME (2009) The role of antioxidants and antioxidant-related enzymes in protective responses to environmentally induced oxidative stress. Mutat Res 674:137–147
230. Dalle-Donne I, Scaloni A, Giustarini D et al (2005) Proteins as biomarkers of oxidative/ nitrosative stress in diseases: the contribution of redox proteomics. Mass Spectrom Rev 24:55–99
231. Dalle-Donne I, Rossi R, Colombo R, Giustarini D, Milzani A (2006) Biomarkers of oxidative damage in human disease. Clin Chem 52:601–623
232. Tsukahara H (2007) Biomarkers for oxidative stress: clinical application in pediatric medicine. Curr Med Chem 14:339–351
233. Celi P (2010) Biomarkers of oxidative stress in ruminant medicine. Immunopharmacol Immunotoxicol. 2010 Sep 18 [Epub ahead of print]

Chapter 2
Methods for Detection of ROS in the Female Reproductive System

Rakesh K. Sharma, Nathan Reynolds, Mitali Rakhit and Ashok Agarwal

Abstract The role of reactive oxygen species (ROS) within the female reproductive system is complex and can contribute to multiple gynecological diseases including infertility. This chapter will describe the various methods available to measure both ROS and other markers of oxidative stress in female infertility. Methods including chemiluminescence, flow cytometry, ELISA, metabolomics that utilize various markers of oxidative stress will be discussed. The effects of these markers in various female diseases are also briefly described.

Keywords Methods for detection · Reactive oxygen species (ROS) · Female Reproductive System · Chemiluminescence · Flow Cytometry · ELISA

R. K. Sharma (✉) · M. Rakhit
Center For Reproductive Medicine, Cleveland Clinic, Desk A19.1
9500 Euclid Avenue, Cleveland, OH 44195, USA
e-mail: sharmar@ccf.org

M. Rakhit
e-mail: mitali.rakhit@gmail.com

N. Reynolds
Biological Sciences, Western Michigan University, 1903 W. Michigan Avenue,
Kalamazoo, MI 49009, USA
e-mail: nathan.m.reynolds@wmich.edu

A. Agarwal
 Center for Reproductive Medicine, Lerner College of Medicine,
Cleveland Clinic, 9500 Euclid Avenue, Cleveland, OH 44195, USA
e-mail: agarwaa@ccf.org

2.1 Introduction

Free radicals are molecules that contain one or more unpaired electrons in their outer shell. They are very unstable and reactive, and attempt to acquire an electron from the surrounding molecule and trigger a chain reaction. Oxygen radicals, such as the superoxide anion (O_2^-), the hydroxyl radical (OH·), hydrogen peroxide (H_2O_2), hypochlorite radical (OHCl·), and the peroxyl radical (ROO·) comprise the highly reactive group of oxygen species called reactive oxygen species (ROS). Also included in this group are the reactive nitrogen radicals, such as nitric oxide (NO·.) and nitric dioxide (NO_2), peroxynitrite anion, and nitroxyl ion [1–5].

ROS are involved in various signal transduction pathways and act as second messengers. Physiological amounts of ROS are necessary for healthy body function. A delicate balance exists between ROS and antioxidants throughout the female reproductive system. Antioxidants are able to break the cycle by donating an electron to the free radical and stabilizing it without destabilizing itself. When this balance is disrupted due to excessive production of ROS or the inability of the antioxidants to neutralize ROS, it results in oxidative stress (OS). OS can damage fats, lipids, nucleic acids, and proteins which are the fundamental building blocks of life. OS is involved in the pathophysiology of a number of diseases of the female reproductive system such as endometriosis, polycystic ovary syndrome, preeclampsia maternal diabetes, and recurrent pregnancy loss [2, 6–12].

2.2 Measurement of Reactive Oxygen Species and Oxidative Stress Markers

A number of OS biomarkers have been investigated in various fluids such as peritoneal fluid, follicular fluid, amniotic fluid, and hydrosalpingeal fluid. ROS is the initial marker and a number of other markers are available to measure the end product of ROS-induced damage such as lipid peroxidation, oxidation of proteins, and DNA damage. Some of the more common methods are described below.

2.3 Measurement of ROS by Chemiluminescence Assay

In this assay, a probe such as luminol (5-amino-2,3 dihydro-1,4 phthalazinedione) is added to the target fluid that reacts with any ROS present in the biofluids. The byproduct of this reaction is the production of photons of light [3, 12]. Luminol reacts with free radicals within the fluid including O_2^-, OH and hydrogen peroxide. Luminol reacts quickly with both intracellular and intercellular free radicals. However, it does not differentiate between the types of free radicals and therefore measures global ROS. The results are expressed in relative light units (RLU), counted photons per minute (cpm), or milliVolts/sec [12].

Herein, we describe the details of measurement of ROS by chemiluminescence assay in unprocessed or neat sample i.e. without any further processing of the sample.

1. Equipment and material

    (a) Disposable polystyrene tubes with caps (12 × 75 mm)
    (b) Eppendorf pipettes (5, 10, 50 and 1000 μL)
    (c) Serological Pipettes (1, 2 mL)
    (d) Desktop centrifuge
    (e) Dimethyl Sulfoxide (DMSO; Catalog # D8779, Sigma Chemical Co., St. Louis, MO)
    (f) Luminol (5-amino-2,3 dihydro-1,4 phthalazinedione; Catalog # A8511, Sigma Chemical Co., St. Louis, MO)
    (g) Dulbecco's Phosphate buffered saline solution 1X (PBS-1X; Catalog # 9235, Irvine Scientific, Santa Ana, CA)
    (h) Luminometer (Model: Berthold Technologies, Autolumat plus LB 953, Oakridge, TN)

2. Preparation of reagents

    (a) *Stock Luminol (100 mM)* Weigh 177.09 mg of luminol and add it to 10 mL of DMSO solution in a polystyrene tube. The tube must be covered in an aluminum foil due to the sensitivity of the luminol to light. It can be stored at room temperature in the dark.
    (b) *Working Luminol (5 mM)* Mix 20 μL luminol stock solution with 380 μL DMSO in a foil-covered polystyrene tube. This must be done prior to every use. Store the solution at room temperature in the dark.
    (c) *DMSO solution* Provided ready to use; store at room temperature until the expiration date.

3. Specimen preparation

    Upon arrival of the specimen, complete record of patient's name, allocated identity number, date, and time of collection is noted.

4. ROS measurement by luminometer

    The luminometer is attached to the desktop computer and a monitor. The software used is Tube master (Berthold Technologies, Autolumat plus LB 953, Oakridge, TN). This procedure is performed in a dark room.

1. Label 11 tubes (12 × 75 mm) in duplicates and add the following reagents as indicated in Table 2.1. Blank (tubes labeled 1–3), Negative Control (tubes labeled 4–6), Patient Sample (tubes labeled 7–8), and Positive Control (tubes labeled 9–11).

    Note: To avoid contamination, change pipette tips after each addition.

2. Gently vortex the tubes to mix the aliquots uniformly.

Table 2.1 Set up for the measurement of ROS

No.	Labeled tubes (no.)	PBS–1 X (μL)	Specimen volume (μL)	Probe luminol (5 mM) (μL)	Hydrogen peroxide (μL)
1	Blank (tubes 1–3)	400	–	–	–
2	Negative control (tubes 4–6)	400	–	10	–
3	Patient (tubes 7–8)	–	400	10	–
4	Positive Control (tubes 9–11)	400	–	10	50

3. Place all the labeled tubes in the luminometer in the following order: Blank (tubes labeled 1–3), Negative Control (tubes labeled 4–6), Patient Sample (tubes labeled 7–8), and Positive Control (tubes labeled 9–11).

Instrument set up

1. Turn on the instrument and the computer. From the desktop, click on 'Berthold tube' master icon to start the program.
2. From the 'Setup menu' select 'Measurement Definition' and then 'New Measurement'. You will be prompted to the following:

    a. 'Measurement Name' (Initials, Date, Analyte, and Measurement).
    b. It will show 'Measurement Definition' on the 'Tool bar'
    c. Click 'Luminometer Measurement' protocol and from the drop menu click on 'Rep. assay'.
    d. Next define each 'Parameters' as follows:

       i. Read time 1 s
       ii. Background read time 0 s.
       iii. Total time 900 s.
       iv. Cycle time 30 s
       v. Delay 'Inj M read (s)' 0 s.
       vi. 'Injector M (μL)' 0 s
       vii. 'Temperature (°C)' 37 °C
       viii. 'Temperature control (0 = OFF) 1 = ON

    e. Press 'Save'

   Note: steps 'v and vi' are used for a large number of samples and the reagents can be added by means of injectors.

3. From the 'Setup' menu select 'Assay Definition' and then 'New Assay': It will ask for the following:

    a. 'Assay Name' (Initials, Date, Analyte, Assay). Click 'OK'
    b. Select 'Measurement Method' and from the drop down menu select the measurement from Step 2a above.
    c. Go to 'Column Menu' Hide everything except the following:

    i. Sample ID
    ii. Status
    iii. RLU mean
    iv. Read date
    v. Read time

  d. Go to 'Sample Type' menu and select 'Normal'
  e. Press 'OK'
  f. Go to file, 'New' click 'Workload' Press 'OK.'

4. Save the' Work Load' (Date, Initial, Sample or experiment ID) in 'Work Load' file.
5. Click 'File name'
6. After saving the 'Work Load' the name of the file will show in the 'Title Bar'
7. The specimens are ready to be analyzed

VIII. Analyzing the samples

1. Load the tubes into the instrument and click 'Start'. It will start scanning for tubes.
2. After scanning it will show how many tubes are detected by the instrument in each batch, press 'Next'.
3. Select the 'Assay Type' and type file name and then click 'Finish'.
4. The 'Excel spreadsheet' will open measurement of the tubes will start.
5. Do not touch or change the screen, wait (3–5 min) to make sure everything is working fine.
6. After finishing measurements, it will ask for 'Save Excel Spread Sheet', save it in the 'My Document'under 'ExcelSheet' folder.
7. Select 'Excel Sheet' name the file and save type as ''Measurement Files'(*.txr). Save the 'Excel Sheet'.

IX. Printing ROS results

1. Print Excel as well as the 'chart'.
2. Close the 'Excel sheet'.
3. Print the 'Work Load' sheet, save, and close it.

X. Calculating Results and Quality Control

1. Calculate the 'average RLU' for Negative control, Samples, and Positive control.
2. Calculate sample ROS by subtracting its average from negative control average.
3. Sample ROS = Average 'RLU mean' for sample—Average 'RLU mean' for negative control.

   Calculated sample ROS = XX (RLU/sec)
   Reference values are established for each lab by testing a number of samples both from healthy women and infertile subjects. It is important to perform a regular quality control of the instrument as well as the reagents.

2.4 Factors Affecting Chemiluminescent Reaction

Having a reliable method of detection of ROS is important [13]. The luminol assay is robust; however, there are various factors that can affect ROS detection by the chemiluminescent reactions [3, 12, 14, 15]. Some of these confounding factors are:

1. The luminometer instrumentation, its calibration, determination of sensitivity, dynamic range, and units used.
2. The volume of the sample, use of reagent, and temperature of the luminometer.
3. Repeated centrifugation: Artificial increase in chemiluminescent signal because of the shearing forces generated by centrifugation [16].
4. Medium pH: Luminol is sensitive to pH changes.

2.5 Types of Luminometers

Several types of luminometers, ranging in features, design, and pricing, can be used in measuring the emitted light from the chemiluminescence assay reaction [17]. There are two different kinds of processing designs for luminometers. While direct current luminometers measure electric current, photon counting luminometers count individual photons [12]. Currently, there are three types of luminometers available for commercial uses. Single and double tube luminometers, which measure one or two samples at a time, are inexpensive and are typically used by small research laboratories. Multiple tube luminometers measure several tubes at a time, and as they are more expensive than single and double tube luminometers. Lastly, plate luminometers measure multiple samples (96 or 360) at a given time, and are typically used by commercial enterprises.

2.6 Measurement of ROS by Other Methods

2.6.1 Measurement of ROS by Nitroblue tetrazolium staining

Nitroblue tetrazolium (NBT) is an electron acceptor that becomes reduced in the presence of ROS to form a blue-black compound, formazan. This simple histochemical staining method can help target cells generating ROS. Cells generating ROS are prepared at a concentration of $1-5 \times 10^6$/mL in Kreb's buffer and about 10 µL is loaded in a glass slide and placed in an incubator for 20 min at 37 °C. It is gently rinsed with 0.154 M NaCl, and the adherent cells are over-layered with an equal volume of NBT (0.4 %) and phorbol 12-myristate 13-acetate (0.20–1 µg/mL) in Kreb's buffer with glucose (5 mM). After 15 min incubation, the slides are washed with 0.154 M NaCl, fixed for 1 min in absolute methanol, and counter stained with 1 or 2 % Safranin. Total 100 consecutive cells on each slide are observed microscopically under oil

($\times 100$) and scored as: cells filled with formazan granules (+++); intermediate formazan density (++); scattered or few formazan granules (+) or no formazan detectable (−). The data are represented as percentage of NBT-positive cells.

2.7 Epifluorescence Microscopy

This technique detects the presence of ROS using a fluorescent end product of an oxidation reaction. The product is generated by the reaction of hydroethidine and a O_2^-, which yields ethidium bromide. This end product emits a red fluorescent light that can be seen through an epifluorescence microscope. Because this technique requires a less expensive piece of equipment, it is more common and easily available to many labs for ROS detection.

2.8 Measurement of ROS by Flow Cytometry

This method utilizes the use of fluorescent probes to detect ROS within the cells [18]. The single cells are required to be suspended at a density of 10^5–10^7 cells/mL and 10,000 events are measured. Individual intracellular ROS radicals can be detected by flow cytometry. Oxidation of 2, 7 dichlorofluorescein diacetate (DCFH-DA) by ROS, which is generated within the cell, makes them highly fluorescent and can be used to measure formation of intracellular levels of hydrogen peroxide. Hydroethidine (HE) can be used for measurement of intracellular levels of superoxide. It is a substance that is oxidized by the O_2^- to become ethidium bromide with red florescence emission. Flow cytometry method has a higher specificity, accuracy, sensitivity, and reproducibility as compared to chemiluminescence for intracellular ROS [18].

2.9 Measurement of Enzymatic Antioxidants

Enzymatic activities can be measured using commercially available assay kits (Cayman Chemical, Ann Arbor, MI) following the methodology described by the manufacturer. Superoxide dismutase (SOD) activity is measured using a tetrazolium salt for detection of superoxide radicals generated by xanthine oxidase and hypoxanthine. The chromophore produced has a maximal absorbance at 525 nm. One unit of SOD is defined as the amount of enzyme needed to exhibit 50 % dismutation of the superoxide radical. Glutathione peroxidase (GPx) activity is measured using a kinetic colorimetric assay that measures activity indirectly by a coupled reaction with glutathione reductase. Glutathione reductase and NADPH reduce oxidized glutathione. NADPH oxidation is accompanied by a decrease in absorbance at 340 nm, and the decrease is directly proportional to the GPx activity in the sample. Similarly, Catalase (CAT) is involved in the detoxification of H_2O_2, which is a toxic product of both normal aerobic metabolism and pathogenic ROS production. This enzyme catalyzes the conversion of two molecules of H_2O_2 to molecular oxygen and two molecules of water (catalytic activity). CAT assay kit utilizes the peroxidatic function of CAT for determination of enzyme activity. This method is based on the reaction of the enzyme with methanol in the presence of an optimal concentration of H_2O_2. The formaldehyde produced is measured spectrophotometrically with 4-amino-3-hydrazino-5-mercapto-1,2,4-triazole (Purpald) as the chromogen. Purpald specifically forms a bicyclic heterocycle with aldehydes, which upon oxidation changes from colorless to a purple color [19, 20].

2.10 Measurement of Total Antioxidants

The sum of endogenous and food-derived antioxidants represents the total antioxidant activity of the extracellular fluid. Measuring the overall antioxidant capacity may give more relevant biological information compared to that obtained by the measurement of individual components, as it considers the cumulative effect of all antioxidants present in plasma and body fluids. The total antioxidant assay relies on the ability of antioxidants in the sample to inhibit the oxidation of 2,2'-azino-di-3-ethylbenzthiazoline sulphonate (ABTS) to $ABTS^+$ by metmyoglobin. Under the reaction conditions used, the antioxidants in the seminal plasma cause suppression of the absorbance at 750 nm to a degree which is proportional to their concentration. The capacity of the antioxidants present in the sample to prevent ABTS oxidation is compared with that of standard—Trolox, a water-soluble tocopherol analog. Results are reported as micromoles of Trolox equivalent. This assay measures the combined antioxidant activities of all its constituents including vitamins, proteins, lipids, glutathione and uric acid.

2.10.1 Measurement of Oxidative DNA Adducts

Immunohistochemistry (IHC) or western blot analysis has been used to study oxidative DNA adducts 8-hydroxy 2-deoxyguanosine [21]. IHC staining can be performed on paraffin sections (5 μ) using mouse monoclonal antibodies specific against 8-hydroxy-2'-deoxyguanosine clone N45.1 using indirect methods. Briefly, after sections are deparaffinized and rehydrated, tissue sections are incubated for 120 min at 37 °C with the primary antibody at 1:50 dilution. Slides are then incubated with a secondary antibody using horse anti-mouse dilutions (1:200) for 30 min at 37 °C. Finally, sections are stained with diaminobenzidine (DAB).

2.10.2 Measurement of Nitric Oxide

Metabolites of NO such as nitrite and nitrate can be determined by nitrate reductase and the Griess reaction [22, 23]. Total NO (nitrite and nitrate) levels in the serum and NO can also be measured using a rapid response chemiluminescence assay [24].

2.10.3 Enzyme-Linked Immunosorbent Assay

Enzyme-linked immunosorbent assay (ELISA) is based on the ability for antigens to detect proteins that have been damaged by OS. This method can detect the presence of a specific antibody or antigen within a fluid sample making it a useful tool to detect the end products of ROS interactions. This is accomplished by exposing the biofluids to the desired assay and analyzing the fluid for the amount of the targeted molecules using a spectrophotometer [12]. Utilizing appropriate antibodies, activities of various antioxidant enzymes such as total superoxide dismutase, CAT and GPx can be measured within the tissue in order to evaluate the OS [12].

2.10.4 Measurement of Lipid Peroxides

Thiobarbituric acid-reacting substances (TBARS) measure primarily malondialdehyde derived from lipid peroxidation, as well as other breakdown products from oxidatively modified proteins, carbohydrates, and nucleic acids [25]. Various commercially available kits can be used. In this assay, formation of malondialdehyde is estimated by the thiobarbituric acid method. About 1 mL of each sample is mixed with 2 mL of trichloroacetic acid (15 %), thiobarbituric acid (0.375 %) and hydrochloric acid (0.25 N) and then heated in a boiling water bath for 15 min. After cooling, the precipitate is centrifuged at 1,000 g for 10 min. The absorbance of the supernatant is measured at a 535 nm wavelength by using a spectrophotometer. The concentration of thiobarbituric acid–reactive substances is determined by considering the coefficient of molar absorptivity of the product. A standard curve is constructed by using a stock solution of 10 mM MDA prepared from tetramethoxypropane (Sigma Chemical Co., St. Louis, MO). The assay is performed in duplicate, and the results are reported as nmol MDA [26]. Sometimes a lipid peroxidation promoter such as ferrous sulfate (2.5 mM) and sodium ascorbate (12.5 mM) is also used [27, 28].

2.10.5 Measurement of Protein Oxidation

The most commonly used marker of protein oxidation is protein carbonyl content [29]. Redox cycling cations such as Fe^{2+} or Cu^{2+} can bind to cation binding locations on proteins and with the aid of further attack by H_2O_2 or O_2 can transform side-chain amine groups on several amino acids (i.e., lysine, arginine, proline, or histidine) into carbonyls. The most convenient procedure is the reaction between 2, 4-dinitrophenylhydrazine (DNPH) and protein carbonyls. DNPH reacts with protein carbonyls, forming a Schiff base to produce the corresponding hydrazone, which can be analyzed spectrophotometrically at 360–385 nm [30]. This assay is available as a kit (Catalog No. 10005020, Cayman Chemical, Ann Arbor, MI). Protein carbonyl content can be measured both in fluids and tissue and is expressed as protein carbonyl content (nmol/ml) or as protein carbonyl content (ng/mg of protein).

2.10.6 Total Plasma Lipid Hydroperoxides

Total plasma lipid hydroperoxides (LHP) are determined using the ferrous oxidation in Xylenol Orange (FOX) assay [31]. The method is based on the principle of the rapid peroxide-mediated oxidation of Fe^{2+} to Fe^{3+} under acidic conditions. The latter, in the presence of xylenol orange, forms a Fe^{3+}-xylenol orange complex that can be measured spectrophotometrically at 560 nm.

2.10.7 Measurement of Total (Free and Esterified) 8-F2-Isoprostane

8-F2-isoprostane measures stable end product of oxidized lipids derived from arachidonic acid [32–34]. Total (free and esterified) 8-F2-isoprostane can be measured using a commercial kit (cat. No. 516351; Cayman Chemical, Ann Arbor, MI). Samples (250 mL) are treated with potassium hydroxide are extracted with ethanol and purified through an affinity column (Catalog No. 416358, Cayman Chemical, Ann Arbor, MI) eluted and analyzed by enzyme immunoassay technology and concentration is expressed in pg/mL.

2.10.8 Measurement of Fat-Soluble Antioxidants

Fat-soluble antioxidants (vitamin A, vitamin E, beta-carotene, and lycopene) confer micronutrient antioxidant protection. These can be measured by HPLC. Samples (300 mL) are mixed with equal volumes of ethyl alcohol containing an internal standard (a-tocopherol acetate) and extracted twice with 2 mL of hexane. The upper organic phase is removed, evaporated to dryness under nitrogen, reconstituted in 300 mL of mobile phase containing 60 % acetonitrile–25 % methanol and 15 % ethylene chloride, and sonicated. Sixty mL is next injected onto a Supelco C18 column with Supelco guard precolumn and the vitamins separated [35]. Absorbance data are obtained from a photodiode array spectrometer set to simultaneously record at 292 nm for a-tocopherol, 326 nm for retinol and 452 nm for carotenoids and then quantified and concentration is expressed in mg/mL.

2.10.9 Measurement of Oxidative Stress by Metabolomics

The study of metabolomics involves quantifying the composition of small molecules within a tissue or biofluid sample to identify the metabolites present within a cellular system and a metabolite profile is generated [36, 37]. Multiple methods of metabolically analyzing fluids have been developed. Morphological indicators alone are not efficient in determining the viability of an embryo. Determining a metabolic profile of embryonic culture media for IVF embryos can be used to assess the metabolic differences between those that implant and those that do not implant [36–39]. Metabolomic analysis can accurately determine if an embryo is likely to be viable. The presence of these metabolites can be indicative of not only the viability of embryos, as is the case when testing embryonic culture media, but also can be used as a diagnostic tool when testing other fluids within the female reproductive system.

Depending on the type of fluid being analyzed, some methods are preferred over others. Some of the platforms for metabolic analysis are as follows: nuclear magnetic resonance (NMR) spectroscopy, liquid chromatography–mass spectrometry (LC–MS) capillary electrophoresis–mass spectrometry (CE–MS), and gas chromatography–mass spectrometry (GC–MS) [36]. NMR technique is useful when analyzing tissue extracts or fluids and requires no fluid separation, but it has a low sensitivity. LC–MS is useful when analyzing thermolabile and polar compounds while CE–MS is useful when analyzing volatile and nonvolatile compounds [36]. CE–MS is most useful when the concentration or the volume of the fluid/tissue is very small.

2.11 Near-Infrared Spectroscopy

Biological fluids can also be subjected to near infrared (IR) spectroscopy. This procedure analyzes the sample with a reflective spectrograph that has a large dynamic range photodiode detector. Results are generally in the range of 700–1,050 nm [37]. The sample is subjected to IR signaling and the machine records the sensitivity of each molecule. The results of IR spectroscopy can show the presence of—thiol or sulfhydryl groups, (−SH), hydroxyl groups (−OH), carbogens (−CH), and amines (−NH) groups based on their corresponding vibrational sensitivities that are indicative of OS [38]. Using the NMR method, Seli et al. [38] observed that cultures with higher concentrations of glutamate and lower levels of pyruvate had embryos that were significantly more likely to result in pregnancy. Glutamate is important in protecting the embryo from damage associated with exposure to concentrations of ammonium [39].

2.12 Raman Analysis

Raman analysis allows to accurately 'fingerprint' biological species within a fluid based on the vibration and molecular motion of biological/chemical species when subjected to spectroscopy [40]. The composition of the material can be determined using Raman analysis. This analysis records the cells sensitivity to the exposure of certain wavelengths and records the results from 50 to 3,450 cm^{-1} [38].Although this method conducts signals at lower intensities compared to infra red spectroscopy. Raman analysis is useful to detect the presence of fatty acids, lactate, and glucose in addition to the presence of sulfhydryls [38].

Utilizing Raman and near-IR spectroscopy method, an increase in −SH and a simultaneous decrease in −CH, and −NH was seen. All these groups are associated with OS [38].

2.13 Proton Nuclear Magnetic Resonance

The broad application of Proton NMR in gynecology and obstetrics allows the in vivo detection of ROS in the tissue [39, 41]. This method requires a tissue or fluid sample to be analyzed with a spectrometer in which the reabsorbance of the magnetic energy is used to determine the molecules present. Typically, the values are in a range of 0–10 ppm.

2.14 Oxidative Stress in Reproductive Diseases

OS either due to the production of excessive ROS or due to the limited ability of antioxidants to scavenge these ROS results in oxidative stress. OS affects a variety of body fluids, tissues, organs; as well, a host of female infertility diseases are attributed to the production of pathological ROS levels. Here we briefly describe the more important effects of ROS and/or its end products on various body fluids, tissues, and certain diseases attributed to OS [7, 8, 11, 42].

2.15 Ovaries

2.15.1 Oocytes and Follicles

OS is involved in the physiological aspects of ovarian function [43, 44]. Various studies have confirmed the role of ROS in follicular maturation, folliculogenesis, function of the corpus luteum, as well as ovulation. Parenchymal steroidogenic cells, endothelial cells, and phagocytic macrophages are among the producers of ROS within the ovaries [7, 8, 45]. In addition, normal ovaries often express many of the common biomarkers of OS [46]. Markers of OS such as Cu-SOD, Zn-SOD, and Mn-SOD, GPx γ glutamyl synthetase, and lipid peroxides have been measured by immunohistochemical staining, mRNA expression, and thiobarbituric acid methods [44, 47]. A decrease was reported in the antioxidant levels, specifically GPx, and these had a negative impact on the fertilization rates of gametes [48]. Another study reported that GPx and Mn-SOD can actually be used as markers to detect oocyte maturation, thereby implying exposure to OS [49]. NO also contributes to ovarian function. High levels of NO have adverse effect on the rate of cleavage, overall embryo quality, and implantation rates. Physiological levels of NO in follicular fluid were reported to be beneficial and these correlated with follicles containing mature oocytes that eventually fertilized [50]. A study by Bedaiwy et al. [51] found that patients presenting with peritoneal factor infertility also had elevated levels of serum NO.

2.15.2 Follicular Fluid and ART Outcomes

The environment of the follicular fluid is thought to play a critical role in oocyte maturation and the eventual development of an embryo [6, 52–54]. The follicular fluid is known to be metabolically active and contains steroid hormones, growth factors, cytokines, granulose cells, and leukocytes [6, 28, 55]. Additionally, there are many antioxidants found in follicular fluid, including vitamin E, carotene, ascorbate, cysteamine, taurine, hypotaurine, transferrin, thioredoxin, and dithiothreitol, which promote healthy oocyte maturation, and oocyte viability, however, the results are conflicting [48, 54–57]. Increased ROS levels have been associated with poor oocyte quality, low fertilization rate, and impaired embryo development [6, 58, 59]. ROS levels of the follicular fluid in women who had undergone a successful IVF treatment were significantly higher than the ROS levels of the fluid in women who did not have success in their IVF treatment [59–61]. This indicates that while an imbalance of prooxidants and antioxidants can cause a disturbance in natural female reproductive tendencies, ROS within the follicular fluid is essential for different phases of oocyte development and maturation, although the exact function remains unknown [28, 53, 54, 62, 63].

Current knowledge of the origin of ROS in follicular fluid remains unclear. It is speculated that ROS may be generated from the metabolic environment surrounding the embryo and that oxidative phosphorylation, NADPH oxidase and xanthine oxidase are sources of ROS from within the oocyte [28]. Steroid hormones, growth factors, cytokines granulose cells, and leukocytes are present components of this environment that are known to increase ROS production [6, 60]. It is speculated that the intra-follicular microenvironment and the condition therein have an important role in the development of the oocyte. Reports suggest that a decrease in ROS within the follicular fluid and an increase in total antioxidant capacity—will increase the viability of the embryo prior to implantation [53, 54, 58, 61]. Futhermore, detecting ROS within the follicular fluid could help predict the chances of pregnancy or explain some of the underlying causes of some female reproductive diseases.

2.16 Amniotic/Placental Fluid and ART Outcomes

During pregnancy, the mother and fetus can be exposed to high levels of OS [64]. It is important to analyze the composition of the amniotic fluid to determine the effect of ROS in fetal development. Amniotic fluid reaches its highest volume in the second and third trimester of gestation. It is composed mostly of fetal urine but also of fetal lung secretions [65]. While the exact source of OS during pregnancy is not well understood, it is possible that the increasing volume of amniotic fluid as the gestation period progresses plays a role in its production. However, additional sources are speculated to contribute to the increased total antioxidant capacity (TAC) noticed in amniotic fluid. These sources are the intramembranous exchange of nutrients between the chorionic plate and amniotic-chorionic membranes, as well as the transmembranous exchange of nutrients between the amniotic-chorionic membranes and the uterine wall [65]. Measuring ROS in amniotic fluid may help study the common conditions of OS such as fetal growth restriction and preeclampsia that result in increased levels of ROS possibly caused by the increased volume of amniotic fluid.

2.17 ROS and IVF Culture Media

Generally, when a couple undergoes IVF, the embryos chosen for implanting are selected by morphological criteria. The current morphological assessment examines the rate of cleavage, fragmentation, inclusion bodies, and the allocation of inner cell mass [36]. It is important to test the culture media because eight of ten transferred embryos will not result in a successful implantation and two of three ART cycles will not result in pregnancy suggesting that morphological criteria alone do not provide enough information when determining the viability of an embryo [38].

Metabolomic analysis provides a method to assess and search for products within the culture media, including metabolites resulting from OS, which may help signal the fertilization potential of the embryos examined [38]. The in vitro culture environment contains a higher concentration of oxygen as opposed to the conditions of in vivo embryos [53]. Assessing the concentration of pyruvate within the culture media can be a helpful in determining viability and the presence of ROS. Pyruvate is thought to be important because of its metabolic significance and implications during the Kreb's cycle in addition to balancing the oxidation and reduction interactions [66].

2.18 Fallopian Tubes

2.18.1 Tubal Factors

The fallopian tubes contain an internal NO system. NO is a vasodilator and acts to improve tubal contractions and motility. A decrease in the production of NO can hamper tubal motility and subsequently cause sperm transport delay, ovum retention, as well as general infertility problems [4, 8]. A significant increase in the levels of NO within the fallopian tubes can be toxic to any invading microbes, including human spermatozoa [8, 67]. The presence of NO synthase has been reported in human tubal cells.

2.18.2 Hydrosalpinx

Hydrosalpinx is defined as a blocked, dilated, and fluid filled fallopian tube that has usually also been affected by a previous tubal infection. This disease is commonly associated with female factor infertility as it is correlated with lower pregnancy and implantation rates as well as an increased rate of miscarriages. OS has often been associated as an underlying factor causing characteristic embryotoxicity in hydrosalpingeal fluid.

Low levels of ROS are beneficial for blastocyst development and may denote the physiological amounts that are present in a normal endosalpinx, high levels are pathological, and result in deleterious effects on the embryo [6, 67, 68]. However, a complete lack of detection of ROS in hydrosalpingeal fluid has a higher correlation with endosalpinges that have undergone a greater amount of damage and deterioration. Embryotoxicity normally associated with the presence of hydrosalpingeal fluid is not necessarily caused by an excess of ROS. ROS levels detected in hydrosalpingeal fluid may originate as byproducts from other natural body processes such as cellular respiration. Higher levels of ROS may also be generated by the inflammatory response resulting from chronic salpingitis [69].

Few studies have been conducted on human subjects with regard to hydrosalpingeal fluid and the ART outcomes. In a study using mouse as a model, Bedaiwy et al. [68] reported that exposure to hydrosalpingeal fluid containing higher concentrations of ROS resulted blastocyst with an increased rate of development. This correlation suggests that in healthier subjects, the granulosa cells will have an increased metabolic activity, which will produce more ROS. However, these authors also noted that the level of ROS detected was most likely not high enough to be considered detrimental to the developing embryo [68].

The removal of the damaged tube via salpingectomy was also shown to improve the implantation rates of the embryo [70]. This suggests that a healthy salpingeal tubal environment correlates with healthy implantation rates. There have been few human studies, which have analyzed the affect of ROS on ART outcomes. The adverse effects of hydrosalpinges have been shown to be reversible by salpingectomy prior to IVF [68].

2.18.3 Amniotic Sac/Placenta

The placenta is an organ that connects the developing fetus to the uterine wall of its mother in order to facilitate nutrient uptake, excretion, and gas exchange. Sufficient uteroplacental circulation is necessary in order to ensure a healthy pregnancy. Early in the pregnancy, the ends of the spiral arteries are blocked by endovascular trophoblast cells, however between weeks 10 and 12, presence of trophoblasts opens up the arteries allowing circulation thereby significantly increasing the oxygen tension. At this point, there is surge in placental OS. However, when the trophoblasts cells are prematurely dislodged from the arteries due to thinning, fragmentation, or reduced endovascular invasion, it results in OS that has been linked to spontaneous abortions and/or recurrent pregnancy loss [71]. The biomarkers of OS in the placenta include: LHP, intracellular ROS, TAC, and DNA adducts-8-hydroxy 2-deoxyguanosine [72].

2.18.4 Reactive Oxygen Species and Endometrium

ROS appear to play a regulatory role in the endometrial cycle [9]. Various studies have suggested a link among OS and the propagation of the proliferative, secretory, and menstrual phases of the monthly cycle in the endometrium. NO regulates endometrial microvasculature and significantly increased concentrations of NO have been linked with implantation failure and subsequent decrease in pregnancy rates [73]. NO synthase is secreted by the endothelial cells on the surface of the endometrium and it helps to prepare for implantation by an embryo. Inducible NOS and endothelial NOS expression have been demonstrated in human endometrium and endometrial vessels [5, 74]. An increase in endothelial NO synthase production and ROS end products has also been reported, specifically in lipid peroxidation concentrations towards the end of the secretory cycle, just prior to the onset of menstruation. Also a decrease in SOD levels was reported during the same stage of this cycle suggesting that OS contributes to the breakdown and subsequent shedding of the endometrial lining. Reduced expression of SOD leads to failed pregnancy [47, 75]. Studies have also suggested that ROS can activate nuclear factor kappa B that promotes the production of prostaglandin F2 alpha and COX-2 mRNA, which further promotes the shedding of the endometrium [71, 76–78].

2.18.5 Reactive Oxygen Species and Peritoneal Fluid

Peritoneal fluid is located in the peritoneal cavity and lubricates the pelvic cavity, uterus, ovaries and fallopian tubes, and contains hormones secreted from the ovarian follicles and corpus luteum [6, 79]. Leukocytes (macrophages) within the peritoneal tissue secrete cytokines, such as interleukins (IL) and tumor necrosis factor (TNF)-α, in a healthy peritoneal environment [80]. However, the macrophages are activated in response to inflammatory diseases, such as endometriosis, and secrete increased amounts of the cytokines [81]. Elevated levels of macrophages induce OS within the peritoneal cavity, thus altering the delicate environment of the peritoneal fluid and promoting infertility [32]. While the exact mechanism of the idiopathic infertility remains unclear, it is thought that there is a correlation between elevated numbers of macrophages present in the peritoneal fluid and infertility [82].

OS and antioxidant biomarkers are present in both serum and peritoneal fluid [83]. In peritoneal fluid, OS is initiated in inflammatory cells and the cellular debris serves as a substrate. In the serum/plasma, the oxidized metabolites such as ox-LDL are incorporated into carriers and they modify lipids, proteins, and carbohydrates in the peripheral circulation.

2.18.6 Reactive Oxygen Species and Endometriosis

Endometriosis is associated with chronic inflammation, and ROS are proinflammatory mediators that modulate cell proliferation [84, 85]. Development of OS in the local peritoneal environment may be one of the links in the chain of events leading to endometriosis-associated infertility [86]. Elevated ROS are produced from erythrocytes and apoptotic endometrioma cells, as well as the activated polymorphonuclear leukocytes and macrophages that are recruited to phagocytize the apoptotic cells [2, 87].

Markers of OS have been found to be elevated both in serum and peritoneal fluid of patients with endometriosis, [88–90].

2.18.7 Levels of Antioxidants

Peritoneal fluid is more susceptible to OS than serum. Markers in peritoneal fluid provide a more localized measure of OS related to endometriosis. Significantly, lower levels of vitamin E were reported in the peritoneal fluid than in plasma. Murphy et al. [83], suggesting that the peritoneal cavity has less antioxidant protection than serum. High amounts of enzymatic antioxidants such as GPx, SOD, Cu, Zn, and Mn-SOD and xanthine oxidase have been reported in women with endometriosis, suggesting ROS generation, and OS activity. [32, 74, 90].

2.18.8 Lipid Peroxides

Women with endometriosis had significantly higher levels of lipid peroxides such as malondialdehyde-modified low-density lipoprotein, and oxidized low-density lipoprotein, 8-F2-isoprostane, paraoxonase activity, 8-hydroxy 1-deoxyguanosine, 8-hydroxy-2-nonenal, and 4-hydroxy-2-nonenal than women without endometriosis [32, 89, 91].

2.18.9 Nitric Oxides

Increased levels of NO and NOS have been reported in the endometrium and in the peritoneal fluid of patients with endometriosis [22, 23, 92].

2.18.10 ROS and Sperm in Peritoneal Fluid

A link has been suggested between elevated ROS levels and the toxicity of the peritoneal fluid to the sperm, although evidence suggests that this is an unlikely cause for women with mild endometriosis [93]. Another hypothesis as to why elevated ROS within the peritoneal fluid impacts fertility is that women with endometriosis are more likely to have elevated levels of proinflammatory cytokines (macrophages) within their peritoneal fluid which could be a contributing factor to their infertility, however, the pathway remains unclear [93]. It is hypothesized that peritoneal fluid diffuses into the fallopian tubes where it may cause damage to sperm [91, 94, 95]. Higher concentrations of ROS were reported in the peritoneal fluid of women with idiopathic infertility compared with fertile controls but were not different from women with endometriosis [81]. However, the sample size was small in this study. Although there have been many reports in support of the hypothesis that elevated ROS levels contribute to idiopathic infertility and infertility related to endometriosis, there have also been some studies suggest that there is no correlation between the two pointing out the continuing controversy in the medical literature [52, 96, 97].

2.18.11 ROS and Immune System

An increase in the production of ROS in endometriosis is attributed to the activation of the immune system [98, 99]. ROS may play a role in the regulation of the expression of genes encoding some immunoregulators, cytokines, and cell adhesion molecules which are involved in the pathogenesis of endometriosis [100, 101]. Women with endometriosis exhibit an increased titer of autoantibodies related to OS that result in an increase in serum autoantibody titers to oxidatively modified low-density lipoproteins [83, 89].

It is not clear whether endometriosis associated OS is due to a lack in antioxidants or an increase in the production of ROS. However, there appears to be an increased amount of enzymatic antioxidant expression by the endometrial cells of women developing endometriosis. Even so, the amount of ROS present in these patients seems to outnumber the defense mechanisms available to prevent them from causing damage [101].

2.18.12 Polycystic Ovary Syndrome

Polycystic ovary syndrome (PCOS) affects 5–10 % of the reproductive aged female population [11]. Women with PCOS may also suffer from metabolic syndrome. Hyperinsulinemia is a characteristic of this disorder and increased presence of insulin circulating in the blood stream as well as TNF-α induced OS on theca cell proliferation [102]. While the exact role of ROS in folliculogenesis remains unclear, it is speculated that the origin of ROS is from the mononuclear cells of hyperglycemic patients [103, 104]. Folliculogenesis is a complex process in which ovarian follicles and the oocyte within develop to maturity. In normal women, the process of folliculogenesis involves many complex endocrine and intra ovarian paracrine reactions, which sustain a suitable intra follicular environment for the developing oocyte [105]. On a molecular level, the women experience androgen and LH hypersecretion, frequent insulin resistance because of hyperinsulinemia, and polycystic ovaries. OS may cause further proliferation of ovarian mesenchymal cells in patients with polycystic ovarian syndrome [105]. Because ROS are known to damage the lipids within cell membranes by lipid peroxidation, they can permanently damage the follicle and oocyte. For this reason, ROS are considered an integral factor in the etiology of PCOS [12]. Higher DNA damage (strand breaks) is seen in PCOS subjects compared to controls [106]. The susceptibility of DNA to OS in these patients may help explain the link between PCOS and ovarian cancer. Higher levels of protein carbonyl (a biomarker of protein oxidation) and C- reactive protein (biomarker of inflammation) as well as increase in levels of malondialdehyde have been reported in PCOS women [72].

2.18.13 Pre-Eclampsia

Pre-eclampsia is usually identified by presence of endothelial cell dysfunction and large amounts of lipid peroxidation activity [107]. Although the etiologic origin of pre-eclampsia is unknown, it continues to one of the leading causes of maternal and neonatal mortality [108, 109].

Women who are diagnosed with pre-eclampsia during pregnancy often have abnormally large amounts of ROS production with respect to NO and superoxide specifically as well as increased levels of antioxidants and high placental lipid peroxidation [107, 110]. An increase in the amount of activated neutrophils in women with pre-eclampsia contributes to the production of ROS. This also contributes to the typical vascular endothelial damage usually associated with the disease [111]. A study by Lee et al. [24] found that neutrophils isolated from pregnant women in their third trimester who were diagnosed with pre-eclampsia produced significantly more ROS than those isolated from healthy mothers. Neutrophil activation in pre-eclampsia may also lead to a greater amount of lipid peroxidation from the excessive production of reactive oxygen species. ROS have also been implicated in the pathogenesis of pre-eclampsia as seen by increased presence of endothelium derived NO levels [112, 113]. High levels of peroxynitrite formation have been associated with reports of increased endothelial NO synthase and decreased SOD, an enzymatic antioxidant, in pre-eclamptic patients [114, 115]. Molecules known as advanced glycation end products (AGE) are generators of ROS and can simultaneously cause vascular dysfunction through an association with cell surface receptors. Elevated levels of these molecules are present in women with pre-eclampsia [114]

A possible explanation for the etiologic origin of pre-eclampsia could involve the perfusion of the placenta in expectant mothers. Poorly perfused placental tissue may trigger free radical process and initiate lipid peroxidation, which is one of the main identifiable characteristics of pre-eclampsia [116]. O_2^- has been reported to initiate lipid peroxidation [117]. This combined with an oxidative imbalance in the blood are mechanisms leading to endothelial cell injury. An increase in plasma thiobarbituric acid reactive substances and decrease in the activities of both enzymatic and nonenzymatic antioxidants is seen in subjects during pregnancy and pre-eclampsia [118]. OS has been suggested as a link between the two-stage model of the pre-eclampsia syndrome; reduction of placental perfusion caused by maternal factors (stage 1) and activation of the maternal endothelium with multi-system disorders (stage 2).

2.18.14 Maternal Diabetes

Pregnant mothers who are diabetic face a higher risk of embryopathies, spontaneous abortions, and perinatal mortality that is associated with excessive OS [119]. Diabetic mothers give birth to a higher number of offspring with congenital malformations compared to healthy mothers. Increased rates of lipid peroxidation and protein carbonylation were reported in experimentally induced diabetic pregnancy [120, 121]. An excess of ROS has been observed when diabetes-induced embryopathy was blocked by antioxidants in vivo and in vitro [2, 8, 42] indicating that addition of antioxidants reduced the potential amount of damage caused by ROS. Siman et al. [122] showed the involvement of ROS in diabetes-induced malformations. In this study, antioxidant treatment with butylated hydroxytoluene, vitamin E or C resulted in a reduction of the appearance of congenital malformations from approximately 25 % to less than 8 %. Diabetes itself is classified as a state of OS and low-density lipoproteins from pregnant diabetic women are highly susceptible to oxidation. In a study by Trocino et al. [123] embryos that have been cultured under hyperglycemic conditions showed increased production levels of ROS and significant reduction in glutathione synthesis and as well as in currently existing levels of the antioxidants. The presence of ROS is increased in embryos that have been exposed to high levels of glucose, possibly due to an increase in O_2^- generation or oxidative metabolism. The prevalence of congenital malformations preceded by embryonic dysmorphogenesis in vitro can be minimized by reducing the exposure of oxidative substrates to embryonic mitochondria or by improving ROS scavenging ability [124].

2.18.15 Recurrent Pregnancy Loss

Recurrent pregnancy loss (RPL) affects 0.5–3 % of reproductive age women and is defined as three or more consecutive spontaneous abortions occurring before 5 months of gestation [125]. Although the etiology of RPL can be attributed to many factors including gynecologic disease, anatomic abnormalities of the uterus, genetic anomalies in the fetus, and sperm DNA damage, almost 50–60 % of cases involving RPL have been observed to implicate the presence of OS [67, 70, 72, 126]. There is an increase in OS during the earlier parts of pregnancy from the trophoblast inside the placenta, as measured by various biomarkers such as TAC and ROS [72]. If there is an excess of OS and a lack of adequate antioxidant levels to protect the surrounding lipids, DNA, and proteins from damage, significant harm could be caused to the developing embryo [127, 128]. Women who are suffering from RPL have been found to possess weakened antioxidant defense systems as well as increased levels of OS biomarkers; therefore, it seems logical to assume that an excess in the presence of ROS may contribute to the pathological basis of the disease [129, 130]. Male partners with large amounts of leukocytes in their semen, morphologically abnormal sperm, or sperm with significant amounts of DNA damage can also contribute to ROS production levels and subsequent spontaneous abortions [131–134]. It is important to be able to detect the various biomarkers of OS in pregnant women in order to effectively predetermine chances for developing RPL and to combat the disease using appropriate treatment methods.

2.19 Conclusions

The detection of ROS in the female reproductive system is important for a variety of reasons. We can analyze the different fluids and determine that lower levels of ROS can be indicative of healthy women, such as in hydrosalpingeal fluid, or that the elevated levels indicate that something is wrong such as in amniotic, peritoneal, and follicular fluid. Additionally, the ROS levels can be used to determine if there is a link between gynecological diseases and OS as well as how likely an oocyte will get fertilized or if an embryo will implant. In short, it is important to detect OS for the continuing understanding of the female reproductive cycle and health.

References

1. Sharma RK, Agarwal A (1996) Role of reactive oxygen species in male infertility. Urology 48(6):835–850
2. Sharma RK, Agarwal A (2004) Role of reactive oxygen species in gynecologic diseases. Reprod Med Biol 3(4):177–199
3. Agarwal A, Cocuzza M, Abdelrazik H, Sharma R (2008) Oxidative stress measurement in patients with male or female factor infertility. Network 2008(2):195–218
4. Rosselli M, Dubey RK, Imthurn B, Macas E, Keller PJ (1995) Effects of nitric oxide on human spermatozoa: evidence that nitric oxide decreases sperm motility and induces sperm toxicity. Hum Reprod 10:1786–1790
5. Rosselli M, Keller PJ, Dubey RK (1998) Role of nitric oxide in the biology, physiology and pathophysiology of reproduction. Hum Reprod Update 4(1):3–24
6. Agarwal A, Saleh RA, Bedaiwy MA (2003) Role of reactive oxygen species in the pathophysiology of human reproduction. Fertil Steril 79:829–843
7. Agarwal A, Gupta S, Sharma R (2005) Oxidative stress and its implications in female infertility—a clinician's perspective. Reprod Biomed Online 11(5):641–650
8. Agarwal A, Gupta S, Sharma RK (2005) Role of oxidative stress in female reproduction. Reprod Biol Endocrinol 14(3):28
9. Agarwal A, Gupta S, Sekhon L, Shah R (2008) Redox considerations in female reproductive function and assisted reproduction: from molecular mechanisms to health implications. Antioxid Redox Signal 10(8):1375–1403
10. de Matos DG, Furnus CC (2000) The importance of having high glutathione (GSH) level after bovine in vitro maturation on embryo development effect of beta-mercaptoethanol, cysteine and cystine. Theriogenology 53(3):761–771
11. Agarwal A, Allamaneni SS (2004) Role of free radicals in female reproductive diseases and assisted reproduction. Reprod Biomed Online 9(3):338–347
12. Agarwal A, Allamaneni SS, Said TM (2004) Chemiluminescence technique for measuring reactive oxygen species. Reprod Biomed Online 9(4):466–468
13. Kobayashi H, Gil-Guzman E, Mahran AM, Sharma RK, Nelson DR, Thomas AJ Jr, Agarwal A (2001) Quality control of reactive oxygen species measurement by luminol-dependent chemiluminescence assay. J Androl 22:568–574
14. Berthold F, Herick K, Siewe RM (2000) Luminometer design and low light detection. Methods Enzymol 305:62–87
15. Aitken RJ, Baker MA, O'Bryan M (2004) Shedding light on chemiluminescence: the application of chemiluminescence in diagnostic andrology. J Androl 25(4):455–65. Review
16. Shekarriz M, DeWire DM, Thomas AJ Jr, Agarwal A (1995) A method of human semen centrifugation to minimize the iatrogenic sperm injuries caused by reactive oxygen species. Eur Urol 28(1):31–35
17. Stanley PE (1999) Commercially available fluorometers, luminometers and imaging devices for low-light level measurements and allied kits and reagents: survey update 6. Luminescence 14(4):201–213
18. Mahfouz RZ, Sharma RK, Said TM, Erenpreiss J, Agarwal A (2009) Association of sperm apoptosis and DNA ploidy with sperm chromatin quality in human spermatozoa. Fertil Steril 91(4):1110–1118

19. Wheeler CR, Salzman JA, Elsayed NM et al (1990) Automated assays for superoxide dismutase, catalase, glutathione peroxidase, and glutathione reductase activity. Anal Biochem 184:193–199
20. Johansson LH, Borg LAH (1988) A spectrophotometric method for determination of catalase activity in small tissue samples. Anal Biochem 174:331–336
21. Takagi Y, Nikaido T, Toki T, Kita N, Kanai M, Ashida T, Ohira S, Konishi I (2004) Levels of oxidative stress and redox-related molecules in the placenta in preeclampsia and fetal growth restriction. Virchows Arch 444:49–55
22. Dong M, Shi Y, Cheng Q, Hao M (2001) Increased nitric oxide in peritoneal fluid from women with idiopathic infertility and endometriosis. J Reprod Med 46:887–891
23. Osborn BH, Haney AF, Misukonis MA, Weinberg JB (2002) Inducible nitric oxide synthase expression by peritoneal macrophages in endometriosis-associated infertility. Fertil Steril 77:46–51
24. Lee KS, Joo BS, Na YJ, Yoon MS, Choi OH, Kim WW (2000) Relationships between concentrations of tumor necrosis factor-alpha and nitric oxide in follicular fluid and oocyte quality. J Assist Reprod Genet 17(4):222–228
25. Guichardant M, Chantegrel B, Deshayes C, Doutheau A, Moliere P, Lagarde M (2004) Specific markers of lipid peroxidation issued from n-3 and n-6 fatty acids. Biochem Soc Trans 32:139–140
26. Campos PC, Ferriani RA, dos Reis RM, de Moura MD, Jordão AA Jr, Navarro PA (2008) Lipid peroxidation and vitamin E in serum and follicular fluid of infertile women with peritoneal endometriosis submitted to controlled ovarian hyperstimulation: a pilot study. Fertil Steril 90:2080–2085
27. Wang Y, Sharma RK, Agarwal A (1997) Effect of cryopreservation and sperm concentration on lipid peroxidation in human semen. Urology 50:409–413
28. Pasqualotto EB, Agarwal A, Sharma RK, Izzo VM, Pinotti JA, Joshi NJ et al (2004) Effect of oxidative stress in follicular fluid on the outcome of assisted reproductive procedures. Fertil Steril 81:973–976
29. Stadtman ER, Oliver CN (1991) Metal-catalyzed oxidation of proteins. Physiological consequences. J Biol Chem 266(4):2005–2008
30. Levine RL, Williams JA, Stadtman ER, Shacter E (1994) Carbonyl assays for determination of oxidatively modified proteins. Methods Enzymol 233:346–357
31. Jiang ZY, Hunt JV, Wolf SP (1992) Detection of lipid hydroperoxides using fox method. Anal Biochem 202:384–389
32. Jackson LW, Schisterman EF, Dey-Rao R, Browne R, Armstrong D (2005) Oxidative stress and endometriosis. Hum Reprod 20(7):2014–2020
33. Fam SS, Morrow JD (2003) The isoprostanes: unique products of arachidonic acid oxidation—a review. Curr Med Chem 10:1723–1740
34. Rokach J, Kim S, Bellone S, Lawson JA, Pratico D, Powell WS, FitzGerald GA (2004) Total synthesis of isoprostanes: discovery and quantitation in biological systems. Chem Phys Lipids 128:35–56
35. Browne RW, Armstrong D (1998) Simultaneous determination of serum retinol, tocopherols, and carotenoids by HPLC. Methods Mol Biol 108:269–275
36. Singh R, Sinclair KD (2007) Metabolomics: approaches to assessing oocyte and embryo quality. Theriogenology 68(Suppl 1):S56–S62
37. Wu H, Southam AD, Hines A, Viant MR (2008) High-throughput tissue extraction protocol for NMR- and MS-based metabolomics. Anal Biochem 372(2):204–212
38. Seli E, Sakkas D, Scott R, Kwok SC, Rosendahl SM, Burns DH (2007) Noninvasive metabolomic profiling of embryo culture media using Raman and near-infrared spectroscopy correlates with reproductive potential of embryos in women undergoing in vitro fertilization. Fertil Steril 88(5):1350–1357
39. Seli E, Botros L, Sakkas D, Burns DH (2008) Noninvasive metabolomic profiling of embryo culture media using proton nuclear magnetic resonance correlates with reproductive potential of embryos in women undergoing in vitro fertilization. Fertil Steril 90(6):2183–2189

40. Scott R, Seli E, Miller K, Sakkas D, Scott K, Burns DH (2008) Noninvasive metabolomic profiling of human embryo culture media using Raman spectroscopy predicts embryonic reproductive potential: a prospective blinded pilot study. Fertil Steril 90(1):77–83
41. Fenton BW, Lin CS, Macedonia C, Schellinger D, Ascher S (2001) The fetus at term: in utero volume-selected proton MR spectroscopy with a breath-hold technique–a feasibility study. Radiology 219(2):563–566
42. Gupta S, Sekhon L, Kim Y, Agarwal A (2010) The role of oxidative stress and antioxidants in assisted reproduction. In: Gupta S, Agarwal A (eds) Current concepts in assisted reproduction and fertility preservation, Current Women's Health Reviews, vol 6(3) pp. 227–238
43. Shiotani M, Noda Y, Narimoto K, Imai K, Mori T, Fujimoto K, Ogawa K (1991) Immunohistochemical localization of superoxide dismutase in the human ovary. Hum Reprod 6:1349–1353
44. Suzuki T, Sugino N, Fukaya T, Sugiyama S, Uda T, Takaya R, Yajima A, Sasano H (1999) Superoxide dismutase in normal cycling human ovaries: immunohistochemical localization and characterization. Fertil Steril 72:720–726
45. Halliwell B, Gutteridge JM (1988) Free radicals and antioxidant protection: mechanisms and significance in toxicology and disease. Hum Toxicol 7(1):7–13
46. Tamate K, Sengoku K, Ishikawa M (1995) The role of superoxide dismutase in the human ovary and fallopian tube. J Obstet Gynaecol 21(4):401–409
47. Sugino N, Nakata M, Kashida S, Karube A, Takiguchi S, Kato H (2000) Decreased superoxide dismutase expression and increased concentrations of lipid peroxide and prostaglandin F(2alpha) in the decidua of failed pregnancy. Mol Hum Reprod 6:642–647
48. Paszkowski T, Traub AI, Robinson SY, McMaster D (1995) Selenium dependent glutathione peroxidase activity in human follicular fluid. Clin Chim Acta 236(2):173–180
49. El Mouatassim S, Guérin P, Ménézo Y (1999) Expression of genes encoding antioxidant enzymes in human and mouse oocytes during the final stages of maturation. Mol Hum Reprod 5(8):720–725
50. Barrionuevo MJ, Schwandt RA, Rao PS, Graham LB, Maisel LP, Yeko TR (2000) Nitric oxide (NO) and interleukin-1beta (IL-1beta) in follicular fluid and their correlation with fertilization and embryo cleavage. Am J Reprod Immunol 44(6):359–364
51. Bedaiwy MA, Falcone T (2004) Laboratory testing for endometriosis. Clin Chim Acta 340(1–2):41–56
52. Ho HN, Wu MY, Chen SU, Chao KH, Chen CD, Yang YS (1997) Total antioxidant status and nitric oxide do not increase in peritoneal fluids from women with endometriosis. Hum Reprod 12(12):2810–2815
53. Yang HW, Hwang KJ, Kwon HC, Kim HS, Choi KW, Oh KS (1998) Detection of reactive oxygen species (ROS) and apoptosis in human fragmented embryos. Hum Reprod 13(4):998–1002
54. Oyawoye O, Abdel Gadir A, Garner A, Constantinovici N, Perrett C, Hardiman P (2003) Antioxidants and reactive oxygen species in follicular fluid of women undergoing IVF: relationship to outcome. Hum Reprod 18(11):2270–2274
55. Pasqualotto EB, Lara LV, Salvador M, Sobreiro BP, Borges E, Pasqualotto FF (2009) The role of enzymatic antioxidants detected in the follicular fluid and semen of infertile couples undergoing assisted reproduction. Hum Fertil (Camb) 12(3):166–171
56. Sabatini L, Wilson C, Lower A, Al-Shawaf T, Grudzinskas JG (1999) Superoxide dismutase activity in human follicular fluid after controlled ovarian hyperstimulation in women undergoing in vitro fertilization. Fertil Steril 72:1027–1034
57. Appasamy M, Jauniaux E, Serhal P, Al-Qahtani A, Grome NP, Muttukrishna S (2008) Evaluation of the relationship between follicular fluid oxidative stress, ovarian hormones, and response to gonadotropin stimulation. Fertil Steril 89:912–921
58. Revelli A, Delle Piane L, Casano S, Molinari E, Massobrio M, Rinaudo P (2009) Follicular fluid content and oocyte quality: from single biochemical markers to metabolomics. Reprod Biol Endocrinol 7:40

59. Das S, Chattopadhyay R, Ghosh S, Ghosh S, Goswami SK, Chakravarty BN et al (2006) Reactive oxygen species level in follicular fluid–embryo quality marker in IVF? Hum Reprod 21:2403–2407
60. Attaran M, Pasqualotto E, Falcone T, Goldberg JM, Miller KF, Agarwal A, Sharma RK (2000) The effect of follicular fluid reactive oxygen species on the outcome of in vitro fertilization. Int J Fertil Womens Med 45:314–320
61. Jana SK, K NB, Chattopadhyay R, Chakravarty B, Chaudhury K (2010) Upper control limit of reactive oxygen species in follicular fluid beyond which viable embryo formation is not favorable.Reprod Toxicol 29:447–51
62. Wiener-Megnazi Z, Vardi L, Lissak A, Shnizer S, Teznick AZ, Ishai D, Lahav-Baratz S, Shiloh H, Koifman M, Dirnfeld M (2004) Oxidative stress indices in follicular fluid as measured by thermochemiluminescence assay correlate with outcome parameters in vitro fertilization. Fertil Steril 82:1171–1176
63. Paszkowski T, Clarke RN (1996) Antioxidant capacity of preimplantation embryo culture medium declines following the incubation of poor quality embryos. Hum Reprod 11: 2493–2495
64. Raicević S, Cubrilo D, Arsenijević S, Vukcević G, Zivković V, Vuletić M, Barudzić N, Andjelković N, Antonović O, Jakovljević V (2010) Oxidative stress in fetal distress: potential prospects for diagnosis. Oxid Med Cell Longev 3(3):214–218
65. Burlingame JM, Esfandiari N, Sharma RK, Mascha E, Falcone T (2003) Total antioxidant capacity and reactive oxygen species in amniotic fluid. Obstet Gynecol 101(4):756–761
66. Morales H, Tilquin P, Rees JF, Massip A, Dessy F, Van Langendonckt A (1999) Pyruvate prevents peroxide-induced injury of in vitro preimplantation bovine embryos. Mol Reprod Dev 52(2):149–157
67. Gupta, S, Surti, N, Metterle, L, Chandra, A, Agarwal, A (2009) Antioxidants and female reproductive pathologies. Review Article. Arch of Med Sci 5(1A):S151–S173
68. Bedaiwy MA, Goldberg JM, Falcone T, Singh M, Nelson D, Azab H et al (2002) Relationship between oxidative stress and embryotoxicity of hydrosalpingeal fluid. Hum Reprod 17:601–604
69. Kubo A, Sasada M, Nishimura T, Moriguchi T, Kakita T, Yamamoto K, Uchino H (1987) Oxygen radical generation by polymorphonuclear leucocytes of beige mice. Clin Exp Immunol 70(3):658–663
70. Gupta S, Agarwal A, Banerjee J, Alvarez JG (2007) The role of oxidative stress in spontaneous abortion and recurrent pregnancy loss: a systematic review. Obstet Gynecol Surv 62(5):335–347
71. Myatt L (2010) Review: reactive oxygen and nitrogen species and functional adaptation of the placenta. Placenta (31Suppl):S66–S69
72. Palacio JR, Iborra A, Ulcova-Gallova Z, Badia R, Martinez P (2006) The presence of antibodies to oxidative modified proteins in serum from polycystic ovary syndrome patients. Clin Exp Immunol 144:217–222
73. Sugino N (2007) The role of oxygen radical-mediated signaling pathways in endometrial function. Placenta 28:S133–S136
74. Ota H, Igarashi S, Tanaka T (2001) Xanthine oxidase in eutopic and ectopic endometrium in endometriosis and adenomyosis. Fertil Steril 75:785–790
75. Sugino N, Takiguchi S, Kashida S, Karube A, Nakamura Y, Kato H (2000) Superoxide dismutase expression in the human corpus luteum during the menstrual cycle and in early pregnancy. Mol Hum Reprod 6:19–25
76. Sugino N, Karube-Harada A, Sakata A, Takiguchi S, Kato H (2002) Nuclear factor-kappa B is required for tumor necrosis factor-alpha-induced manganese superoxide dismutase expression in human endometrial stromal cells. J Clin Endocrinol Metab 87(8):3845–3850
77. Sugino N (2004) Withdrawal of ovarian steroids stimulates prostaglandin F2alpha production through nuclear factor-kappaB activation via oxygen radicals in human endometrial stromal cells: potential relevance to menstruation. J Reprod Dev 50:215–225

78. Sugino N (2007) The role of oxygen radical-mediated signaling pathways in endometrial function. Placenta 28:S133–S136
79. Bedaiwy MA, Falcone T (2003) Peritoneal fluid environment in endometriosis. Clinicopathological implications. Minerva Ginecol 55(4):333–345
80. Berbic M, Schulke L, Markham R, Tokushige N, Russell P, Fraser IS (2009) Macrophage expression in endometrium of women with and without endometriosis. Hum Reprod 24(2):325–332
81. Bedaiwy MA, Falcone T, Sharma RK, Goldberg JM, Attaran M, Nelson DR, Agarwal A (2002) Prediction of endometriosis with serum and peritoneal fluid markers: a prospective controlled trial. Hum Reprod 17:426–431
82. Montagna P, Capellino S, Villaggio B, Remorgida V, Ragni N, Cutolo M, Ferrero S (2008) Peritoneal fluid macrophages in endometriosis: correlation between the expression of estrogen receptors and inflammation. Fertil Steril 90(1):156–164
83. Murphy AA, Palinski W, Rankin S, Morales AJ, Parthasarathy S (1998) Evidence for oxidatively modified lipid-protein complexes in endometrium and endometriosis. Fertil Steril 69:1092–1094
84. Ngô C, Chéreau C, Nicco C, Weill B, Chapron C, Batteux F (2009) Reactive oxygen species controls endometriosis progression. Am J Pathol 175(1):225–234
85. Alexandre J, Nicco C, Chereau C, Laurent A, Weill B, Goldwasser F, Batteux F (2006) Improvement of the therapeutic index of anticancer drugs by the superoxide dismutase mimic mangafodipir. J Natl Cancer Inst 98:236–244
86. Gupta S, Agarwal A, Krajcir N, Alvarez JG (2006) Role of oxidative stress in endometriosis. Reprod Biomed Online 13:126–134
87. Zeller JM, Henig I, Radwanska E, Dmowski WP (1987) Enhancement of human monocyte and peritoneal macrophage chemiluminescence activities in women with endometriosis. Am J Reprod Immunol Microbiol 13(3):78–82
88. Arumugam K, Dip YC (1995) Endometriosis and infertility: the role of exogenous lipid peroxides in the peritoneal fluid. Fertil Steril 63:198–199
89. Shanti A, Santanam N, Morales AJ, Parthasarathy S, Murphy AA (1999) Autoantibodies to markers of oxidative stress are elevated in women with endometriosis. Fertil Steril 71:1115–1118
90. Szczepańska M, Koźlik J, Skrzypczak J, Mikołajczyk M (2003) Oxidative stress may be a piece in the endometriosis puzzle. Fertil Steril 79(6):1288–1293
91. Saito H, Seino T, Kaneko T, Nakahara K, Toya M, Kurachi H (2002) Endometriosis and oocyte quality. GynecolObstet Invest 53(Suppl 1):46–51
92. Khorram O, Lessey BA (2002) Alterations in expression of endometrial endothelial nitric oxide synthase and alpha(v)beta(3) integrin in women with endometriosis. Fertil Steril 78:860–864
93. Oak MK, Chantler EN, Williams CA, Elstein M (1985) Sperm survival studies in peritoneal fluid from infertile women with endometriosis and unexplained infertility. Clin Reprod Fertil 3(4):297–303
94. Polak G, Koziol-Montewka M, Gogacz M, Blaszkowska I, Kotarski J (2001) Total antioxidant status of peritoneal fluid in infertile women. Eur J Obstet Gynecol Reprod Biol 94:261–263
95. Storey BT (1997) Biochemistry of the induction and prevention of lipoperoxidative damage in human spermatozoa. Mol Hum Reprod 3:203–213
96. do Amaral VF, Bydlowski SP, Peranovich TC, Navarro PA, Subbiah MT, Ferriani RA (2005) Lipid peroxidation in the peritoneal fluid of infertile women with peritoneal endometriosis. Eur J Obstet Gynecol Reprod Biol 1:119(1):72–5
97. Wang Y, Sharma RK, Falcone T, Goldberg J, Agarwal A (1997) Importance of reactive oxygen species in the peritoneal fluid of women with endometriosis or idiopathic infertility. Fertil Steril 68(5):826–830
98. Eskenazi B, Warner ML (1997) Epidemiology of endometriosis. Obstet Gynecol Clin North Am 24(2):235–258

99. Gleicher N, el-Roeiy A, Confino E, Friberg J (1987) Is endometriosis an autoimmune disease? Obstet Gynecol 70(1):115–122
100. Gleicher N (1987) A potential animal model for autoimmune reproductive failure. Am J Reprod Immunol Microbiol. 14(4):122
101. Van Langendonckt A, Casanas-Roux F, Donnez J (2002) Oxidative stress and peritoneal endometriosis. Fertil Steril 77(5):861–870
102. Kodaman PH, Duleba AJ (2008) Statins in the treatment of polycystic ovary syndrome. Semin Reprod Med 26(1):127–138
103. Gonzalez F, Rote NS, Minium J, Kirwan JP (2006) Reactive oxygen species-induced oxidative stress in the development of insulin resistance and hyperandrogenism in polycystic ovary syndrome. J Clin Endocrinol Metab 91:336–340
104. González F, Rote NS, Minium J, Weaver AL, Kirwan JP (2010) Elevated circulating levels of macrophage migration inhibitory factor in polycystic ovary syndrome. Cytokine 51(3):240–244
105. Dumesic DA, Padmanabhan V, Abbott DH (2008) Polycystic ovary syndrome and oocyte developmental competence. Obstet Gynecol Surv 63(1):39–48
106. Dincer Y, Akcay T, Erdem T, Ilker Saygili E, Gundogdu S (2005) DNA damage, DNA susceptibility to oxidation and glutathione level in women with polycystic ovary syndrome. Scand J Clin Lab Invest 65:721–728
107. Ishihara O, Hayashi M, Osawa H, Kobayashi K, Takeda S, Vessby B, Basu S (2004) Isoprostanes, prostaglandins and tocopherols in pre-eclampsia, normal pregnancy and non-pregnancy. Free Radic Res 38:913–918
108. Sidiqui IA, Jaleel A, Tamimi W, Al Kadri HM (2010) Role of oxidative stress in the pathogenesis of preeclampsia. Arch Gynecol Obstet 282(5):469–474
109. Wang Y, Walsh SW (1996) Antioxidant activities and mRNA expression of superoxide dismutase, catalase, and glutathione peroxidase in normal and preeclamptic placentas. J Soc Gynecol Investig 3:179–184
110. Hubel CA (1999) Oxidative stress in the pathogenesis of preeclampsia. Proc Soc Exp Biol Med 222(3):222–235
111. Holthe MR, Staff AC, Berge LN, Lyberg T (2004) Leukocyte adhesion molecules and reactive oxygen species in preeclampsia. Obstet Gynecol 103(5 Pt 1):913–922
112. Davidge ST, Hubel CA, Brayden RD, Capeless EC, McLaughlin MK (1992) Sera antioxidant activity in uncomplicated and preeclamptic pregnancies. Obstet Gynecol 79(6):897–901
113. Matsubara K, Matsubara Y, Hyodo S, Katayama T, Ito MJ (2010) Role of nitric oxide and reactive oxygen species in the pathogenesis of preeclampsia. Obstet Gynaecol Res 36(2):239–247
114. Cooke CL, Brockelsby JC, Baker PN, Davidge ST (2003) The receptor for advanced glycation end products (RAGE) is elevated in women with preeclampsia. Hypertens Pregnancy 22(2):173–184
115. Roggensack AM, Zhang Y, Davidge ST (1999) Evidence for peroxynitrite formation in the vasculature of women with preeclampsia. Hypertension 33(1):83–89
116. Diamant S, Kissilevitz R, Diamant Y (1980) Lipid peroxidation system in human placental tissue: general properties and the influence of gestational age. Biol Reprod 23(4):776–781
117. Radi R, Beckman JS, Bush KM, Freeman BA (1991) Peroxynitrite-induced membrane lipid peroxidation: the cytotoxic potential of superoxide and nitric oxide. Arch Biochem Biophys 288(2):481–487
118. Pasupathi P, Manivannan U, Manivannan P, Deepa M (2010) Cardiac troponins and oxidative stress markers in non-pregnant, pregnant and preeclampsia women. Bangladesh Med Res Counc Bull 36(1):4–9
119. Djordjevic A, Spasic S, Jovanovic-Galovic A, Djordjevic R, Grubor-Lajsic G (2004) Oxidative stress in diabetic pregnancy: SOD, CAT and GSH-Px activity and lipid peroxidation products. J Matern Fetal Neonatal Med 16:367–372
120. Cederberg J, Basu S, Eriksson UJ (2001) Increased rate of lipid peroxidation and protein carbonylation in experimental diabetic pregnancy. Diabetologia 44(6):766–774

121. Pedersen L, Tygstrup I, Pedersen J (1964) Congenital malformations in newborn infants of diabetic women correlation with maternal diabetic vascular complications. Lancet 1(7343):1124–1126
122. Simán CM, Eriksson UJ (1997) Vitamin C supplementation of the maternal diet reduces the rate of malformation in the offspring of diabetic rats. Diabetologia 40(12):1416–1424
123. Trocino RA, Akazawa S, Ishibashi M, Matsumoto K, Matsuo H, Yamamoto H, Goto S, Urata Y, Kondo T, Nagataki S (1995) Significance of glutathione depletion and oxidative stress in early embryogenesis in glucose-induced rat embryo culture. Diabetes 44(8): 992–998
124. Eriksson UJ, Borg LA (1991) Protection by free oxygen radical scavenging enzymes against glucose-induced embryonic malformations in vitro. Diabetologia 34(5):325–331
125. Falcone T, Hurd W (2007) Clinical Reproductive Medicine and Surgery. Mosby Elsevier, Philadelphia, p.123
126. Burton GJ, Hempstock J, Jauniaux E (2003) Oxygen, early embryonic metabolism and free radical-mediated embryopathies. Reprod Biomed Online 6(1):84–96
127. Hempstock J, Jauniaux E, Greenwold N, Burton GJ (2003) The contribution of placental oxidative stress to early pregnancy failure. Hum Pathol 34(12):1265–1275
128. Jauniaux E, Watson AL, Hempstock J, Bao YP, Skepper JN, Burton GJ (2000) Onset of maternal arterial blood flow and placental oxidative stress. A possible factor in human early pregnancy failure. Am J Pathol 157:2111–2122
129. Sane AS, Chokshi SA, Mishra VV, Barad DP, Shah VC, Nagpal S (1991) Serum lipoperoxides in induced and spontaneous abortions. Gynecol Obstet Invest 31(3):172–175
130. Simşek M, Naziroğlu M, Simşek H, Cay M, Aksakal M, Kumru S (1998) Blood plasma levels of lipoperoxides, glutathione peroxidase, beta carotene, vitamin A and E in women with habitual abortion. Cell Biochem Funct 16(4):227–231
131. Gil-Villa AM, Cardona-Maya W, Agarwal A, Sharma R, Cadavid A (2010) Assessment of sperm factors possibly involved in early recurrent pregnancy loss. Fertil Steril 94(4): 1465–1472
132. Gil-Villa AM, Cardona-Maya W, Agarwal A, Sharma R, Cadavid A (2009) Role of male factor in early recurrent embryo loss: do antioxidants have any effect? Case report. Fertil Steril 92:565–571
133. Gomes FM, Navarro PA, de Abreu LG, Ferriani RA, dos Reis RM, de Moura MD (2008) Effect of peritoneal fluid from patients with minimal/mild endometriosis on progesterone release by human granulosa-lutein cells obtained from infertile patients without endometriosis: a pilot study. Eur J Obstet Gynecol Reprod Biol 138(1):60–65
134. Larson KL, DeJonge CJ, Barnes AM, Jost LK, Evenson DP (2000) Sperm chromatin structure assay parameters as predictors of failed pregnancy following assisted reproductive techniques. Hum Reprod 15(8):1717–1722

Chapter 3
Oxidative Stress and The Endometrium

Botros Rizk, Marwa Badr and Christina Talerico

Abstract The endometrium plays a key role in successful implantation and pregnancy. A delicate balance exists between the antioxidant mechanisms and the reactive oxygen species in the endometrium. When pregnancy occurs, successful implantation results in the production of Human Chorionic Gonadotropin (HCG) which maintains progesterone levels. Progesterone increases the activity of superoxide dismutase in the endometrium which in turn suppresses the production of reactive oxygen species and prostaglandin F2α. (PGF2α). On the other hand, in the absence of pregnancy, progesterone levels drop and superoxide dismutase activity declines. As a result, cyclooxygenase enzyme 2 (COX2) and PGF2α increase, resulting in menstruation. HCG might improve the uterine environment prior to implantation by suppressing the apoptotic response to oxidative stress in the maternal decidua.

B. Rizk
Division of Reproductive Endocrinology and Infertility, University of South Alabama In Vitro Fertilization Laboratory, 251 Cox Street, Suite A, Mobile, AL 36604, USA
e-mail: botros4@gmail.com

M. Badr
Division of Reproductive Endocrinology and Infertility, Department of Obstetrics and Gynecology, University of South Alabama, 251 Cox Street, Suite A, Mobile, AL 36604, USA
e-mail: mbadr2004@gmail.com

M. Badr
Ain Shams University, Cairo, Egypt

C. Talerico (✉)
University of South Alabama College of Medicine, 5851 USA Drive North, Mobile, AL 36688, USA
e-mail: ctm601@jaguar1.usouthal.edu

Keywords Endometrium · Oxidative stress · Maternal decidua · Antioxidant defense mechanisms · Decidualization

3.1 Introduction

The endometrium is a highly organized structure which is cyclically controlled by the ovarian steroids and pituitary hormones. The endometrium varies in appearance in health and disease [1, 2] from a pink, shiny, and glistening appearance to a polypoid structure and possible endometrial hyperplasia and carcinoma (Fig. 3.1). Rizk and Sallam [3] discussed the role of the endometrium in implantation and the physiological changes in pregnancy. De Ziegler et al.[4] elegantly discussed the impact of endometritis on implantation. They argued that an increase in aromatase enzyme activity may be a part of the explanation of the negative impact of chronic endometritis on endometrial receptor activity [4]. Jauniaux and Rizk [5] presented the impact of abnormalities in the implantation process on the outcome of pregnancy after assisted reproduction. In this chapter, we discuss the relationship between the endometrium and oxidative stress in health as well as different disease states such as early pregnancy loss, endometriosis, and cancer [5–9].

3.2 The Balance Between Oxidative Stress and Antioxidant Defense Mechanisms

A delicate balance exists between reactive oxidative stress and the various defense systems against oxidative stress in the endometrium. Optimum physiologic levels of oxygen are essential. However, toxic reactive oxygen species such as the superoxide radical, hydrogen peroxide, and the hydroxyl radical are generated from oxygen. These oxygen radicals could potentially have harmful effects including protein damage, lipid peroxidation, and possibly DNA damage [10–14]. These may impact membrane structure and function. Under aerobic conditions, superoxide radicals are scavenged by metallo enzymes [10]. Copper–zinc superoxide dismutase is located in the cytosol, whereas manganese superoxide is located in the mitochondria. Both superoxide dismutases metabolize superoxides. Hydrogen peroxide is formed which is metabolized by glutathione peroxidase and catalase to water and oxygen. Under physiologic, aerobic conditions, the endometrium is a perfect example of the balance between reactive oxygen species and the successful antioxidant system. Disruption of the balance may result in abnormalities of normal reproductive physiological functions such as menstruation and decidualization. Endometrial stromal cells serve as a potential source of reactive oxygen species. The endoplasmic reticulum and the electron transport systems both in the mitochondria and the nucleus are the intracellular sources of

3 Oxidative Stress and the Endometrium

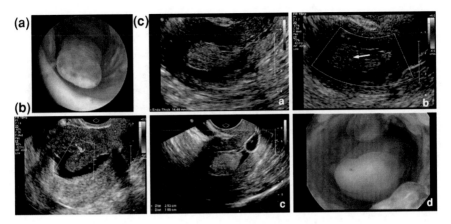

Fig. 3.1 a Endometrial polyp by hysteroscopy. Reproduced from deCherney [1]; Chap. 1, p. 2. b, c Endometrial polyp by ultrasonography, sonohysterography, and hysteroscopy. Modified from Brown [2]; Chap. 5, p. 48

reactive oxygen species. Superoxide dismutase is highly expressed in glandular epithelial as well as stromal cells in the endometrium. The two systems, namely reactive oxygen species and superoxide dismutase balance each other and play very important roles in reproductive physiology such as menstruation and decidualization of the endometrium.

3.3 Oxidative Stress and Decidualization

Successful implantation relies upon decidualization of stromal cells [5]. Decidualization is a characterization of the endometrium of the pregnant uterus. It is a response of maternal cells to progesterone. Progesterone action on the endometrium increases glandular epithelial secretion, stimulates the accumulation of glycogen in stroma, and promotes the stromal vascularity of the spiral arterioles.

The antioxidant system is essential for the successful transformation of the endometrium to become the decidua. Superoxide dismutases such as copper–zinc as well as manganese are important in the process of decidualization [10–14]. The human endometrium has been extensively studied both during the menstrual cycle as well as the early stages of pregnancy [10]. In the endometrial glandular epithelium, constant immunostaining of copper–zinc as well as manganese superoxide dismutase throughout the menstrual cycle is demonstrated. Stromal cells demonstrate no immunostaining for either of these superoxide dismutases during the proliferative phase and only moderate immunostaining in stromal cells showing morphologically decidualized changes after the mid-secretory phase [11]. On the other hand, there is intensive immunostaining in decidualized stromal cells of early pregnancy. There is strong evidence that superoxide dismutase expression in the stromal cells is associated with decidualization.

Sugino [10] studied the relation between ovarian steroids and superoxide dismutases. Estrogen and progesterone increased copper–zinc superoxide dismutase activities as well as mRNA levels. Cyclic AMP had no effect. In contradistinction, manganese superoxide dismutase activities as well as mRNA levels were increased by both estrogen and progesterone as well as cAMP. Those stimulatory effects were inhibited by protein kinase-A inhibitor [14]. Sugino [10] proposed that estrogen and progesterone induces manganese superoxide dismutase by cAMP. However, copper–zinc superoxide dismutase is regulated by a separate pathway. Potentially, there are different roles for the two types of superoxide dismutase in the endometrial stromal cells undergoing decidualization.

3.4 Oxidative Stress and Menstruation

Menstruation is an orchestrated series of events that involves prostaglandins, cytokines, and proteinases. Prostaglandin F2-α (PGF2-α) is responsible for endometrial shedding by causing vasoconstriction of the spiral blood vessels. This is followed by vasodilatation in the endometrium prior to menstruation. PGF2-α concentrations increase in the endometrium toward the late secretory phase and reach their peak just before menstruation.

In the human endometrium during the late secretory phase, superoxide dismutase activity is decreased and lipid peroxide levels increase just before menstruation [11, 12]. In the human endometrial stromal cells, hydrogen peroxide stimulates the production of PGF2-α [12, 13]. The intricate relationship between superoxide dismutase, the active oxygen species, and PGF2-α is essential in the regulation of menstruation. When the level of ovarian steroids decrease, menstruation occurs. This is possibly mediated through the production of PGF2-α as a result of the withdrawal of ovarian steroids (Fig. 3.2). Sugino [10] elegantly discusses implantation and the balance in oxidative and antioxidative forces in the endometrium. As a result of successful implantation, progesterone levels are well-maintained, copper–zinc superoxide dismutase activity in the endometrium is high and therefore, the production of reactive oxygen species and subsequently PGF2-α levels are suppressed. They also demonstrated that in the decidua of early pregnancy, copper–zinc superoxide dismutase levels are high and reactive oxygen species levels are low [11].

In addition to causing endometrial cell apoptosis, reactive oxygen species may also regulate cellular functions by the production of substances that have biological activities. Withdrawal of ovarian steroids activates a transcription factor known as nuclear factor-κB by production of reactive oxygen species. This in turn stimulates the production of cyclooxygenase enzyme-2 (COX-2) and PGF2-α in the human endometrial stromal cells. PGF2-α in its turn initiates the menstrual process. Sugino [10] presented the intracellular signaling pathway for the increase in PGF2-α production in the stromal cells by reactive oxygen species. The mechanism by which the withdrawal of ovarian steroids cause the decline in

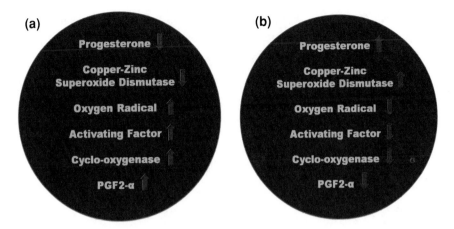

Fig. 3.2 a In the absence of pregnancy, progesterone levels drop resulting in a decrease in superoxide dismutase and increase in oxygen radicals and finally PGF2-α levels increase and menstruation starts. **b** When pregnancy occurs, progesterone levels are maintained and superoxide dismutase levels increase and oxygen radicals decrease and finally PGF2- α levels remain low and menstruation does not start

the expression of superoxide dismutase could be as a result of the drop in progesterone or by some other factors that are switched on by the withdrawal of ovarian steroids (Fig. 3.2).

3.5 Human Chorionic Gonadotropin and Oxidative Stress

Human chorionic gonadotropin (HCG) is routinely used to trigger ovulation and complete the oocyte meiotic division. Furthermore, it is frequently used for luteal support. Rizk [15–17] stated that while HCG is as effective as progesterone for luteal phase support, it carries a higher risk of ovarian hyperstimulation syndrome. HCG favorably impacts both the ovary and the endometrium. The main role of HCG is assumed to be the release of ovarian steroids, namely estradiol and progesterone as well as others. A very interesting issue is whether HCG confers resistance to oxidative stress-induced apoptosis in decidualizing human endometrial stromal cells.

Kajihara et al. [18] investigated the effect of HCG on the expression of genes that are involved in the resistance to oxidative stress in decidualizing human endometrial stromal cells. Human endometrial stromal cells were isolated and incubated with adenosine monophosphate and medroxyprogesterone acetate in the presence or absence of recombinant HCG at various concentrations. Recombinant HCG conferred additional protection to decidualizing human endometrial stromal cells against stress-induced apoptosis. HCG might improve the uterine environment prior to implantation by suppressing the apoptotic response in the maternal decidua under oxidative stress [18].

3.6 Progesterone Receptor Activity and Oxidative Stress

Survival of early pregnancy is dependent on progesterone signaling in the maternal decidua. Progesterone is produced by the corpus luteum of pregnancy. Progesterone supplementation is the standard of practice. In most assisted reproductive cycles, progesterone signaling could be affected by oxidative stress [19]. The mechanism by which progesterone withdrawal results in increased PGF2-α has been elegantly demonstrated by Sugino [10] (Fig. 3.2). The Japanese investigators postulated that progesterone withdrawal impacts the balance between oxidants and antioxidants resulting in increased expression of cyclooxygenase enzyme 2 (COX2) which results in increased PGF2-α production.

3.7 Early Pregnancy Loss and Oxidative Stress

Early pregnancy loss is a complicated issue that could result from a myriad of problems. Oxidative stress is a common pathology for different etiologies of early pregnancy loss. It has been questioned whether endoplasmic reticulum stress is a cofactor of oxidative stress in decidual cells from patients with early pregnancy loss [20]. Liu et al. [20] performed proteomic analysis from women who suffered from early pregnancy loss and they analyzed markers for endoplasmic reticulum stress, redox status, apoptotic features, and cell viability. They demonstrated that the cell survival of the decidual cells was dose-dependently reduced by hydrogen peroxide and could be reversed by the presence of vitamin E. The authors suggested that sustained endoplasmic reticulum stress occurs in decidual cells resulting in early pregnancy loss [20]

In early pregnancy, the levels of reactive oxygen species are low in the decidua [11]. The synthetic capacity is very low after conception in human decidua. These two features are essential for the maintenance of pregnancy. Sugino et al.[12] explained the possible mechanism for uterine contractions in spontaneous abortions and the absence of contractions in missed abortions. The PGF2-α and lipid peroxide concentrations in the decidua of missed abortions were found to be similar to that of normal pregnancy [13]. However, these levels are much lower than those of incomplete abortion where uterine bleeding occurs and uterine contractions follow. Sugino et al. [12] suggest that lipid peroxides in the decidua could be responsible for the increased PGF2-α production. This may lead to strong uterine contractions and possible expulsion of the uterine contents.

Brosens et al. [21] noted that the cyclic endometrial decidualization followed by menstrual shedding is confined to only a few species. Placental formation starts by deep trophoblastic invasion into maternal tissues and its access to the vascular system. Both pregnancy and menstruation are inflammatory conditions that involve a degree of physiologic ischemia followed by reperfusion. The emergence of cyclic menstruation may not have been an evolutionary coincidence [21]. Menstruation may serve to protect the uterine tissues from oxidative stress associated with deep placentation.

3.7.1 The Fallopian Tube and Oxidative Stress

The fallopian tubes play a role in the success and maintenance of early pregnancy. Hydrosalpinges have a detrimental outcome on pregnancy outcome even when IVF is performed [8]. Somatic cell embryo coculture enhances embryo development in vitro by producing embryotrophic factors as well as removing potentially harmful substances. The mechanisms by which somatic cells remove the toxic substances remain a subject for investigation. Cheong et al., [22] demonstrated that in the mouse model, the oviductal microsomal epoxide hydrolase reduces active oxygen species level and thereby enhances preimplantation mouse embryo development.

3.7.2 Uterine Binding Proteins and Oxidative stress

Uterine binding protein FKBP52 is a steroid receptor associated protein. It belongs to the large family of immunophilins and is associated with the cytoskeleton during mitosis. FKBP52 deficiency might be implicated in implantation failure as a result in reduction of progesterone responsiveness [23]. Most clinicians believe in progesterone supplementation as a tool to diminish the odds of increased implantation failure, however, progesterone responsiveness is not taken into account. Treatment with antioxidants such as α-tocopherol and N-acetylcysteine may also reduce implantation failure [23].

3.8 Oxidative Stress and Placental Development

The development of the placenta is affected by the uterine environments. During the first trimester, the development of the placenta occurs in a low oxygen environment supported by histiotrophic nutrition from the endometrial glands. The growth of the chorionic sac is remarkably uniform between individuals. A major switch to hemotrophic nutrition occurs toward the end of the first trimester. The rise in intraplacental oxygen concentration poses a major challenge to placental tissues. At this time, extensive villus remodeling occurs. Later in pregnancy, nutrition deprivation and vascular compromise play an important role. Oxidative stress and particularly its effect on endoplasmic reticulum stress could be the reason for vascular compromise and hence play a role in intrauterine growth restriction [24–26].

Successful human pregnancy depends on trophoblastic invasion of the maternal spiral arteries during the process of trophoblastic invasion. There is loss of smooth muscle and elastic lamina from the vessel wall as well as the inner third of the myometrium. There is significant vasodilation that accompanies this process.

Failure of trophoblastic invasion accompanies several complications of pregnancy such as early onset preeclampsia and fetal growth restriction [25, 26].

3.9 Oxidative Stress and Carcinogenesis

Oxidative stress and an impaired antioxidant system are significant in the pathophysiology of endometrial disease including endometrial hyperplasia and endometrial cancer. Pejic et al. [27] explored the lipid peroxidation levels and antioxidant enzyme activities in women with different forms of gynecological disease. Endometrial tissues from patients with different diagnoses were subject to assays for superoxide dismutase, catalase, glutathione peroxidase, glutathione reductase, and lipid hydroperoxides. Patients with hyperplastic and malignant lesions such as adenocarcinoma, had enhanced lipid peroxidation and altered uterine antioxidant activities compared with patients with benign uterine disease such as polyps and myoma. Lipid peroxide levels and antioxidant enzyme activities in the uterus could be clinically useful in the evaluation of gynecological disorders [27].

Oxidative stress simply arises when the production of reactive oxygen species exceeds the intrinsic antioxidant defenses. Reactive oxygen species function as second messengers in the intracellular signaling cascades that control the cell homeostasis. Disruption of function at the cellular level and even cell death may occur. Oxidative stress and an impaired antioxidant system have been proposed as a potential factor involved in the pathophysiology of carcinogenesis. Superoxide dismutase activity is significantly decreased in hyperplasia and adenocarcinoma patients. Glutathione peroxidase and reductase are increased in hyperplastic patients and only glutathione reductase is increased in adenocarcinoma patients. Patients with premalignant and malignant uterine disease have advanced lipid peroxidation and increased antioxidant activity compared to patients with benign disease such as uterine polyps and myomas [27]

3.10 Oxidative Stress and Endometriosis

Endometriosis represents the ectopic implantation of the endometrium in the pelvic peritoneum and the ovaries (Fig. 3.3a, b). It is thought that the endometrium in women with endometriosis has different abnormalities at the cellular level which predisposes it to its implantation in the pelvis and its persistence [7]. The question of whether oxidative stress is part of the reason for the successful implantation and persistence of the endometrium in the pelvis has been discussed by Gupta et al., [28]. Several hypotheses have been proposed as to why oxidative stress may occur in women with endometriosis. Menstrual reflux transplants cell debris in the peritoneal cavity which may be associated with the development of endometriosis (Fig. 3.3). Erythrocytes release hemoglobin which acts as a

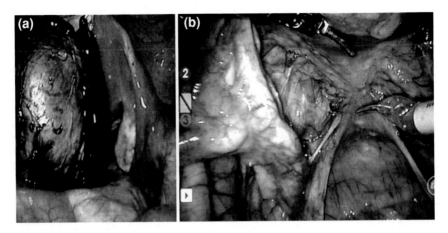

Fig. 3.3 a Ovarian endometrioma (Courtesy Garcia-Velasco and Rizk). **b** Pelvic endometriosis in the cul-de-sac demonstrating vesicles and inflammatory reaction (Courtesy of Botros Rizk, MD)

proinflammatory factor. It contains the redox generating iron molecule. Gupta et al. [28] proposed that the presence of iron macrophages and environmental contaminants such as polychlorinated biphenyls in the peritoneal cavity could cause an imbalance between antioxidants and reactive oxygen species (Figs. 3.4 and 3.5).

Nitric oxide and nitric oxide synthase are detected in greater amounts in the endometrium of women with endometriosis. Increased expression is particularly evident in the glandular epithelium. Rizk and Sallam [3] suggested that the variations in the expression of the endothelial nitric oxide synthase gene may be involved in endometrial angiogenesis. They may modulate the process of endometriosis and may have implications for implantation.

The endometrium of patients with endometriosis may exhibit increased lipid protein complex modification. On one hand, the lipid peroxide concentrations are elevated and on the other hand, reduced levels of antioxidant enzymes such as superoxide dismutase may be observed.

3.10.1 Glutathione and Endometrial Stromal Cells

Glutathione is one of the cornerstones in the intracellular antioxidant system. It plays a key role in endometrial detoxification and also plays a key role in the pathogenesis of endometriosis. Estrogen is well known as a major risk factor in the development and progression of endometriosis. Glutathione levels were significantly increased following the in vitro culture and treatment of endometrial stromal cells with estradiol, tumor necrosis factor, and interleukin1-β. Increased production of estradiol and proinflammatory interleukins in the peritoneal cavity

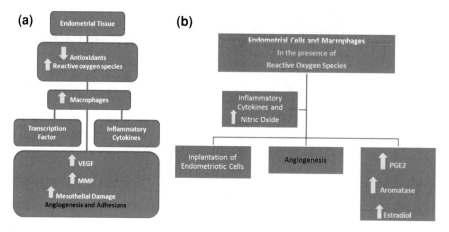

Fig. 3.4 a The role of reactive oxygen species, inflammatory cytokines, and vascular endothelial growth factor in the pathogenesis of endometriosis. b The role of reactive oxygen species, inflammatory cytokines, and prostaglandins in the pathogenesis of endometriosis

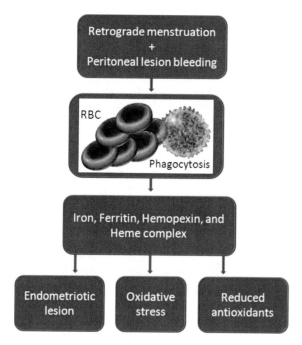

Fig. 3.5 The role of retrograde menstruation, iron, and oxidative stress in the pathogenesis of endometriosis

may result in the establishment of endometriosis through increased levels of glutathione [29].

3.10.2 The Role of Iron in the Pathogenesis of Endometriosis

Peritoneal endometriosis is portrayed by Garcia-Velasco and Rizk [6] as a chronic inflammatory disease characterized by increased numbers of peritoneal macrophages and their secreted products. Inflammation plays a major role in pain and infertility associated with endometriosis, but is also extensively involved in the molecular processes that lead to peritoneal lesion development. Peritoneal oxidative stress is important for the endometriosis-associated inflammatory pathogenesis (Fig. 3.5). Excessive production of reactive oxygen species, secondary to peritoneal influx of pro-oxidants such as iron during retrograde menstruation, may initiate cellular damage and activate nuclear factor-kappa B, which increases proinflammatory gene expression (Fig. 3.5). This transcription factor increases prostaglandin biosynthesis which may be implicated in the peritoneal proinflammatory process in endometriosis [30].

3.10.3 Angiogenesis in Endometriosis

Angiogenesis is an integral component of the implantation of the ectopic endometrium and establishment of pelvic endometriosis. VEGF is the main mechanism by which angiogenesis is established. Oxidative stress may contribute to angiogenesis in ectopic endometrial implants and helps the progression of endometriosis by increasing VEGF production [28, 31]. Glycodelin, a glycoprotein whose expression in increased by oxidative stress, may possibly function as an autocrine factor that increases VEGF expression in the ectopic endometrial tissue [32].

3.10.4 The Roles of Thioredoxin and Thioredoxin-Binding Protein-2 in Endometriosis

Oxidative stress has been implicated in the pathogenesis of endometriosis [27, 28, 31]. Thioredoxin protects cells against oxidative stress, and thioredoxin-binding protein-2 (TBP-2) is a suppressor factor of thioredoxin in the biological cellular function. Patients with endometriosis show lower levels of TBP-2 with no significant changes of TRX mRNA levels leading to a high TRX to TBP-2 ratio in comparison to patients without endometriosis. Seo et al. [33] suggested that the high TRX/TBP-2 ratio may be associated with the establishment of endometriosis [33].

3.11 Conclusion

In this chapter, we have reviewed the role of oxidative stress in menstruation, decidualization, and implantation. Successful implantation maintains progesterone levels and increased superoxide dismutase activity, while reactive oxygen species, and PGF2α levels are suppressed. HCG and progesterone are useful for the improvement of the endometrium for implantation.

3.12 Keypoints

- Oxidative stress and antioxidant defense mechanisms play an important role in menstruation, decidualization, and implantation.
- Successful implantation results in maintenance of progesterone levels and increased superoxide dismutase activity in the endometrium.
- Superoxide dismutase activity suppresses the production of reactive oxygen species and subsequently PGF2-α production.
- Menstruation is an orchestrated series of events that involves prostaglandins, cytokines, and proteinases.
- In the absence of pregnancy, progesterone levels drop, superoxide dismutase activity drops, and COX2 and PGF2-α increase resulting in menstruation
- HCG might improve the uterine environment prior to implantation by suppressing the apoptotic response to oxidative stress in the maternal deciduas.
- PGF2α and lipid peroxide concentrations in the decidua of missed abortions are similar to those of normal pregnancy.
- PGF2α and lipid peroxide levels in missed abortion and normal pregnancy are significantly lower than incomplete abortion where uterine bleeding occurs and uterine contractions follow.
- Oxidative stress may be involved in the pathogenesis of endometriosis by several mechanisms.
- The presence of macrophages, iron, and environmental contaminants such as polychlorinated biphenyls in the peritoneal cavity of endometriosis patients could cause an imbalance between antioxidants and reactive oxygen species.

References

1. DeCherney A, Hill MJ (2010) The future of imaging and assisted reproduction. In: Rizk B (ed) Ultrasonography in reproductive medicine and infertility. Cambridge University Press, Cambridge pp. 1–10 (Chapter 1)
2. Brown WW III (2010) Sonohysterography. In: Rizk B (ed) Ultrasonography in reproductive medicine and infertility. Cambridge University Press, Cambridge, pp 42–53 (Chapter 5)

3. Rizk B, Sallam HN (2012) The uterine factor in infertility. In: Rizk B, Sallam HN (eds) Clinical infertility and in vitro fertilization. Jaypee Brothers Medical Publishers, St. Louis, pp 84–96 (Chapter 7)
4. de Zeigler D, Fraisse T et al (2008) Endometrial receptivity. In: Rizk B, Garcia-Velasco J, Makrigiannakis A (eds) Infertility and assisted reproduction. Cambridge University Press, New York, pp 38–45 (Chapter 4)
5. Jauniaux E, Rizk B (eds) (2012) Pregnancy after reproductive technology. Cambridge University Press, Cambridge
6. Garcia-Velasco JA, Rizk B (eds) (2010) Endometriosis: current management and future trends. Jaypee Medical Publishers, St. Louis
7. Gardner DK, Rizk B, Falcone T (eds) (2011) Human assisted reproductive technology: future trends in laboratory and clinical practice (Cambridge Medicine). Cambridge University Press, Cambridge
8. Aboulghar M, Rizk B (eds) (2011) Ovarian stimulation. Cambridge University Press, Cambridge
9. Dickey RP (2010) The endometrium. In: Rizk B (ed) Ultrasonography in reproductive medicine and infertility. Cambridge University Press, Cambridge, pp 97–102 (Chapter 12)
10. Sugino N (2007) The role of oxygen radical-mediated signaling pathways in endometrial function. Placenta 28(Suppl A), Trophoblast Res 21:S133–S136
11. Sugino N, Shimamura K, Takiguchi S et al (1996) Changes in activity of superoxide dismutase in the human endometrium throughout the menstrual cycle and in early pregnancy. Hum Reprod 11:1073–1078
12. Sugino N, Karube-Harada A, Kashida S et al (2001) Reactive oxygen species stimulate prostaglandin F2α production in human endometrial stromal cells in vitro. Hum Reprod 16:1797–1801
13. Sugino N, Nakata M, Kashida S et al (2000) Decreased superoxide dismutase expression and increased concentrations of lipid peroxide and prostaglandin F2-α in the decidua of failed pregnancy. Mol Hum Reprod 6:642–647
14. Sugino N, Karube-Harada A, Sakata A et al (2002) Different mechanisms for the induction of copper-zinc superoxide dismutase and manganese superoxide dismutase by progesterone human endometrial stromal cells. Hum Reprod 17:1709–1714
15. Rizk B (2006) Ovarian hyperstimulation syndrome: epidemiology, pathophysiology, prevention and management. Cambridge University Press, Cambridge
16. Rizk B (2009) Genetics of ovarian hyperstimulation syndrome. Reprod Biomed Online 19:14–27
17. Rizk B, Smitz J (1992) Ovarian hyperstimulation syndrome after superovulation using GnRH agonists for IVF and related procedures. Hum Reprod 7:320–327
18. Kajihara T, Uchino S, Itakura A (2011) Human chorionic gonadotropin confers resistance to oxidative stress-induced apoptosis in decidualizing human endometrial stromal cells. Fertil Steril 95(4):1302–1307 Mar 15
19. Leitao B, Jones MC, Fusi L et al (2010) Silencing of the JNK pathway maintains progesterone receptor activity in decidualizing human endometrial stromal cells exposed to oxidative stress signals. FASEB J 24(5):1541–1551 May
20. Ax Liu, He WH, Yin LJ, Lv PP et al (2011) Sustained endoplasmic reticulum stress as a cofactor of oxidative stress in decidual cells from patients with early pregnancy loss. J Clin Endocrinol Metab 96(3):E493–E497
21. Brosens JJ, Wilson MS, Lam EW (2009) FOXO transcription factors: from cell fate decisions to regulation of human female reproduction. Adv Exp Med Biol 665:227–241
22. Cheong AW, Lee YL, Liu WM et al (2009) Oviductal microsomal epoxide hydrolase (EPHX1) reduce reactive oxygen species (ROS) level and enhances preimplantation mouse embryo development. Biol Reprod 81(1):126–132
23. Hirota Y, Acar N, Tranguch S et al (2010) Uterine FK506-binding protein 52 (FKBP52)-peroxiredoxin (PRDX6) signaling protects pregnancy from overt oxidative stress. Proc Natl Acad Sci USA 107(35):15577–15582

24. Burton GJ, Jauniaux E, Charnock-Jones DS (2010) The influence of the intrauterine environment on human placental development. Int J Dev Biol 54(2–3):303–312
25. Burton GJ, Woods AW, Jauniaux E et al (2009) Rheological and physiological consequences of conversion of the maternal spiral arteries for uteroplacental blood flow during human pregnancy. Placenta 30(6):473–482
26. Burton GJ, Yung HW, Cindrova-Davies T, Charnock-Jones DS (2009) Placental endoplasmic reticulum stress and oxidative stress in the pathophysiology of unexplained intrauterine growth restriction and early onset preeclampsia. Placenta 30(Suppl A):S43–S48
27. Pejic S, Todorovic A, Stojiljokvic V (2009) Antioxidant enzymes and lipid peroxidation in endometrium of patients with polyps, myoma, hyperplasia and adenocarcinoma. Reprod Biol Endocrinol 23(7):149 Dec
28. Gupta S, Sekhon L, Aziz N, Agarwal A (2008) The impact of oxidative stress on female reproduction and ART: an evidence-based review. In: Rizk B, Garcia-Velasco J, Makrigiannakis A (eds) Infertility and assisted reproduction. Cambridge University Press, New York, pp 629–642 (Chapter 64)
29. Lee SR, Kim SH, Lee HW et al (2009) Increased expression of glutathione by estradiol, tumor necrosis factor-alpha, and interleukin 1-beta in endometrial stromal cells. Am J Reprod Immunol 62(6):352–356 Dec
30. Kobayashi H, Yamada Y, Kanayama S et al (2009) The role of iron in the pathogenesis of endometriosis. Gynecol Endocrinol 25(1):39–52 Jan
31. Gupta S (2012) Oxidative stress and and its role in endometriosis-mechanistic and therapeutic implications. In: Rizk B, Sallam H (eds) Clinical infertility and in vitro fertilization. Jaypee Brothers Medical Publishers, New Delhi, pp 316–325 (Chapter 37)
32. Park JK, Song M, Dominguez CE et al (2006) Glycodelin mediates the increase in vascular endothelial growth factor in response to oxidative stress in the endometrium. Am J Obstet Gynecol 195(6):1772–1777
33. Seo SK, Yang HI, Lee KE et al (2010) The roles of thioredoxin and thioredoxin-binding protein-2 in endometriosis. Hum Reprod 25(5):1251–1258

Chapter 4
Oxidative Stress, Oogenesis and Folliculogenesis

Malcolm A. Paine, Elizabeth H. Ruder, Terryl J. Hartman, Jeffrey Blumberg and Marlene B. Goldman

Abstract There is increasing evidence that nutrition and other lifestyle factors are important in the etiology of reproductive failure. Diet and its constituent antioxidants as well as oxidative stress in females may be influential in the timing and continuation of viable pregnancies. This chapter reviews oxidative stress with particular reference to folliculogenesis and oogenesis. The evidence suggests that there are threshold levels for oxidative stress that may depend on anatomic site and stage of preconception. Oxidative stress, may influence oogenesis, folliculogenesis and the establishment of a viable early pregnancy. Reactive oxygen species are continuously formed in cells of the reproductive tract, secondary to biochemical

M. A. Paine
Obstetrics and Gynecology, Dartmouth-Hitchcock Medical Center, 1 Medical Center Drive, Lebanon, NH 03756, USA
e-mail: Malcolm.a.paine@Hitchcock.org

E. H. Ruder
National Cancer Institute, NIH Division of Cancer Epidemiology and Genetics, 6120 Executive Blvd, Suite 320, Rockville, MD 20852, USA
e-mail: rudereh@mail.nih.gov

T. J. Hartman
Nutritional Sciences, Penn State University, 110 Chandlee Laboratory, University Park, PA 16802, USA
e-mail: tjh9@psu.edu

JeffreyBlumberg
Antioxidants Research Laboratory, Jean Mayer USDA Human Nutrition Research Center on Aging, Tufts University, 711 Washington Street, Boston, MA 02111, USA
e-mail: jeffrey.blumberg@tufts.edu

M. B. Goldman (✉)
Dartmouth-Hitchcock Medical Center Obstetrics & Gynecology, One Medical Center Drive, Lebanon, NH 03756, USA
e-mail: Marlene.B.Goldman@Dartmouth.edu

reactions, such as those within the mitochondrial respiratory chain, and also as a result of external factors.

Keywords Oxidative stress · Antioxidant vitamins · Oogenesis · Folliculogenesis · Reproductive failure · Secondary oocyte quality · Corpus luteum function

4.1 Introduction

Oxidative stress (OS), may influence oogenesis, folliculogenesis and the establishment of a viable early pregnancy (Fig. 4.1). Reactive oxygen species (ROS) are continuously formed in cells of the reproductive tract, secondary to biochemical reactions, such as those within the mitochondrial respiratory chain, and also as a result of external factors (Fig. 4.2). OS induces lipid peroxidation, alters DNA and protein in a structural and functional manner, promotes apoptosis, and causes cumulative oxidative damage. Antioxidants (e.g., vitamins C and E and β-carotene) and trace elements (e.g., selenium, zinc, and copper) functioning as essential cofactors for antioxidant enzymes are capable of disposing, scavenging, or suppressing the formation of ROS. Inadequate production or intake of antioxidants or excessive generation of ROS results in OS. Evidence from in vitro, animal model, and clinical studies suggests that OS plays a role in the etiology of adverse reproductive events in both women and men [1–8]. In vitro and animal studies point to several mechanisms by which OS may affect fertility; however, no study has directly addressed the effects of OS on fertility in women.

Dietary Reference Intakes (DRIs) for some antioxidants are increased during pregnancy [9]. For example, vitamin C requirements are elevated in pregnancy due to both hemodilution and active placental transfer to the fetus; increased requirements are also noted for zinc, selenium, copper, and manganese. However, sufficient scientific evidence is not available to support a change in the requirement for vitamin E during pregnancy [9].

Pregnancy is a period of increased OS due to heightened metabolic activity. Decreased plasma thiols [10, 11] and increased placental lipid peroxides and decreased expression of antioxidant enzymes have been reported during pregnancy [12, 13]. This chapter reviews the impact of OS on oogenesis and folliculogenesis.

4.2 Oogenesis

Gametogenesis in the female, beginning with ovarian differentiation, is first seen at 6–8 weeks of embryo development, with rapid mitotic division of germ cells. The maximum oogonal content of the gonad of 6–7 million is reached by 16–20 weeks gestation. Commencing around 12 weeks, the oogonia are transformed to oocytes,

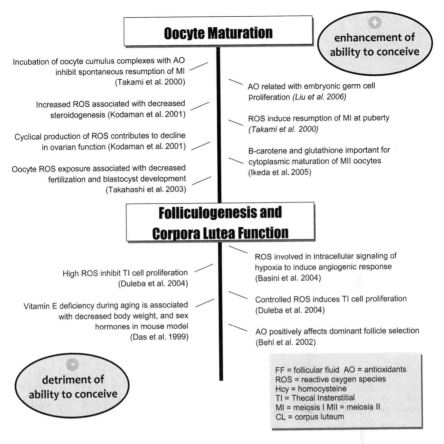

Fig. 4.1 Selected studies on reactive oxygen species and antioxidants and oocyte maturation, folliculogenesis, and corpora lutea function, by influence on conception

entering the first meiotic division, arresting in prophase I. Ova are formed from the two meiotic divisions of the oocyte. The first is completed just before ovulation, stimulated by follicle stimulating hormone (FSH), and the second at the time of sperm penetration.

The mechanisms linking OS to impaired oogenesis are not well known. OS may disturb spindle organization leading to aneuploidy [14], although high follicular oxygen and antioxidant concentration may aid in correct chromosomal alignment [15].

4.2.1 Ovarian Germ Cells to Secondary Oocytes: The Role of Oxidative Stress

Liu and colleagues [16] investigated the antioxidant effects of daidzen, an isoflavone found principally in soybeans, on germ cell proliferation in ovarian cells from chicken embryos. Daidzen exposure increased proliferation of germ

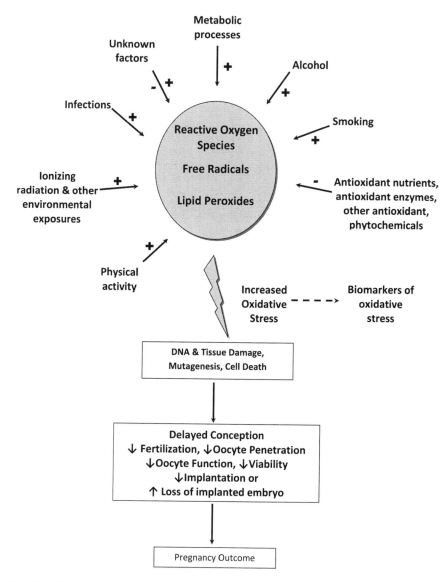

Fig. 4.2 Role of reactive oxygen species and oxidative stress in oocyte function and fertility

cells ($p < 0.05$) and helped restore overall antioxidant levels after hypoxanthine/xanthine oxidase (HX/XO) system challenge.

Subsequent to pubertal hormonal changes, a number of primary oocytes begin to grow each month. One primary oocyte becomes dominant and resumes meiosis I (MI). The resumption of MI is induced by elevated ROS and inhibited by antioxidants [17, 18], perhaps indicating that the preovulatory follicle is an important promoter of the ovulatory sequence via regulated generation of ROS. However, it has been suggested that cyclical ROS production may, over time, contribute to oophoritis associated with autoimmune premature ovarian failure [19] and exacerbated by diminished antioxidant status.

Elevated preovulatory luteinizing hormone (LH) results in oocyte maturation with the second meiotic division (MII) [20]. The oocyte remains suspended in metaphase until fertilization. Both human and rat granulosa and luteal cells respond negatively to ROS and adversely affect MII progression, leading to diminished gonadotropin and anti-steroidogenic actions, DNA damage, and inhibited protein ATP production [19]. A recent study revealing a positive association among follicular fluid (FF), estradiol (E2) and total antioxidant capacity (TAC) suggests that E2 may also play a role in the ovarian antioxidant–oxidant balance in granulosa cells [21].

Glutathione (GSH) is a nonprotein sulfhydryl tripeptidyl molecule and may be the major source of redox potential in the oocyte [22]. Elevation of glutathione reductase (GSHR) activity appears to be responsible for maintaining gamete viability [23]. GSH is also critical for oocyte maturation, particularly in the cytoplasmic maturation required for preimplantation development and formation of the male sperm pronucleus [24, 25].

β-carotene has been recognized in bovine models for its ability to enhance oocyte cytoplasmic maturation [26]. The contrasting relationship of antioxidants, deleterious for the progression of MI, but beneficial for MII, implies a complex role for ROS in the ovarian environment. Such findings require an appreciation of ROS as multifunctional agents in which their effects may vary over the continuum of concentration and developmental stages [27].

4.2.2 Impact of ROS on the Aging Oocyte

Human follicular fluid free radical activity and apoptosis of human granulosa and cumulus cells increase with age [28–30]. It has been hypothesized that prolonged exposure of aged oocytes to ROS negatively affects calcium homeostasis and impairs Ca^{2+} oscillation-dependent signaling, thereby causing a decline in oocyte developmental ability [31]. At the time of sperm penetration, drastic changes occur in intracellular oocyte free calcium concentration ($[Ca^{2+}]_i$). Initially, there is a single long-lasting increase in $[Ca^{2+}]_i$, followed by short, repetitive transient $[Ca^{2+}]_i$, lasting for several hours. These changes in $[Ca^{2+}]_i$ trigger the release of the cortical granule thereby preventing penetration of the ovum by additional sperm and stimulate the resumption of meiosis [31]. Follicular fluid aspirates from younger (aged 27–32 years) and older (aged 39–45 years) women undergoing in vitro fertilization (IVF) treatment indicated lower specific activity of catalase, lower concentration of glutathione transferase (GST) and higher superoxide dismutase (SOD) activity in the older women relative to the younger women. No differences were observed between the two groups for glutathione peroxidase

(GSHPx) or GSHR activity or for protein expression of catalase or SOD. The age-related changes caused a reduction in the catalase/SOD ratio and a slight reduction in the GSHPx/SOD ratio, suggesting an overall decrease in ROS scavenging ability with aging. However, the indication for IVF treatment and the cause of the infertility were not described and may have influenced FF antioxidant and ROS activity.

Whole ovaries from reproductive aging rats (aged 8–9 months) and control animals (aged 26 days) were removed and homogenized following corpus luteum (CL) induction and prostaglandin F2α (PGF2α) administration in the midluteal phase to induce luteal regression [32]. Analysis of homogenized ovaries at baseline 2 and 24 hours post-PGF2α administration indicated alterations in antioxidant defense with age. Diminished antioxidant status may induce apoptosis during luteal regression and lead to decreased progesterone synthesis. The aged ovaries had elevated vitamin E content at 0, 2, and 24 hours ($p < 0.05$), and lower GSHR levels at 2 and 24 hours post-PGF2α administration ($p < 0.01$) compared to control ovaries. No significant differences in GSHPx, catalase, or thiobarbituric acid-reacting substances (TBARS), an index of lipid peroxidation, were detected. Thus, a shift toward a higher concentration of vitamin E may occur to help protect the aging ovary during luteolysis and compensate for the decline in the luteal cell ability to quench ROS, as evidenced by lower GSHR.

4.3 Folliculogenesis

The increase in steroid hormone production of developing follicles is accompanied by an increase in the activity of cytochrome P450, which, in turn, generates ROS (e.g., H_2O_2) [33]. A study of ROS generation during the ovulatory cascade in response to LH indicated that a gonadotropin stimulated, protein kinase C activated, NADPH/NADH oxidase-type superoxide generator in the preovulatory follicle exists and may be a regulating factor in ROS production during ovulation [18]. Behl and Pandey [34] investigated whether changes in the antioxidant enzyme catalase (which converts H_2O_2 to H_2O and O_2) and estradiol activity of ovarian follicular cells in various stages of development varied with FSH. Granulosa cells isolated from dissected goat ovarian follicles indicated that large follicles (>6 mm) exhibited greater catalase activity than granulosa cells from small (<3 mm) or medium (3–6 mm) sized follicles. After a uniform dose of FSH (200 ng/mL), both catalase activity and estradiol release were greater in large follicles than in medium or small follicles. Since the dominant follicle will be the follicle with the highest estrogen concentration, the concomitant increases in catalase and estradiol in response to FSH suggest a role for catalase in follicular selection and prevention of apoptosis.

ROS-dependent oxidation of lipoproteins into oxidized low-density lipoprotein (oxLDL) and the up-regulation of its lectin-like oxLDL receptor-1 (LOX-1) occurs in endothelial cells [35]. Recently, Bausenwein and colleagues [36] established that oxLDL and LOX-1 are present in follicular fluid or human granulosa cells and are elevated in obese women.

Detoxification of ROS (produced by active metabolism) by aldo–keto reductase appears to contribute to the maintenance of the genital tract [22]. In fact, granulosa cells and the epithelia of the genital tract produce high levels of aldose reductase and aldehyde reductase [37].

In a rat model, maternal under-nutrition resulted in reduced offspring ovarian follicle numbers and mRNA levels of regulatory genes and may be mediated by increased ovarian OS coupled with a decreased ability to repair the resultant oxidative damage [38]. OS may induce DNA damage and trigger apoptosis of the follicle [39]. An in vitro crossover study of temporal effects of sow FF suggested that a major role of FF is to provide protection from OS [40]. Maturation with FF relative to the control medium for all or part of in vitro maturation (IVM) increased cumulus expansion and progesterone production and decreased the incidence of cumulus cell apoptosis ($p < 0.05$). Granulosa cell hypoxia is physiological during follicular growth [41]. Low oxygen tension stimulates follicular angiogenesis, needed for follicular growth and development. Impairment of angiogenesis within ovarian follicles contributes to follicular atresia [42]. ROS may act as signal transducers [43] or intracellular messengers [44] of the angiogenic response.

SOD catalyzes the dismutation of superoxide radicals (O_2^-) to O_2 and H_2O_2 and plays an essential role in the defense against oxygen toxicity [45]. In human tissues, there are two forms of intra-cellular SOD: mitochondrial manganese SOD (MnSOD) and cytoplasmic copper/zinc SOD (Cu/ZnSOD) [46]. Mitochondria are the major consumers of cellular oxygen, thereby providing support to the hypothesis that ROS are involved in intracellular signaling between tissue hypoxia and angiogenic response. Basini et al. [47] studied the response of antioxidant enzymes (SOD, catalase, peroxidase) and ROS in granulosa cells isolated from swine follicles and held in normoxic, hypoxic, and anoxic environments. ROS were decreased ($p < 0.05$) and SOD and peroxidase activities were significantly increased ($p < 0.05$) by hypoxic and anoxic compared to normoxic conditions, but the difference in activity between the two test conditions was not statistically significant. Catalase activity was unaffected by hypoxic or anoxic conditions. Catalase may exhibit lack of change as it is located in peroxisomes [48], whereas SOD and peroxidase are found in the cytoplasm or the mitochondria. High SOD concentrations in the FF have been associated with oocytes that failed to fertilize [49]. SOD levels have been positively correlated with ovarian steroidogenesis [50], and preovulatory antral follicles demonstrate lipid peroxidation [2] and regulation of follicular hydroperoxides may be mediated by GSHPx [51].

In a series of 56 subjects undergoing IVF and 13 age-matched healthy controls, FF and serum samples were compared for levels of lipid peroxidation, reduced glutathione, GSHPx, and vitamins A, E, and C [52]. Lipid peroxidation levels in serum and FF of patients undergoing IVF were lower and vitamins A and C and GSHPx concentrations in FF were higher than among controls. After 45 days of multivitamin and mineral supplementation, described in detail in Ozkaya et al.

[53], there were decreases in FF and serum lipid peroxidation levels along with increases in serum GSH, vitamin C and E concentrations, and GSHPx and vitamin C concentrations in FF. These results suggest that multivitamin and mineral supplementation in women undergoing IVF may strengthen the antioxidant defense system by decreasing OS.

4.3.1 Antioxidants in Folliculogenesis

OS and apoptosis are consequences of folliculogenesis, follicular atresia, and luteal regression. However, the ROS increase can be countered (be it desirable or undesirable) by antioxidant status [27]. Estradiol has been shown to have antioxidant properties and at high concentrations, perhaps by an antioxidant mechanism, suppressed basal and H_2O_2-induced apoptosis in sheep ovarian surface epithelial cells [54]. This is in contrast with some recent evidence that estradiol at 0.1 μM for a minimum of 24 hours produced oxidative DNA *damage* in mouse ovarian surface epithelium, and that this damage can be prevented by vitamin E [55]. Follicular ROS initiate apoptosis whereas follicular GSH, in addition to FSH, protects against apoptosis in cultured preovulatory rat follicles [56]. Oocyte GSH synthesis is believed to be stimulated by low-molecular weight thiol compounds including cysteine, cysteamine, and β-mercaptoethanol [57, 58]. Supplementation of cysteamine during IVM of sheep oocytes indicated that a single cysteamine supplement of 200 μmol increased morula and blastocyst development ($p < 0.05$), but no such effect was found with IVM β-mercaptoethanol supplementation. However, both cysteamine and β-mercaptoethanol supplementation were found to decrease intracellular peroxidase content, most likely via increased GSH synthesis [59]. Interestingly, increased serum GSHR was significantly associated with decreased time to pregnancy in 83 female participants in a prospective study with preconception enrollment [60]. No statistically significant associations were found with GSHPx, SOD, catalase, or thiobarbituric acid. There is also evidence that OS could be sensed by cytokeratin-positive (CK +) granulosa cells and that changes in catalase and GSHPx activities in the preovulatory follicular fluid might indicate an ovarian disorder and have clinical significance [61].

Uterine peroxidase and uterine alkaline phosphatase are estrogen-inducible enzymes associated with uterine growth [62]. Histological analysis indicated that ovaries of rats on a 70-days vitamin E deficient diet showed degenerated follicles, follicles with increased diameter, and hypertrophy of the granulosa cells, whereas control animals exhibited healthy, large follicles [63]. In addition to vitamin E, other antioxidants such as manganese (a cofactor for SOD) are known to influence LH secretion in female rats [64] and ascorbic acid (vitamin C) has been shown to stimulate gonadotropin release in male rats [65], thus suggesting that antioxidants stimulate the release of gonadotropins from the anterior pituitary.

Melatonin and its metabolites are free radical scavengers and also modulate gene transcription for antioxidant enzymes [66, 67]. A recent investigation of the link between OS and poor oocyte quality among women undergoing IVF and embryo transfer indicated that the concentration of 8-hydroxydeoxyguanine (8-OHdG), a biomarker of oxidative DNA damage, was significantly greater in the FF of women with a high rate (>30%) of degenerative oocytes [68]. Subsequent fertilization rate was improved in a subset of women who failed to become pregnant in the previous IVF-embryo transfer cycle when they were provided with 3 mg of melatonin per day, 600 mg α-tocopherol (vitamin E) per day, or both melatonin and α-tocopherol from the 5th day of the previous menstrual cycle to day of oocyte retrieval. Compared with the previous IVF cycle, administration of any of the three treatments was associated with a significantly reduced intrafollicular concentration of 8-OHdG ($p < 0.05$). Hexanoyl–lysine adduct (a measure of lipid peroxidation) was significantly reduced by treatment with α-tocopherol.

Basini et al. [47] reported that ROS, under moderate concentrations, play a role in signal transduction processes involved in growth and protection from apoptosis. The in vitro effects of antioxidants and OS on proliferation of rat thecal-interstitial (T-I) cells were investigated by Duleba et al. [69]. T-I cells develop in the secondary follicle stage and control follicle growth and atresia, regulate ovarian steroidogenesis, and may provide mechanical support for ovarian follicles [70]. ROS were found to induce a biphasic effect with lower concentrations inducing T-I proliferation ($p < 0.01$) and higher concentrations inhibiting proliferation ($p < 0.01$), suggesting that controlled levels of ROS may be needed to maintain DNA synthesis, T-I cell proliferation, and growth of ovarian mesenchyme. It is uncertain whether these in vitro results can be extrapolated to the in vivo environment.

4.4 Secondary Oocyte Quality

It is hypothesized that ROS are released in connection with follicle rupture and are involved in the process itself, possibly as a result of production by inflammatory cells, such as macrophages and neutrophils [22]. Chao and colleagues [71] investigated murine oocyte competence, ovarian mitochondrial DNA (mtDNA) mutation, and oxidative damage after repeated ovarian hyperstimulation by exogenous gonadotropin. With repeated stimulation, an increase in degenerative oocytes and ovulated immature oocytes was seen, indicating a decrease in oocyte quality. Ovarian oxidative damage increased with repeated cycles of stimulation, with a statistically significant increase in lipid peroxides and an increase in 8-OHdG, between the first and sixth cycles ($p < 0.05$). In addition, an increase in mtDNA large-scale deletions was noted with increased ovarian stimulation. Related work by Tarin et al. [72] suggests that the timing of antioxidant administration may affect the number and quality of ovulated oocytes in female mice, as assessed by morphology and chromosome distribution. Mice were administered a mixture of vitamins C and E beginning early (after weaning) or late (beginning at 32 weeks of age), continuing through sacrifice at 40–42, 50–52, or 57–62 weeks after exogenous stimulation [72]. Overall, ovulated oocyte quality was evaluated by summation of the number of retrieved oocytes and total percentage of ovaries exhibiting morphological traits indicative of apoptosis at all three time points of sacrifice from both antioxidant supplementation groups combined and compared to the control. An increased number of normal MII oocytes was observed ($p = 0.039$), and percentage of apoptotic oocytes decreased ($p = 0.041$), in the antioxidant group compared to the control group. The preventive effects of supplementation were greatest when supplementation began after weaning and continued to time of sacrifice.

GSH is thought to be a highly relevant biochemical marker for mammalian oocyte viability [58, 73]. Levels of GSH from samples collected during hamster IVM indicated ovulated oocytes suspended in metaphase of MII, along with their associated cumulus cells, have approximately double the GSH concentration compared to immature germinal vesicle stage oocytes [73]. Increased GSH was also found through the preimplantation stage in bovine oocytes [57].

Nitrogen oxide species (RNOS) are derived primarily from nitric oxide (NO). Nitric oxide synthase (NOS), partially encoded by the genes inducible NOS (NOS II) and endothelial NOS (NOS III), is expressed in the ovary and data suggest that the NOS II-NO-(cyclic GMP) system plays a role in oocyte meiotic maturation [22, 74].

4.5 Corpus Luteum Function

ROS are generated by the corpus luteum (CL) via the monooxygenase reaction as a byproduct during steroid hormone synthesis, when mitochondrial P450 systems can represent an additional burden over the common OS of the respiratory chain [33].

Antioxidants such as β-carotene (giving its bright yellow pigmentation), α-tocopherol, and vitamin C [19, 33] scavenge ROS, protect against oxidative damage of P450 systems and are contained in high concentrations in the CL. The levels of β-carotene and α-tocopherol in bovine CL mirror both the ovarian cycle and plasma progesterone levels [75, 76]. ROS are produced during luteal regression, where they play a role in apoptosis [19]. Activities of SOD, catalase and other antioxidant enzymes in various human and mammalian CL seem to parallel the steroidogenic capacity of the cells [33].

4.6 Steroidogenesis

Overexposure of the ovary to H_2O_2 causes uncoupling of the LH receptor from adenylate cyclase, with consequent impairment of protein synthesis and cholesterol utilization by mitochondrial P450 side chain cleavage ($P450_{scc}$), most likely through impaired production of steroidogenic acute regulatory protein (StAR)

[19]. StAR is responsible for transport of cholesterol to the inner mitochondrial membrane, where $P450_{scc}$ converts cholesterol to pregnenolone [19]. Lecithin-cholesterol acyltransferase (LCAT) plays an important role in reverse cholesterol transport and estrogen synthesis in the follicle. Cigliano and colleagues [77] investigated the estrogen: progesterone ratio and LCAT activity following titration of vitamin C and α-tocopherol in human preovulatory FF. A positive association of high FF LCAT activity was noted with vitamin C and α-tocopherol accumulation, and vice versa—lower activity was associated with consumption of vitamin C and α-tocopherol. The evidence suggests that the mature follicle appears to accumulate these vitamins in the FF to protect LCAT from oxidative damage and promote steroidogenesis.

4.7 Fertilization

Oocyte GSH concentrations fluctuate with the cell cycle, being highest during meiotic metaphase, the critical period for spindle growth and development and for sperm chromatin remodeling [73]. High GSH concentrations may also aid in meiotic spindle formation [78]. Thus, exposure to OS before fertilization appears to disrupt the meiotic spindle and increase risk of abnormal zygote formation. The activity of ROS generated during gamete fusion is inhibited, due to increased production of antioxidants, particularly SOD [27]. As a result, it is unlikely that O_2^- affects gamete fusion when adequate antioxidant defenses are available [79]. High levels of secreted GSHR in the epithelia of the oviducts may indicate involvement in fertilization and embryo development and provide protection for oocytes against excessively produced ROS that occur during ovulation [22, 23].

4.8 The Oocyte and In Vitro Fertilization

During IVF, the FF removed from the ovary has no therapeutic use and has become a "biological window" [30] for understanding the environment of the mature oocyte in infertility. Oyawoye et al. [80] prospectively evaluated TAC with the Ferric Reducing Antioxidant Power (FRAP) assay, using FF collected from 63 women undergoing oocyte retrieval for IVF after controlled ovarian stimulation. Baseline TAC did not differ with oocyte presence in the follicles, although it was significantly higher in FF samples of those oocytes that achieved successful fertilization. However, significantly lower baseline TAC was observed in the FF where the resultant embryo survived to the day of transfer. This discrepancy may be due to the effects of ROS being dependent on the stage of embryo development [80]. The percent of TAC loss 72 hours post-harvest did not differ significantly between follicles containing oocytes and those that did not, nor did it differ by fertilization status or embryo survival to time of transfer. These results suggest that antioxidant consumption in FF may have little value in predicting successful fertilization and embryo viability up to the time of transfer. However, the variation in outcomes may also reflect differential impact on OS by the various causes of infertility and confounding by indication for IVF.

A recent longitudinal study identified that IVF treatment itself induced the production of ROS, reflected by decreased tocopherols, TAC (measured by the 2,2'-azobis-3-ethylbenzthiazoline-6-sulfonic acid decolorization method), and time at which oxidation rate was maximal [81].

ROS and TAC were measured by chemiluminescence in the FF of 53 women undergoing IVF by Attaran et al. [82]. They noted that significantly higher FF ROS levels were present in individuals who became pregnant compared to those who did not, although TAC did not differ by pregnancy status. Significantly, lack of a reference value for healthy women with unstimulated cycles prevents comparison of the study population with a healthy fertile population. Nonetheless, these findings suggest physiologic concentrations of FF ROS may be indicative of healthy developing oocytes, a metabolically active system and a potential marker of IVF success.

In a series of 189 women undergoing IVF, Wiener-Megnazi et al. [30] used a thermochemiluminescence (TCL) assay to measure OS in FF samples. After controlling for age, OS was found to be positively correlated with the number of retrieved mature oocytes ($p < 0.0001$). All pregnancies occurred when the FF TCL amplitude at 50 s was within the range of 347–569 counts per second (cps). With 385 and 569 cps as limits for the occurrence of conception, the negative predictive value beyond this range was 96% and the positive predictive value was 32% ($p < 0.004$). These results suggest a beneficial threshold level for OS. The existence of an acceptable threshold level was also suggested in the evaluation of 208 FF samples from 78 women undergoing controlled ovarian stimulation [83]. Similar to the findings of Pasqualotto et al. [84] who failed to find an association with lipid peroxidation and TAC, oocyte maturity did not vary with the changing levels of ROS and lipid peroxidation, in either grade II or grade III oocytes. However, the Das et al. [83] study also found an overall negative correlation between ROS in FF and embryo quality, similar to the association of lower TAC with decreased fertilization potential shown by Oyawoye et al. [80]. Uniquely, Das et al. [83] reported a favorable effect of ROS on percent embryo formation up to approximately 100 cps in both grade II and grade III oocytes, after which embryo formation declined. More recently, authors from the same institution published a series of 803 FF samples from 128 women demonstrating increased ROS production in FF was associated with an elevated level of lipid peroxidation and decreased TAC, further indicating significant negative correlations between ROS and IVF outcome parameters including oocyte quality, fertilization rate, and embryo quality in tubal factor patients [85]. Pasqualotto et al. [84] reported both lipid peroxidation and TAC to be positively correlated with pregnancy, but not fertilization rate. However, the nonparametric nature of the correlation precludes the ability to detect differences over the continuum of values.

Many thiols are capable of scavenging free radicals. However, homocysteine (Hcy) is a sulfur-containing amino acid produced primarily in vivo during demethylation of methionine during DNA/RNA methylation and appears to possess some prooxidant activity [86, 87]. Plasma Hcy is negatively associated with fruit and vegetable consumption, folate and vitamin B_6 and B_{12} levels, and endurance exercise and positively associated with alcohol consumption, caffeine intake, and tobacco use [88]. Elevated Hcy induces endothelial dysfunction and promotes disease of the vasculature, in part by reducing the availability of NO and activation of protease activated receptors (PARs) to generate ROS [87]. Ebisch et al. [89] measured concentrations of Hcy, GSH, cysteine, and cysteinylglycine in FF of 156 women undergoing infertility treatment. Hcy was negatively associated with embryo quality on culture day three (OR = 0.58, 95% CI 0.35–0.97), suggesting that Hcy is inversely related to fertility outcome. In a separate study of single oocytes at the time of retrieval, FF Hcy levels were negatively correlated with follicular fluid vitamin B_{12}, folate and fertilization rate, and positively correlated with FF malondialdehyde, a biomarker of lipid peroxidation [90].

4.9 The Effects of Exogenous Agents

Diet modulation provides indirect evidence of the importance of OS and its control with antioxidant intake. For example, preconceptional multivitamin supplementation may enhance fertility, perhaps by increasing menstrual cycle regularity [91, 92] or via prevention of ovulatory disorders [93]. In contrast, caffeinated beverages are associated with decreased fertility [94].

Cigarettes and alcohol both cause decreased fertility in women [95], probably, in part, due to an increase in OS. Cigarette smoke is well known to increase systemic OS [96] and ethanol metabolism generates ROS through the electron transport chain. Both cigarette smoking and alcohol consumption may lead to lipid peroxidation, protein oxidation, and DNA damage. However, this may be confounded by differing dietary intakes, including that of vitamin C and α-tocopherol, in smokers [97]. Increased ROS are present in FF of smokers, and they have a lower IVF success rate [98]. Smokers ($n = 17$) have also been found to have lower concentrations of FF β-carotene and decreased IVF success compared to nonsmokers ($n = 43$) [99]. The authors found no smoking-related differences in FF or plasma concentrations of vitamin E or lycopene, which suggests follicular loss of β-carotene occurs in response to OS due to smoking. β-carotene has previously been reported to be one of the micronutrients whose concentration is most strongly influenced by smoking [100].

Environmental pollutants acting as endocrine disruptors, or in an otherwise antagonistic manner, may interfere with ovarian development, folliculogenesis, and steroidogenesis [101]. Many of these contaminants have prooxidant capabilities, but oxidation is not likely to be the mechanism behind the endocrine disruption. Plastics (e.g., phthalates), industrial compounds (e.g., polychlorinated biphenyl, PCB), pesticides (e.g., dichlorodiphenyltrichloroethane, DDT; methoxychlor, MXC; vinclozolin;

and atrazine), detergents and surfactants (e.g., octylphenol, nonylphenol, bisphenol-A) have potential estrogenic, anti-estrogenic, and/or anti-androgenic effects, thus potentiating the ability to mimic endogenous hormone mediated mechanisms and interfere with normal female fertility [101]. There is evidence from in vitro studies in rat ovary and uterus that exposure to mercury and cadmium decreases the quantity of ATP via reduction in the ATP-hydrolyzing enzyme. This reduction may affect mammalian fertility [102], but in a Japanese study, self-reported time to pregnancy was not associated with hair mercury concentration among 193 women [103].

4.10 Conclusion

The role of OS in female oogenesis and folliculogenesis is an area requiring continued research. The scientific evidence suggests that OS is an important mediator of conception, although there appears to be a threshold for the benefit or harm from OS, depending on anatomic location and stage of preconception. For example, resumption of MI is induced by an increase in ROS and inhibited by a high antioxidant status, and low FF ROS are associated with successful IVF procedures, perhaps as an indication of a healthy, metabolically active follicle. In addition, care must be given to acknowledge any undesirable effects of excessive vitamin supplementation [104]. Human research investigating OS can be especially challenging. In vitro studies of humans and other mammals offer promising insight, but the nature of these experiments can introduce additional OS not found in vivo. In addition, most studies of FF composition are conducted in women undergoing IVF where follicle maturation is stimulated with exogenous hormones, creating a milieu that differs from the FF of women not undergoing ovarian stimulation.

4.11 Bullet points

- The apparent benefit of dietary antioxidants for MII and detriment for MI progression implies a complex, multifunctional role for ROS in the ovarian environment over the continuum of concentration, and developmental stages.
- The resumption of meiosis is induced by elevated ROS and inhibited by antioxidants, suggesting that the preovulatory follicle is an important promoter of the ovulatory sequence via regulated generation of ROS.
- Mature follicles appear to accumulate antioxidant vitamins in the FF to protect lecithin-cholesterol acyltransferase from oxidative damage and promote steroidogenesis.
- Oxidative stress and apoptosis are consequences of folliculogenesis, follicular atresia and luteal regression and can be modulated by increases in antioxidant defenses.

- A positive association among FF, E2, and TAC suggests E2 may also play a role in the ovarian antioxidant–oxidant balance in granulosa cells.
- Physiologic concentrations of ROS in FF may reflect healthy developing oocytes and serve as a potential biomarker of IVF success.
- Most studies of FF composition are conducted in women undergoing IVF where follicle maturation is stimulated with exogenous hormones, creating a milieu that differs from the FF of women not undergoing ovarian stimulation.
- The corpus luteum contains high levels of both ROS and antioxidant enzymes.
- Activities of SOD, catalase, and other antioxidant enzymes in corpora lutea appear to parallel cellular steroidogenic capacity.
- Age-related decreases in antioxidant status may induce apoptosis during luteal regression and lead to decreased progesterone synthesis.
- Preconceptual multivitamin supplementation may enhance fertility by increasing menstrual cycle regularity and/or preventing ovulatory disorders.
- Multivitamin and mineral supplementation in women undergoing IVF may strengthen the antioxidant defense system by decreasing OS.
- OS is an important mediator of conception, although there appears to be a threshold for its benefit or harm depending on anatomic location and stage of preconception.

References

1. Sharma RK, Agarwal A (1996) Role of reactive oxygen species in male infertility. Urology 48(6):835–850
2. Jozwik M, Wolczynski S, Jozwik M, Szamatowicz M (1999) Oxidative stress markers in preovulatory follicular fluid in humans. Mol Hum Reprod 5(5):409–413
3. Duru NK, Morshedi M, Schuffner A, Oehninger S (2000) Semen treatment with progesterone and/or acetyl-L-carnitine does not improve sperm motility or membrane damage after cryopreservation-thawing. Fertil Steril 74(4):715–720
4. Shen H, Ong C (2000) Detection of oxidative DNA damage in human sperm and its association with sperm function and male infertility. Free Radic Biol Med 28(4):529–536
5. Vural P, Akgul C, Yildirim A, Canbaz M (2000) Antioxidant defence in recurrent abortion. Clin Chim Acta 295(1–2):169–177
6. Walsh SW, Vaughan JE, Wang Y, Roberts LJ (2000) 2nd Placental isoprostane is significantly increased in preeclampsia. FASEB J 14(10):1289–1296
7. Acevedo CG, Carrasco G, Burotto M, Rojas S, Bravo I (2001) Ethanol inhibits L-arginine uptake and enhances NO formation in human placenta. Life Sci 68(26):2893–2903
8. Sikka SC (2001) Relative impact of oxidative stress on male reproductive function. Curr Med Chem 8(7):851–862
9. A Report of the Panel on Macronutrients, Subcommittees on Upper Reference Levels of Nutrients and Interpretation and Uses of Dietary Reference Intakes, and the Standing Committee on the Scientific Evaluation of Dietary Reference Intakes Dietary reference intakes for energy, carbohydrate, fiber, fat, fatty acids, cholesterol, protein, and amino acids (macronutrients) (2005) The National Academies Press

10. Franke C, Demmelmair H, Decsi T, Campoy C, Cruz M, Molina-Font JA et al (2010) Influence of fish oil or folate supplementation on the time course of plasma redox markers during pregnancy. Br J Nutr 103(11):1648–1656
11. Raijmakers MT, Roes EM, Steegers EA, van Der Wildt B, Peters WH (2001) Umbilical cord and maternal plasma thiol concentrations in normal pregnancy. Clin Chem 47(4):749–751
12. Wisdom SJ, Wilson R, McKillop JH, Walker JJ (1991) Antioxidant systems in normal pregnancy and in pregnancy-induced hypertension. Am J Obstet Gynecol 165(6 Pt 1):1701–1704
13. Myatt L, Cui X (2004) Oxidative stress in the placenta. Histochem Cell Biol 122(4):369–382
14. Tarin JJ, Vendrell FJ, Ten J, Blanes R, van Blerkom J, Cano A (1996) The oxidizing agent tertiary butyl hydroperoxide induces disturbances in spindle organization, c-meiosis, and aneuploidy in mouse oocytes. Mol Hum Reprod 2(12):895–901
15. Hu Y, Betzendahl I, Cortvrindt R, Smitz J, Eichenlaub-Ritter U (2001) Effects of low O2 and ageing on spindles and chromosomes in mouse oocytes from pre-antral follicle culture. Hum Reprod 16(4):737–748
16. Liu H, Zhang C, Zeng W (2006) Estrogenic and antioxidant effects of a phytoestrogen daidzein on ovarian germ cells in embryonic chickens. Domest Anim Endocrinol 31(3):258–268
17. Takami M, Preston SL, Behrman HR (2000) Eicosatetraynoic and eicosatriynoic acids, lipoxygenase inhibitors, block meiosis via antioxidant action. Am J Physiol Cell Physiol 278(4):C646–C650
18. Kodaman PH, Behrman HR (2001) Endocrine-regulated and protein kinase C-dependent generation of superoxide by rat preovulatory follicles. Endocrinology 142(2):687–693
19. Behrman HR, Kodaman PH, Preston SL, Gao S (2001) Oxidative stress and the ovary. J Soc Gynecol Investig 8(1):S40–2, Jan–Feb 2001
20. Thibault C, Szollosi D, Gerard M (1987) Mammalian oocyte maturation. Reprod Nutr Dev 27(5):865–896
21. Appasamy M, Jauniaux E, Serhal P, Al-Qahtani A, Groome NP, Muttukrishna S (2008) Evaluation of the relationship between follicular fluid oxidative stress, ovarian hormones, and response to gonadotropin stimulation. Fertil Steril 89(4):912–921
22. Fujii J, Iuchi Y, Okada F (2005) Fundamental roles of reactive oxygen species and protective mechanisms in the female reproductive system. Reprod Biol Endocrinol. 2(3):43
23. Kaneko T, Iuchi Y, Kawachiya S, Fujii T, Saito H, Kurachi H et al (2001) Alteration of glutathione reductase expression in the female reproductive organs during the estrous cycle. Biol Reprod 65(5):1410–1416
24. Yoshida M, Ishigaki K, Nagai T, Chikyu M, Pursel VG (1993) Glutathione concentration during maturation and after fertilization in pig oocytes: relevance to the ability of oocytes to form male pronucleus. Biol Reprod 49(1):89–94
25. Eppig JJ (1996) Coordination of nuclear and cytoplasmic oocyte maturation in eutherian mammals. Reprod Fertil Dev 8(4):485–489
26. Ikeda S, Kitagawa M, Imai H, Yamada M (2005) The roles of vitamin A for cytoplasmic maturation of bovine oocytes. J Reprod Dev 51(1):23–35
27. Ruder EH, Hartman TJ, Blumberg J, Goldman MB (2008) Oxidative stress and antioxidants: Exposure and impact on female fertility. Hum Reprod Update 14(4):345–357
28. Sadraie SH, Saito H, Kaneko T, Saito T, Hiroi M (2000) Effects of aging on ovarian fecundity in terms of the incidence of apoptotic granulosa cells. J Assist Reprod Genet 17(3):168–173
29. Moffatt O, Drury S, Tomlinson M, Afnan M, Sakkas D (2002) The apoptotic profile of human cumulus cells changes with patient age and after exposure to sperm but not in relation to oocyte maturity. Fertil Steril 77(5):1006–1011
30. Wiener-Megnazi Z, Vardi L, Lissak A, Shnizer S, Reznick AZ, Ishai D et al (2004) Oxidative stress indices in follicular fluid as measured by the thermochemiluminescence assay correlate with outcome parameters in in vitro fertilization. Fertil Steril 82(Suppl 3):1171–1176

31. Takahashi T, Takahashi E, Igarashi H, Tezuka N, Kurachi H (2003) Impact of oxidative stress in aged mouse oocytes on calcium oscillations at fertilization. Mol Reprod Dev 66(2):143–152
32. Yeh J, Bowman MJ, Browne RW, Chen N (2005) Reproductive aging results in a reconfigured ovarian antioxidant defense profile in rats. Fertil Steril 84(Suppl 2):1109–1113
33. Hanukoglu I (2006) Antioxidant protective mechanisms against reactive oxygen species (ROS) generated by mitochondrial P450 systems in steroidogenic cells. Drug Metab Rev 38(1–2):171–196
34. Behl R, Pandey RS (2002) FSH induced stimulation of catalase activity in goat granulosa cells in vitro. Anim Reprod Sci 70(3–4):215–221
35. Chen J, Mehta JL, Haider N, Zhang X, Narula J, Li D (2004) Role of caspases in ox-LDL-induced apoptotic cascade in human coronary artery endothelial cells. Circ Res 94(3):370–376
36. Bausenwein J, Serke H, Eberle K, Hirrlinger J, Jogschies P, Hmeidan FA et al (2010) Elevated levels of oxidized low-density lipoprotein and of catalase activity in follicular fluid of obese women. Mol Hum Reprod 16(2):117–124
37. Kaneko T, Iuchi Y, Takahashi M, Fujii J (2003) Colocalization of polyol-metabolizing enzymes and immunological detection of fructated proteins in the female reproductive system of the rat. Histochem Cell Biol 119(4):309–315
38. Bernal AB, Vickers MH, Hampton MB, Poynton RA, Sloboda DM (2010) Maternal undernutrition significantly impacts ovarian follicle number and increases ovarian oxidative stress in adult rat offspring. PLoS ONE 5(12):e15558
39. Revelli A (2009) Follicular fluid content and oocyte quality: from single biochemical markers to metabolomics. Reprod Biol Endocrinol 4(7):40
40. Grupen CG, Armstrong DT (2010) Relationship between cumulus cell apoptosis, progesterone production and porcine oocyte developmental competence: temporal effects of follicular fluid during IVM. Reprod Fertil Dev 22(7):1100–1109
41. Tropea A, Miceli F, Minici F, Tiberi F, Orlando M, Gangale MF et al (2006) Regulation of vascular endothelial growth factor synthesis and release by human luteal cells in vitro. J Clin Endocrinol Metab 91(6):2303–2309
42. Greenwald GS, Terranova PF (1988) Follicular selection and its control. In: The physiology of reproduction. Raven Press, New York 387–445
43. Schroedl C, McClintock DS, Budinger GR, Chandel NS (2002) Hypoxic but not anoxic stabilization of HIF-1alpha requires mitochondrial reactive oxygen species. Am J Physiol Lung Cell Mol Physiol 283(5):L922–L931
44. Pearlstein DP, Ali MH, Mungai PT, Hynes KL, Gewertz BL, Schumacker PT (2002) Role of mitochondrial oxidant generation in endothelial cell responses to hypoxia. Arterioscler Thromb Vasc Biol 22(4):566–573
45. McCord JM, Fridovich I (1969) Superoxide dismutase. an enzymic function for erythrocuprein (hemocuprein). J Biol Chem 244(22):6049–6055
46. Iwase K, Nagasaka A, Kato K, Itoh A, Jimbo S, Hibi Y et al (2006) Cu/Zn- and mn-superoxide dismutase distribution and concentration in adrenal tumors. J Surg Res 135(1):150–155
47. Basini G, Grasselli F, Bianco F, Tirelli M, Tamanini C (2004) Effect of reduced oxygen tension on reactive oxygen species production and activity of antioxidant enzymes in swine granulosa cells. BioFactors 20(2):61–69
48. Del Rio LA (2010) Peroxisomes as a cellular source of reactive nitrogen species signal molecules. Arch Biochem Biophys, 3 Nov 2010
49. Sabatini L, Wilson C, Lower A, Al-Shawaf T, Grudzinskas JG (1999) Superoxide dismutase activity in human follicular fluid after controlled ovarian hyperstimulation in women undergoing in vitro fertilization. Fertil Steril 72(6):1027–1034
50. Suzuki T, Sugino N, Fukaya T, Sugiyama S, Uda T, Takaya R et al (1999) Superoxide dismutase in normal cycling human ovaries: Immunohistochemical localization and characterization. Fertil Steril 72(4):720–726

51. Paszkowski T, Traub AI, Robinson SY, McMaster D (1995) Selenium dependent glutathione peroxidase activity in human follicular fluid. Clin Chim Acta 236(2):173–180
52. Ozkaya MO, Naziroglu M (2010) Multivitamin and mineral supplementation modulates oxidative stress and antioxidant vitamin levels in serum and follicular fluid of women undergoing in vitro fertilization. Fertil Steril 94(6):2465–2466
53. Ozkaya MO, Naziroglu M, Barak C, Berkkanoglu M (2011) Effects of multivitamin/mineral supplementation on trace element levels in serum and follicular fluid of women undergoing in vitro fertilization (IVF). Biol Trace Elem Res 139(1):1–9
54. Murdoch WJ, Van Kirk EA (2002) Steroid hormonal regulation of proliferative, p53 tumor suppressor, and apoptotic responses of sheep ovarian surface epithelial cells. Mol Cell Endocrinol 186(1):61–67
55. Symonds DA, Merchenthaler I, Flaws JA (2008) Methoxychlor and estradiol induce oxidative stress DNA damage in the mouse ovarian surface epithelium. Toxicol Sci 105(1):182–187
56. Tsai-Turton M, Luderer U (2006) Opposing effects of glutathione depletion and follicle-stimulating hormone on reactive oxygen species and apoptosis in cultured preovulatory rat follicles. Endocrinology 147(3):1224–1236
57. de Matos DG, Furnus CC (2000) The importance of having high glutathione (GSH) level after bovine in vitro maturation on embryo development effect of beta-mercaptoethanol, cysteine and cystine. Theriogenology 53(3):761–771
58. Luberda Z (2005) The role of glutathione in mammalian gametes. Reprod Biol 5(1):5–17
59. de Matos DG, Gasparrini B, Pasqualini SR, Thompson JG (2002) Effect of glutathione synthesis stimulation during in vitro maturation of ovine oocytes on embryo development and intracellular peroxide content. Theriogenology 57(5):1443–1451
60. Jackson LW, Schisterman EF, Browne RW (2005) Oxidative stress and female fecundity [abstract 106]. Society of Pediatric and Perinatal Reproductive Epidemiologic Research
61. Serke H, Bausenwein J, Hirrlinger J, Nowicki M, Vilser C, Jogschies P et al (2010) Granulosa cell subtypes vary in response to oxidized low-density lipoprotein as regards specific lipoprotein receptors and antioxidant enzyme activity. J Clin Endocrinol Metab 95(7):3480–3490
62. Manning JP, Steinetz BG, Giannina T (1969) Decidual alkaline phosphatase activity in the pregnant and pseudopregnant rat. Ann N Y Acad Sci 166(2):482–509
63. Das P, Chowdhury M (1999) Vitamin E-deficiency induced changes in ovary and uterus. Mol Cell Biochem 198(1–2):151–156
64. Pine M, Lee B, Dearth R, Hiney JK, Dees WL (2005) Manganese acts centrally to stimulate luteinizing hormone secretion: A potential influence on female pubertal development. Toxicol Sci 85(2):880–885
65. Karanth S, Yu WH, Walczewska A, Mastronardi CA, McCann SM (2001) Ascorbic acid stimulates gonadotropin release by autocrine action by means of NO. Proc Natl Acad Sci U S A. 98(20):11783–11788
66. Tan DX, Manchester LC, Terron MP, Flores LJ, Reiter RJ (2007) One molecule, many derivatives: a never-ending interaction of melatonin with reactive oxygen and nitrogen species? J Pineal Res 42(1):28–42
67. Tomas-Zapico C, Coto-Montes A (2005) A proposed mechanism to explain the stimulatory effect of melatonin on antioxidative enzymes. J Pineal Res 39(2):99–104
68. Tamura H, Takasaki A, Miwa I, Taniguchi K, Maekawa R, Asada H et al (2008) Oxidative stress impairs oocyte quality and melatonin protects oocytes from free radical damage and improves fertilization rate. J Pineal Res 44(3):280–287
69. Duleba AJ, Foyouzi N, Karaca M, Pehlivan T, Kwintkiewicz J, Behrman HR (2004) Proliferation of ovarian theca-interstitial cells is modulated by antioxidants and oxidative stress. Hum Reprod 19(7):1519–1524
70. Erickson GF, Magoffin DA, Dyer CA, Hofeditz C (1985) The ovarian androgen producing cells: A review of structure/function relationships. Endocr Rev Summer 6(3):371–399

71. Chao HT, Lee SY, Lee HM, Liao TL, Wei YH, Kao SH (2005) Repeated ovarian stimulations induce oxidative damage and mitochondrial DNA mutations in mouse ovaries. Ann N Y Acad Sci 1042:148–156
72. Tarin JJ, Perez-Albala S, Cano A (2002) Oral antioxidants counteract the negative effects of female aging on oocyte quantity and quality in the mouse. Mol Reprod Dev 61(3):385–397
73. Zuelke KA, Jeffay SC, Zucker RM, Perreault SD (2003) Glutathione (GSH) concentrations vary with the cell cycle in maturing hamster oocytes, zygotes, and pre-implantation stage embryos. Mol Reprod Dev 64(1):106–112
74. Nakamura Y, Yamagata Y, Sugino N, Takayama H, Kato H (2002) Nitric oxide inhibits oocyte meiotic maturation. Biol Reprod 67(5):1588–1592
75. Haliloglu S, Baspinar N, Serpek B, Erdem H, Bulut Z (2002) Vitamin A and beta-carotene levels in plasma, corpus luteum and follicular fluid of cyclic and pregnant cattle. Reprod Domest Anim 37(2):96–99
76. Schweigert FJ (2003) Research note: Changes in the concentration of beta-carotene, alpha-tocopherol and retinol in the bovine corpus luteum during the ovarian cycle. Arch Tierernahr 57(4):307–310
77. Cigliano L, Balestrieri M, Spagnuolo MS, Dale B, Abrescia P (2002) Lecithin-cholesterol acyltransferase activity during maturation of human preovulatory follicles with different concentrations of ascorbate, alpha-tocopherol and nitrotyrosine. Reprod Fertil Dev 14(1–2):15–21
78. Oliver JM, Albertini DF, Berlin RD (1976) Effects of glutathione-oxidizing agents on microtubule assembly and microtubule-dependent surface properties of human neutrophils. J Cell Biol 71(3):921–932
79. Miesel R, Drzejczak PJ, Kurpisz M (1993) Oxidative stress during the interaction of gametes. Biol Reprod 49(5):918–923
80. Oyawoye O (2003) Antioxidants and reactive oxygen species in follicular fluid of women undergoing IVF: Relationship to outcome. Hum Reprod 18(11):2270–2274
81. Aurrekoetxea I, Ruiz-Sanz JI, del Agua AR, Navarro R, Hernandez ML, Matorras R et al (2010) Serum oxidizability and antioxidant status in patients undergoing in vitro fertilization. Fertil Steril 94(4):1279–1286
82. Attaran M, Pasqualotto E, Falcone T, Goldberg JM, Miller KF, Agarwal A et al (2000) The effect of follicular fluid reactive oxygen species on the outcome of in vitro fertilization. Int J Fertil Womens Med 45(5):314–320
83. Das S, Chattopadhyay R, Ghosh S, Ghosh S, Goswami SK, Chakravarty BN et al (2006) Reactive oxygen species level in follicular fluid—embryo quality marker in IVF? Hum Reprod 21(9):2403–2407
84. Pasqualotto EB, Agarwal A, Sharma RK, Izzo VM, Pinotti JA, Joshi NJ et al (2004) Effect of oxidative stress in follicular fluid on the outcome of assisted reproductive procedures. Fertil Steril 81(4):973–976
85. Jana SK (2010) Upper control limit of reactive oxygen species in follicular fluid beyond which viable embryo formation is not favorable. Reprod Toxicol 29(4):447–451
86. Garry PJ, Vellas BJ (1996) Aging and nutrition. In: Present knowledge in nutrition. ILSI Press, Washington p. 414–49
87. Tyagi N, Sedoris KC, Steed M, Ovechkin AV, Moshal KS, Tyagi SC (2005) Mechanisms of homocysteine-induced oxidative stress. Am J Physiol Heart Circ Physiol 289(6):H2649–H2656
88. Chrysohoou C, Panagiotakos DB, Pitsavos C, Zeimbekis A, Zampelas A, Papademetriou L et al (2004) The associations between smoking, physical activity, dietary habits and plasma homocysteine levels in cardiovascular disease-free people: The 'ATTICA' study. Vasc Med 9(2):117–123
89. Ebisch IM, Peters WH, Thomas CM, Wetzels AM, Peer PG, Steegers-Theunissen RP (2006) Homocysteine, glutathione and related thiols affect fertility parameters in the (sub)fertile couple. Hum Reprod 21(7):1725–1733
90. Berker B, Kaya C, Aytac R, Satiroglu H (2009) Homocysteine concentrations in follicular fluid are associated with poor oocyte and embryo qualities in polycystic ovary syndrome patients undergoing assisted reproduction. Hum Reprod 24(9):2293–2302

91. Czeizel AE, Metneki J, Dudas I (1994) Higher rate of multiple births after periconceptional vitamin supplementation. N Engl J Med 330(23):1687–1688
92. Dudas I, Rockenbauer M, Czeizel AE (1995) The effect of preconceptional multivitamin supplementation on the menstrual cycle. Arch Gynecol Obstet 256(3):115–123
93. Chavarro JE, Rich-Edwards JW, Rosner BA, Willett WC (2007) Diet and lifestyle in the prevention of ovulatory disorder infertility. Obstet Gynecol 110(5):1050–1058
94. Christian MS, Brent RL (2001) Teratogen update: evaluation of the reproductive and developmental risks of caffeine. Teratology 64(1):51–78
95. Sadeu JC, Hughes CL, Agarwal S, Foster WG (2010) Alcohol, drugs, caffeine, tobacco, and environmental contaminant exposure: Reproductive health consequences and clinical implications. Crit Rev Toxicol 40(7):633–652
96. Yanbaeva DG, Dentener MA, Creutzberg EC, Wesseling G, Wouters EF (2007) Systemic effects of smoking. Chest 131(5):1557–1566
97. Schroder H, Marrugat J, Elosua R, Covas MI (2002) Tobacco and alcohol consumption: Impact on other cardiovascular and cancer risk factors in a southern european mediterranean population. Br J Nutr 88(3):273–281
98. Paszkowski T, Clarke RN, Hornstein MD (2002) Smoking induces oxidative stress inside the graafian follicle. Hum Reprod 17(4):921–925
99. Tiboni GM, Bucciarelli T, Giampietro F, Sulpizio M, Di Ilio C (2004) Influence of cigarette smoking on vitamin E, vitamin A, beta-carotene and lycopene concentrations in human pre-ovulatory follicular fluid. Int J Immunopathol Pharmacol 17(3):389–393
100. Alberg A (2002) The influence of cigarette smoking on circulating concentrations of antioxidant micronutrients. Toxicology 180(2):121–137
101. Uzumcu M, Zachow R (2007) Developmental exposure to environmental endocrine disruptors: Consequences within the ovary and on female reproductive function. Reprod Toxicol 23(3):337–352
102. Milosevic M, Petrovic S, Demajo M, Horvat A (2005) Effects of metal ions on plasma membrane Mg^{2+}-atpase in rat uterus and ovaries. Ann N Y Acad Sci 1048:445–448
103. Arakawa C, Yoshinaga J, Okamura K, Nakai K, Satoh H (2006) Fish consumption and time to pregnancy in japanese women. Int J Hyg Environ Health 209(4):337–344
104. Tarin JJ, Brines J, Cano A (1998) Antioxidants may protect against infertility. Hum Reprod 13(6):1415–1416

Chapter 5
Placental Vascular Morphogenesis and Oxidative Stress

Amani Shaman, Beena J. Premkumar and Ashok Agarwal

Abstract The placenta is a hemochorial organ, meaning that it is directly bathed by maternal blood. Favorable fetal growth depends on optimal placental evolution and development, as it represents the interface between the maternal and fetal environments. The placenta plays a crucial role in fetal nutrition, respiration, and hormone synthesis. Vasculogenesis and angiogenesis are essential for normal placental development and effective maternal-fetal exchange. The onset of maternal circulation to the placenta is associated with a burst of oxidative stress (OS). This OS can serve at a physiological level to trigger pathways of differentiation in the regulation of villous remodeling, trophoblastic invasion, and production of angiogenic factors. In excess, however, OS can lead to the development of complications involving the placenta, such as fetal loss, preeclampsia, and intrauterine growth restriction.

Keywords Placenta · Vasculogenesis · Angiogenesis · Oxygen · Oxidative stress · Antioxidants · Preeclampsia · Intrauterine growth restriction

---

A. Shaman · B. J. Premkumar
Center for Reproductive Medicine, Cleveland Clinic Foundation,
9500 Euclid Avenue, Desk A19.1, Cleveland, OH 44195, USA
e-mail: shamana@ccf.org

B. J. Premkumar
e-mail: premkub@ccf.org

A. Agarwal (✉)
Center for Reproductive Medicine, Lerner College of Medicine, Cleveland Clinic,
9500 Euclid Avenue, Cleveland, OH 44195, USA
e-mail: agarwaa@ccf.org

5.1 Introduction

The placenta is a hemochorial organ, meaning that it is directly bathed by maternal blood. It derives from trophoectoderm cells originating in the extraembryonic conceptus [1, 2]. The placenta acts as an organ for fetal nutrition, respiration, hormone synthesis [3] through a complex, and highly vascularized capillary network.

The normal progression of placental vasculogenesis and angiogenesis are responsible for establishing maternal–fetal transfer and exchanges. Vessel formation de novo is described as vasculogenesis, while angiogenesis refers to the formation of new vessels from pre-existing vessels [4–6]. By the end of gestation, the placenta will have developed a capillary network that spans up to 550 km in length and 15 m^2 in surface area for effective maternal-fetal exchange [7].

In addition to stimulating placental development, a hypoxic setting is necessary during early pregnancy ($pO_2 < 20$ mm Hg, or $\sim 5\% \ O_2$) to protect the developing embryo from the teratogenic effects of reactive oxygen species (ROS) during critical phases of organogenesis [8–10]. In normal pregnancies, the onset of intraplacental maternal arterial circulation is associated with a degree of oxidative stress (OS). However, above a certain level, OS can impair the development of placental vascular networks and result in pregnancy complications, such as pregnancy loss, preeclampsia, and intrauterine growth restriction (IUGR).

5.2 Placental Development

The placenta is in a constant state of growth and differentiation throughout pregnancy to ensure for effective maternal-fetal exchanges. The development of placental vessels and trophoblasts occurs through a chain of complex and highly regulated processes (Fig. 5.1) [7, 11].

The invasion of the maternal spiral arterioles by the endovascular trophoblast early in pregnancy establishes a conduit of high flow and low resistance to perfuse the intervillous space [2]. The occlusion of the maternal spiral arteries by endovascular trophoblast cell plugs is responsible for the low O_2 tension associated with early pregnancy. At the end of the first trimester, the release of these plugs causes an increase in O_2 tension, which triggers the development of OS in the placenta [12, 13].

5.3 Placental Vascular Morphogenesis

By day 21 post-conception, the first morphological evidence of vasculogenesis can be observed in the mesenchymal villi. Unique pluripotent mesenchymal cells found under the cytotrophoblastic layer differentiate into hemangioblastic cells, which will later give rise to endothelial and hematopoietic cells [14]. As early as 23 days

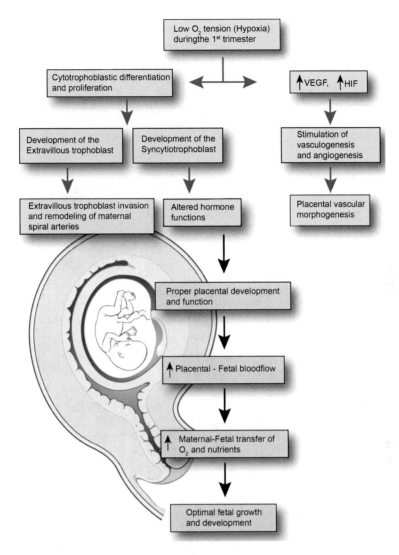

Fig. 5.1 Effects of hypoxia on placental and fetal development Reprinted with permission, Cleveland Clinic Center for Medical Art and Photography © 2004–2011. All Rights Reserved

post-conception, primitive capillaries can be recognized [3]. The anatomical connection to the embryonic circulation develops around 32–35 days post-conception, when the villous capillaries fuse with allantoic vessels [7].

The development of capillary networks during the first and early second trimesters can result from two mechanisms: (1) the elongation of pre-existing endothelial tubes by non-branching angiogenesis and (2) lateral extension of these endothelial tubes, also called sprouting angiogenesis [7, 15]. After 25 weeks of gestation, the pattern of

Fig. 5.2 Vasculosyncytial membrane in the terminal villi Reprinted with permission, Cleveland Clinic Center for Medical Art and Photography © 2004–2011. All Rights Reserved

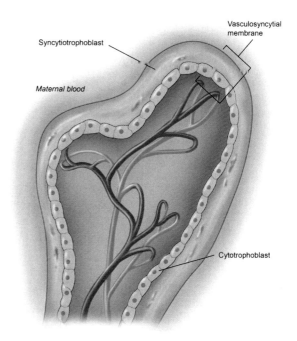

vascular growth shifts towards non-branching angiogenesis, and the total capillary lengths increase until term [16].

Thinning of trophoblastic epithelium between the maternal and fetal circulations produces an entity known as the vasculosyncytial membrane. This membrane provides regions for adequate diffusional exchange between the two circulations [7, 15]. The formation of capillary sinusoids facilitates exchanges between maternal and fetal circulation. In addition, these sinusoids increase the vessel diameter, reducing the flow rate around the vasculosyncytial membrane, which gives more time for maternal-fetal exchanges to take place. Within the fetal capillary network, sinusoid formation can lessen the vascular resistance, allowing for equal distribution of blood flow throughout the villous tree (Fig. 5.2) [3].

5.4 Factors that Regulate Vasculogenesis and Angiogenesis in the Human Placenta

The normal vascular development of the placenta is contingent upon the regulated balancing of actions between angiogenic stimulators and inhibitors to maintain vascular integrity and normal placentation for the advancement of vasculogenesis and angiogenesis. If these interactions are altered, changes within the placenta can occur, potentially leading to negative outcomes. Although these factors have unique functions, their synergistic effects promote optimal progression of these vascular processes, and thus ensure effective vascular development and placental exchange for favorable fetal growth. Placental O_2 tension and concentration are key regulators of placental development and the expressions of several growth factors are mediated by local O_2 concentration [7].

5.5 Vascular Endothelial Growth Factor

The main growth factor involved in vasculogenesis and angiogenesis is vascular endothelial growth factor (VEGF). VEGF is stimulated in response to hypoxia within placental tissue [17–19] and is suppressed by hyperoxia [19].

VEGF is predominantly expressed as VEGFA by cytotrophoblastic cells during early gestation. Later on in pregnancy, VEGFA is expressed by mesenchymal cells and macrophages, called Hofbauer cells [14, 20, 21]. VEGFA promotes the differentiation of mesenchymal cells in the villous core into hemangioblastic cells and stimulates endothelial cell proliferation, migration, apoptosis, and vascular permeability. These processes support early vessels in angiogenic remodeling and capillary network formation within the mesenchymal villous core [7, 20].

VEGFA acts through two receptors: Flt-1 and KDR, which are also referred to as VEGFR-1 and VEGFR-2, respectively [20]. Flt-1 is found in trophoblastic and endothelial cells, while KDR exists in villous endothelial cells [14, 20, 21]. When O_2 levels are low, VEGFA regulation occurs through message stability or transcription [17].

Another receptor, known as soluble Flt-1 (sFlt-1) is expressed by villous endothelium, macrophages, and the trophoblast. It acts as an inhibitor of VEGFA [20], blocking interactions between VEGF and its receptors and is recognized to play an essential role in the regulation of angiogenesis via binding to VEGF and placental growth factor (PlGF) [22]. The activity of sFlt is increased in hypoxia, while under 40 % O_2, its levels have been observed to be significantly decreased [23].

5.6 Pigment Epithelium-Derived Factor

Pigment epithelium-derived factor (PEDF) is considered a potent anti-angiogenic factor that is present in most body tissues [24], including the placental vessels and trophoblasts of normal pregnancies [25]. As an angiogenic inhibitor, PEDF has been demonstrated to counteract [26, 27] the effects of sustained VEGF activity, such as weakening of vascular junctions and leaking of vessels [28].

5.7 Placental Growth Factor

PlGF is considered a member of the VEGF family and is highly expressed in the extravillous trophoblast. PlGF acts by binding to the Flt-1 receptor and is increasingly expressed toward term. The specific role of PlGF in placental vessel development is unclear, however, it has been suggested to participate in angiogenesis, rather than vasculogenesis [7, 29].

The expression of PlGF is down-regulated in hypoxic settings. Premature placental hyperoxia driven by PlGF early on might lead to reduced branching angiogenesis and unsuccessful formation of terminal villi [23].

5.8 Angiopoietins

Angiopoietins are important growth factors that are present in two forms, ANG-1 and 2, and are located in the villous trophoblast [7]. Both ANG-1 and ANG-2 act through the tyrosine kinase receptor, Tie2, in villous endothelial cells and the trophoblast. ANG-1 helps to stabilize newly formed vessels by promoting the survival of endothelial cells. It also plays a role in the final stages of vascular remodeling, leading to the development of a more complex vascular network [30]. ANG-2 acts as functional antagonist to ANG-1 by destabilizing vessels and increasing their susceptibility to the angiogenic effects of VEGFA. In the absence of a stimulus for ANG-2 expression, vessels will regress [29].

5.9 Hypoxia Inducible Factor

As its name suggests, hypoxia inducible factor (HIF) is stimulated in hypoxic settings. In particular, HIF-1 is involved in the maintenance of homeostasis as well as the transcription of certain genes, including VEGF and erythropoietin [31–33], both of which are stimulated by hypoxic conditions [34]. HIF can also be mediated by factors, such as hormones, cytokines, and growth factors [35].

HIF-1α and HIF-2α are expressed in the syncytiotrophoblast, villous cytotrophoblast, and the feto-placental vascular endothelium. During the first trimester of pregnancy, the levels of both HIF-1α and HIF-2α are increased. As pregnancy progresses, their levels decline [35]. Hypoxic conditions up-regulate the expression of HIFs. Thus, the low O_2 tension setting of early pregnancy required for favorable feto-placental development is heavily regulated by HIFs [1, 36].

5.10 Oxidative Stress and Reactive Species

The increase in O_2 tension caused by the onset of maternal circulation into the intervillous spaces leads to transient placental OS [12]. The progressive nature of placental OS from the periphery to the center limits the impact of oxidative damage [37]. During this time, OS helps in placental remodeling, leading to regression of superficial villi by increasing cell death. OS also promotes the transformation of the chorion frondosum into the chorion laeve, giving the placenta its discoid shape [38].

Placental OS develops as a result of a temporary imbalance between the increased production of ROS secondary to elevated O_2 tension and the simultaneous activity of antioxidant defense mechanisms [12].

Reactive oxygen species generated in the placenta most commonly includes the superoxide (SO) anion. Other ROS are the hydroxyl radical, hydrogen peroxide (H_2O_2), and peroxynitrite ($ONOO^-$). Nitric oxide (NO) and carbon monoxide (CO) are other reactive species that can be produced in the placenta.

5.11 Superoxide Anion

Placental production of the SO anion can occur by NADPH oxidase (NOX), which is present in Hofbauer cells in the form of NOX-2. Two other isoforms, NOX-1 and NOX-5, have been identified in the syncytiotrophoblast and vascular endothelium. These isoforms may play an important role in the generation of ROS and in O_2-sensing [39, 40]. The enzyme xanthine oxidase (XO), present in the villous trophoblast, stroma, and endothelial cells, can also generate the SO anion; however, its role in inducing placental OS remains unclear. Production of the SO anion additionally occurs in the mitochondrial electron transport chain, which represents the most important source for SO production, as 2–3 % of O_2 consumed by mitochondria is converted to SO [41].

In the placenta, the SO anion regulates angiogenesis, in addition to transcription factors, generation of antioxidants, proliferation, and matrix remodeling [40].

5.12 Nitric Oxide

Nitric oxide can be generated by either the endothelial isoform of NO synthase (eNOS), located in the villous vascular endothelium and the syncytiotrophoblast [42], or by the inducible isoform (iNOS), expressed in Hofbauer cells. The functions of NO include anti-adhesion and anti-aggregation in the syncytiotrophoblast [43] in addition to regulation of utero-placental and feto-placental vasculature and trophoblast apoptosis. It also influences the flow of blood into the intervillous space through maternal artery dilatation [40].

Together with Flt-1, NO can down-regulate DNA synthesis of the trophoblast and endothelial cells [44], and prevent the proliferation of vascular smooth muscle [45]. In this way, the proliferative functions of VEGF-activated KDR might be opposed by NO-regulated growth in trophoblast and endothelial cells [44].

5.13 Peroxynitrite

The concomitant presence of the SO anion and NO in the placenta fuels formation of the oxidant, ONOO$^-$, which in turn affects NO levels to the extent of altering physiological processes [46]. In the feto-placental vasculature, ONOO$^-$ diminishes levels of the vasodilator NO, leaving the vascular resistance unregulated. Nitration and consequent suppression of placental iNOS could potentially contribute to the vasoconstriction seen in preeclampsia [40]. Furthermore, inhibition of SO dismutase (SOD) activity by ONOO$^-$ could intensify OS [47]. Endoplasmic reticulu, (ER) stress can also be stimulated by ONOO$^-$, as has been demonstrated in trophoblasts [40, 48].

5.14 Carbon Monoxide

In many body tissues including the placenta, the vasodilator, CO [49], is synthesized during the oxidation of heme by the antioxidant enzyme, heme oxygenase (HO). In the placenta, HO is present in three isoforms. HO-1 has been found in villous trophoblastic cells, and the presence of HO-2 has been noted in endothelial and smooth muscle cells of the placental villi vasculature [50], with the protein content of HO-2 being higher than HO-1 in the placenta [51–53]. Compared with HO-1 and HO-2, the HO-3 isoform is relatively inactive [54, 55]. Placental HO has been identified in the syncytiotrophoblast, endothelium, and smooth muscles of the umbilico-placental vasculature [50, 51].

CO has been shown to exhibit NO-like properties including vasodilation [56, 57], anti-aggregation of platelets [58], and is suggested to be an important anti-apoptotic and anti-inflammatory mediator [59–61].

Amounting evidence suggests that CO is essential for the normal development of the placenta and proper execution of placental functions. Sustained vasodilation of the spiral arterioles and placental vasculature was observed with administration of exogenous CO, resulting in reduction of placental vascular resistance [62]. CO has also been reported to exert a cytoprotective effect in placental tissues through the inhibition of apoptosis caused by hypoxia-reoxygenation (H/R) in the syncytiotrophoblast [63]. Moreover, it has been suggested that in the placenta, CO may play a joint role with NO in hemodynamic regulation [51–53].

5.15 Oxidative Stress and Antioxidants

In the placenta, the presence of OS can regulate the concentrations and activities of several essential antioxidant enzymes, allowing it to accommodate to the new high-O_2 environment [37]. Up until 10-12 weeks of gestation, placental antioxidant defenses are absent [64]. After this time, the placental mRNA expressions of antioxidants increase as O_2 tension rises. Both enzymatic and non-enzymatic antioxidant defense systems are present in the placenta. They consist of SOD, including Mn-SOD and Cu/Zn-SOD, catalase, glutathione peroxidase (GPX), glutathione S-transferase, thiol/disulfide oxidoreductase, and vitamins C and E [43].

The expressions and activities of catalase and GPX in the placenta depend on their locations in placental lobule, with higher expressions in the center compared to periphery of the lobule. These changes may reflect O_2 gradients between the center and the periphery of the lobule, with the center being well-oxygenated secondary to the flow of maternal blood. On the other hand, the activity and mRNA concentration of SOD do not seem to differ throughout the placenta [37].

However, failure of the placenta to adapt to increased levels of O_2 can result in clinical consequences, such as miscarriage, preeclampsia, and IUGR.

5.16 Clinical Outcomes Associated with Oxidative Stress and Vascular Dysfunction

5.16.1 Fetal Loss

The degree of OS depends on both the severity of the placental insult and the efficiency of antioxidant defenses [43]. Prior to the development of these defenses, premature, and disorganized flow of maternal blood into the intervillous space is linked to early pregnancy loss (EPL), through the combination of ROS-induced damage and inadequate supply of antioxidant defenses [65, 66].

In addition to the paucity of antioxidant enzymes early in pregnancy, the syncytiotrophoblast is the first villous tissue to be exposed to the increase in O_2 tension as it comes in contact with the maternal circulation [67–69]. As such, EPL and other pregnancy complications [12] could result from insufficiently equipped antioxidant defenses that leave the syncytiotrophoblast particularly sensitive to OS-induced damage.

5.16.2 Preeclampsia and IUGR

The lack of maternal spiral artery conversion leads to the retention of smooth muscle. Spontaneous vasoconstriction can occur in the spiral arteries, predisposing the placenta to ischemic-reperfusion type injuries. The loss or reduction of arterial blood flow to a lobule, whether transient or not, will result in decreased O_2 tension. As the trophoblast extricates blood from the intervillous space, a sharp increase in O_2 tension occurs in association with the restored inflow of maternal arterial blood [70]. During this process, OS has been observed to result from H/R injuries and lead to apoptosis in vitro [7, 71, 72].

In IUGR placentas, the processes of capillary elongation, branching, and dilation as well as formation of terminal villi do not occur sufficiently, significantly impeding feto-placental blood flow and gas exchange. As a result, the fetus is left susceptible to ensuing hypoxia and acidosis [73, 74]. The improper development and villous tree injuries observed in IUGR placentas consequently depletes the syncytiotrophoblast, limiting the maternal-fetal transfer of blood and nutrients [75].

Inadequate development of the placental vasculature may be a shared pathophysiology between preeclampsia (Fig. 5.3) and IUGR, as they are commonly encountered in conjunction with one another.

5.16.2.1 Hypoxia/Reoxygenation

H/R-induced impairment of placental perfusion causes placental OS, as the villi are left exposed to the damaging effects of ROS and RNS [76]. Generalized maternal endothelial dysfunction with subsequent development of preeclampsia symptoms can ensue [77]. Furthermore, chronically under-perfused placental villi could sustain injuries associated with IUGR [70].

The placental villous endothelium and trophoblast of preeclamptic pregnancies has demonstrated higher expressions of xanthine dehydrogenase (XDH)/XO than normal placentas. These increases have been found in concert with the presence of nitrotyrosine residues, which indicates the formation of $ONOO^-$ and thus, OS [78]. In addition to elevated expressions in invasive cytotrophoblasts [78], high levels of nitrotyrosine residues have also been detected in villous stromal cells, and in and around the villous vascular endothelium of preeclamptic patients compared to patients with normal pregnancies [46]. These findings might indicate that increased local generation of the SO anion could be responsible for the formation of $ONOO^-$, along with NO production and insufficient SOD scavenging ability [70].

In the feto-placental vasculature, the formation of $ONOO^-$ depletes the vasodilator NO, leaving the vascular resistance unregulated. Nitration and consequent suppression of placental iNOS could potentially contribute to the vasoconstriction of preeclampsia [40]. Furthermore, inhibition of SOD activity by $ONOO^-$ could intensify OS [47]. Increased eNOS activity has also been observed in the stem villous vasculature of preeclamptic pregnancies compared with normal pregnancies [79].

The maternal and umbilical cord plasma of preeclamptic pregnancies contains markedly increased levels of lipid peroxides [80], along with reduced antioxidant concentrations [81]. Lipid peroxidation is a well-known marker of OS. An increase in the production of lipid peroxides has been demonstrated with hypoxia, in correlation with increased trophoblastic generation of sFlt-1 [82]. Taken together, OS in preeclampsia, evidenced by increased plasma measurements of lipid peroxides, may up-regulate the production of sFlt-1 in the placenta.

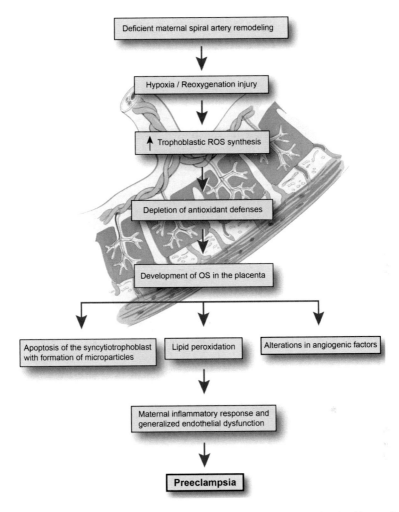

Fig. 5.3 Mechanisms involved in the development of Preeclampsia Reprinted with permission, Cleveland Clinic Center for Medical Art and Photography © 2004–2011. All Rights Reserved

5.16.2.2 Apoptosis

Fluctuating O_2 tension in the placenta can also enhance apoptosis. In particular, increased apoptotic activity has been reported in the syncytiotrophoblast of preeclamptic placentas compared with those of normal pregnancies [71, 72]. Syncytiotrophoblastic apoptosis has been suggested to be induced by OS resulting in trophoblast shedding, termed "trophoblast deportation" [83], and subsequent maternal endothelial dysfunction from up-regulated inflammatory responses [83, 84].

Independently and together, hypoxia and H/R can cause considerable apoptosis [85], potentially leading to IUGR, secondary to the observed increase in activity of p53, a pro-apoptotic protein, in the villous trophoblast [86, 87]. Syncytiotrophoblastic activation of the unfolded protein response indicates ER stress, which also leads to apoptosis [2].

5.16.2.3 Effects of Hypoxia on Angiogenic factors

Preeclamptic conditions and IUGR with preserved end diastolic flow (PED) are considered to be in a state of utero-placental hypoxia. The mother is normoxic, but the placenta and fetus are hypoxic due to impaired utero-placental circulation. Here, placental peripheral villi also show branching angiogenesis, but blood flow to the fetus is normal or decreased [88, 89]. Placentas affected by utero-placental hypoxia have demonstrated up-regulated VEGF and down-regulated PlGF [44]. These results indicate that VEGF was stimulated by placental hypoxia and led to the angiogenic alterations observed in these placentas.

Reported data on VEGFA levels in relation to preeclampsia have been inconsistent. While some studies have recorded higher serum and plasma levels of VEGFA in term preeclamptic patients [90–94], others have failed to produce similar results [95, 96]. In comparison to normotensive controls of similar age, studies have found levels of VEGF and VEGFR-1 to be increased, unchanged, and decreased [97–101]. Likewise, inconsistencies in VEGFA and VEGFR-1 concentrations have also been reported in pregnancies affected by IUGR+PED. Tissue levels of VEGFA in these conditions may be unchanged or decreased [99, 102, 103], with a higher level of VEGFR-1 [100].

In vitro, sFlt 1 is a vasoconstrictor that causes endothelial dysfunction and supporting data from Maynard et al. [104] has demonstrated that overexpression of sFlt1 is a causative factor. Under normoxia, as well as in response to hypoxia, the expressions of sFlt1 and soluble endoglin (sEng) were shown to be increased in cultured placental trophoblasts from preeclamptic patients when compared with normal placental trophoblasts [105]. In pregnant rats, elevated sEng, an antiangiogenic factor, has been observed to exacerbate sFlt1-induced vascular damage, inducing a preeclampsia-like condition [106].

In vitro studies suggest that CO can increase intraplacental fetal perfusion through vasodilation of placental blood vessels and enhanced utero-placental blood flow to the intervillous spaces [107]. Women with preeclampsia have been observed to have decreased expression of HO-1 [108]. Both HO-1 and its metabolite, CO, have been reported to suppress the VEGF-induced expression of sFlt; hence, they have been suggested to regulate elevations of the placental expressions of sFlt and Eng [109, 110]. Pregnant mice lacking HO-1 have been found with disordered placental vasculogenesis, increased circulating sFlt1 levels, and hypertension [111].

The levels of HIF-1α and HIF-2α have been found to be elevated in the placental villi of preeclamptic women compared to those with normal pregnancies [112, 113]. In addition, the HIF target gene, VEGF, is highly expressed in the placentas of preeclamptic and IUGR pregnancies [100], possibly implicating an acclamatory response of the vasculature to the chronically hypoxic environment [35].

5.16.2.4 Effects of Hyperoxia on Angiogenic Factors

It has long been established that the angiogenic factor, VEGF, is crucial for the establishment of an extensive vascular network early in placental development. However, prolonged elevation of VEGF later in gestation could result in unfavorable outcomes.

In fact, IUGR placentae have demonstrated markedly increased PlGF along with hypoxia-induced suppression of PlGF mRNA, implicating placental hyperoxia in the pathogenesis of IUGR [114]. A marked increase of PlGF in placentae affected by IUGR has been observed to stimulate trophoblast proliferation and suppress the growth of endothelial cells. These findings might indicate that early-onset utero-placental hyperoxia favors a shift to premature dominance of PlGF, which accounts for reduced branching angiogenesis and absent non-branching angiogenesis in terminal villi, as well as disrupted trophoblastic growth [23], as observed in the placentas of IUGR pregnancies.

A marked reduction in capillary number and capillary area of terminal villi has been observed in IUGR placentas when compared to normal term placentas [73]. In response to hyperoxia, placental branching angiogenesis ceases with the altered expressions of angiogenic factors in the placental villi [115, 116]. The relative hyperoxia is a possible explanation for the decrease in VEGF, increase in PlGF, and altered angiopoietin expressions encountered in IUGR placentas, since O_2 regulates these factors in different manners [29, 114]. Given that angiogenesis is stimulated in hypoxic settings, it will likely be inhibited by hyperoxia between the intervillous space and placental villi of IUGR pregnancies [117].

5.17 Conclusion

Vasculogenesis and angiogenesis are essential for normal placental development and effective maternal-fetal exchange. Many factors that regulate placental vascular development are regulated by O_2; therefore, maintaining an appropriate level of O_2 throughout gestation is crucial for placental vascular development. Any disturbance in O_2 homeostasis that affects its physiological level can lead to the development of OS secondary to disrupted metabolism from hypoxia or hyperoxia, with subsequent adverse effects on placental development. Failure to adapt to changes in O_2 tension can lead to abnormal placental vascular development, resulting in a range of adverse pregnancy outcomes, such as pregnancy loss, preeclampsia, and IUGR.

Potential treatment strategies may be instrumental in the clinical management of preeclampsia. Restoring the balance between angiogenic (e.g. VEGF, PEDF,

PIGF, ANG), and anti-angiogenic (e.g. sFlt1, sEng) factors in preeclamptic pregnancies could provide a therapeutic basis. The use of angiogenic agonists or inhibitors of anti-angiogenic factors are other potential strategies for therapy.

References

1. Patel J, Landers K et al (2010) Regulation of hypoxia inducible factors (HIF) in hypoxia and normoxia during placental development. Placenta 31(11):951–957
2. Scifres CM, Nelson DM (2009) Intrauterine growth restriction, human placental development and trophoblast cell death. J Physiol 587(Pt 14):3453–3458
3. Charnock-Jones DS, Burton GJ (2000) Placental vascular morphogenesis. Baillieres Best Pract Res Clin Obstet Gynaecol 14(6):953–968
4. Folkman J, Shing Y (1992) Angiogenesis. J Biol Chem 267(16):10931–10934
5. Risau W, Flamme I (1995) Vasculogenesis. Annu Rev Cell Dev Biol 11:73–91
6. Risau W (1997) Mechanisms of angiogenesis. Nature 386(6626):671–674
7. Burton GJ, Charnock-Jones DS et al (2009) Regulation of vascular growth and function in the human placenta. Reproduction 138(6):895–902
8. Burton GJ, Hempstock J et al (2003) Oxygen, early embryonic metabolism and free radical-mediated embryopathies. Reprod Biomed Online 6(1):84–96
9. Burton GJ, Watson AL et al (2002) Uterine glands provide histiotrophic nutrition for the human fetus during the first trimester of pregnancy. J Clin Endocrinol Metab 87(6):2954–2959
10. Jauniaux E, Gulbis B et al (2003) The human first trimester gestational sac limits rather than facilitates oxygen transfer to the foetus—a review. Placenta 24(Suppl A):S86–S93
11. Myatt L (2006) Placental adaptive responses and fetal programming. J Physiol 572(Pt 1):25–30
12. Jauniaux E, Watson AL et al (2000) Onset of maternal arterial blood flow and placental oxidative stress. A possible factor in human early pregnancy failure. Am J Pathol 157(6):2111–2122
13. Huston J (1992) The materno-trophoblastic interface: uteroplacental blood flow. In: Barnea E, Hustin J, Jauniaux E (eds) The first twelve weeks of gestation. Springer, Heidelberg, pp 97–110
14. Charnock-Jones DS, Kaufmann P et al (2004) Aspects of human feto-placental vasculogenesis and angiogenesis. I Mol Regul Placenta 25(2–3):103–113
15. Kaufmann P, Mayhew TM et al (2004) Aspects of human feto-placental vasculogenesis and angiogenesis. II Changes during normal pregnancy. Placenta 25(2–3):114–126
16. Mayhew TM (2002) Feto-placental angiogenesis during gestation is biphasic, longitudinal and occurs by proliferation and remodelling of vascular endothelial cells. Placenta 23(10):742–750
17. Wheeler T, Elcock CL et al (1995) Angiogenesis and the placental environment. Placenta 16(3):289–296
18. Gleadle JM, Ebert BL et al (1995) Regulation of angiogenic growth factor expression by hypoxia, transition metals, and chelating agents. Am J Physiol 268(6 Pt 1):C1362–C1368
19. Shore VH, Wang TH et al (1997) Vascular endothelial growth factor, placenta growth factor and their receptors in isolated human trophoblast. Placenta 18(8):657–665
20. Demir R, Kayisli UA et al (2006) Sequential steps during vasculogenesis and angiogenesis in the very early human placenta. Placenta 27(6–7):535–539
21. Demir R, Seval Y et al (2007) Vasculogenesis and angiogenesis in the early human placenta. Acta Histochem 109(4):257–265
22. Kendall RL, Thomas KA (1993) Inhibition of vascular endothelial cell growth factor activity by an endogenously encoded soluble receptor. Proc Natl Acad Sci U S A. 90(22):10705–10709

23. Ahmed A, Dunk C et al (2000) Regulation of placental vascular endothelial growth factor (VEGF) and placenta growth factor (PIGF) and soluble Flt-1 by oxygen—a review. Placenta 21(Suppl A):S16–S24
24. Dawson DW, Volpert OV et al (1999) Pigment epithelium-derived factor: a potent inhibitor of angiogenesis. Science 285(5425):245–248
25. Plunkett BA, Fitchev P et al (2008) Decreased expression of pigment epithelium derived factor (PEDF), an inhibitor of angiogenesis, in placentas of unexplained stillbirths. Reprod Biol 8(2):107–120
26. Liu H, Ren JG et al (2004) Identification of the antivasopermeability effect of pigment epithelium-derived factor and its active site. Proc Natl Acad Sci U S A 101(17):6605–6610
27. Yamagishi S, Nakamura K et al (2006) Pigment epithelium-derived factor inhibits advanced glycation end product-induced retinal vascular hyperpermeability by blocking reactive oxygen species-mediated vascular endothelial growth factor expression. J Biol Chem 281(29):20213–20220
28. Villasante A, Pacheco A et al (2007) Vascular endothelial cadherin regulates vascular permeability: Implications for ovarian hyperstimulation syndrome. J Clin Endocrinol Metab 92(1):314–321
29. Plaisier M (2011) Decidualisation and angiogenesis. Best Pract Res Clin Obstet Gynaecol 25(3):259–271
30. Zygmunt M, Herr F et al (2003) Angiogenesis and vasculogenesis in pregnancy. Eur J Obstet Gynecol Reprod Biol 110(Suppl 1):S10–S18
31. Chen L, Endler A et al (2009) Hypoxia and angiogenesis: regulation of hypoxia-inducible factors via novel binding factors. Exp Mol Med 41(12):849–857
32. Dunwoodie SL (2009) The role of hypoxia in development of the Mammalian embryo. Dev Cell 17(6):755–773
33. Ietta F, Wu Y et al (2006) Dynamic HIF1A regulation during human placental development. Biol Reprod 75(1):112–121
34. Wang GL, Semenza GL (1995) Purification and characterization of hypoxia-inducible factor 1. J Biol Chem 270(3):1230–1237
35. Pringle KG, Kind KL et al (2010) Beyond oxygen: complex regulation and activity of hypoxia inducible factors in pregnancy. Hum Reprod Update 16(4):415–431
36. Genbacev O, Zhou Y et al (1997) Regulation of human placental development by oxygen tension. Science 277(5332):1669–1672
37. Hempstock J, Bao YP et al (2003) Intralobular differences in antioxidant enzyme expression and activity reflect the pattern of maternal arterial bloodflow within the human placenta. Placenta 24(5):517–523
38. Burton GJ, Jauniaux E et al (2010) The influence of the intrauterine environment on human placental development. Int J Dev Biol 54(2–3):303–312
39. Cui XL, Brockman D et al (2006) Expression of NADPH oxidase isoform 1 (Nox1) in human placenta: involvement in preeclampsia. Placenta 27(4–5):422–431
40. Webster RP, Roberts VH et al (2008) Protein nitration in placenta—functional significance. Placenta 29(12):985–994
41. Chance B, Sies H et al (1979) Hydroperoxide metabolism in mammalian organs. Physiol Rev 59(3):527–605
42. Eis AL, Brockman DE et al (1995) Immunohistochemical localization of endothelial nitric oxide synthase in human villous and extravillous trophoblast populations and expression during syncytiotrophoblast formation in vitro. Placenta 16(2):113–126
43. Myatt L, Cui X (2004) Oxidative stress in the placenta. Histochem Cell Biol 122(4):369–382
44. Ahmed A, Dunk C et al (1997) Role of VEGF receptor-1 (Flt-1) in mediating calcium-dependent nitric oxide release and limiting DNA synthesis in human trophoblast cells. Lab Invest 76(6):779–791
45. Garg UC, Hassid A (1989) Nitric oxide-generating vasodilators and 8-bromo-cyclic guanosine monophosphate inhibit mitogenesis and proliferation of cultured rat vascular smooth muscle cells. J Clin Invest 83(5):1774–1777

46. Myatt L, Rosenfield RB et al (1996) Nitrotyrosine residues in placenta. Evidence of peroxynitrite formation and action. Hypertension 28(3):488–493
47. MacMillan-Crow LA, Crow JP et al (1998) Peroxynitrite-mediated inactivation of manganese superoxide dismutase involves nitration and oxidation of critical tyrosine residues. Biochemistry 37(6):1613–1622
48. Yung HW, Korolchuk S et al (2007) Endoplasmic reticulum stress exacerbates ischemia-reperfusion-induced apoptosis through attenuation of Akt protein synthesis in human choriocarcinoma cells. FASEB J 21(3):872–884
49. Ahmed A, Rahman M et al (2000) Induction of placental heme oxygenase-1 is protective against TNFalpha-induced cytotoxicity and promotes vessel relaxation. Mol Med 6(5):391–409
50. Yoshiki N, Kubota T et al (2000) Expression and localization of heme oxygenase in human placental villi. Biochem Biophys Res Commun 276(3):1136–1142
51. Lyall F, Barber A et al (2000) Hemeoxygenase expression in human placenta and placental bed implies a role in regulation of trophoblast invasion and placental function. FASEB J 14(1):208–219
52. McLean M, Bowman M et al (2000) Expression of the heme oxygenase-carbon monoxide signalling system in human placenta. J Clin Endocrinol Metab 85(6):2345–2349
53. McLaughlin BE, Lash GE et al (2003) Heme oxygenase expression in selected regions of term human placenta. Exp Biol Med (Maywood) 228(5):564–567
54. Maines MD (1997) The heme oxygenase system: a regulator of second messenger gases. Annu Rev Pharmacol Toxicol 37:517–554
55. McCoubrey WK Jr, Huang TJ et al (1997) Isolation and characterization of a cDNA from the rat brain that encodes hemoprotein heme oxygenase-3. Eur J Biochem 247(2):725–732
56. Johnson RA, Kozma F et al (1999) Carbon monoxide: from toxin to endogenous modulator of cardiovascular functions. Braz J Med Biol Res 32(1):1–14
57. Wang R (1998) Resurgence of carbon monoxide: an endogenous gaseous vasorelaxing factor. Can J Physiol Pharmacol 76(1):1–15
58. Wagner CT, Durante W et al (1997) Hemodynamic forces induce the expression of heme oxygenase in cultured vascular smooth muscle cells. J Clin Invest. 100(3):589–596
59. Sass G, Soares MC et al (2003) Heme oxygenase-1 and its reaction product, carbon monoxide, prevent inflammation-related apoptotic liver damage in mice. Hepatology 38(4):909–918
60. Song R, Kubo M et al (2003) Carbon monoxide induces cytoprotection in rat orthotopic lung transplantation via anti-inflammatory and anti-apoptotic effects. Am J Pathol 163(1):231–242
61. Brouard S, Otterbein LE et al (2000) Carbon monoxide generated by heme oxygenase 1 suppresses endothelial cell apoptosis. J Exp Med 192(7):1015–1026
62. Bainbridge SA, Farley AE et al (2002) Carbon monoxide decreases perfusion pressure in isolated human placenta. Placenta 23(8–9):563–569
63. Bainbridge SA, Belkacemi L et al (2006) Carbon monoxide inhibits hypoxia/reoxygenation-induced apoptosis and secondary necrosis in syncytiotrophoblast. Am J Pathol 169(3):774–783
64. Longtine MS, Nelson DM (2011) Placental dysfunction and fetal programming: the importance of placental size, shape, histopathology, and molecular composition. Semin Reprod Med 29(3):187–196
65. Jauniaux E, Hempstock J et al (2003) Trophoblastic oxidative stress in relation to temporal and regional differences in maternal placental blood flow in normal and abnormal early pregnancies. Am J Pathol 162(1):115–125
66. Jauniaux E, Johns J et al (2005) The role of ultrasound imaging in diagnosing and investigating early pregnancy failure. Ultrasound Obstet Gynecol 25(6):613–624
67. Watson AL, Skepper JN et al (1998) Susceptibility of human placental syncytiotrophoblastic mitochondria to oxygen-mediated damage in relation to gestational age. J Clin Endocrinol Metab 83(5):1697–1705

68. Watson AL, Palmer ME et al (1997) Variations in expression of copper/zinc superoxide dismutase in villous trophoblast of the human placenta with gestational age. Placenta 18(4):295–299
69. Watson AL, Skepper JN et al (1998) Changes in concentration, localization and activity of catalase within the human placenta during early gestation. Placenta 19(1):27–34
70. Hung TH, Burton GJ (2006) Hypoxia and reoxygenation: a possible mechanism for placental oxidative stress in preeclampsia. Taiwan J Obstet Gynecol 45(3):189–200
71. Leung DN, Smith SC et al (2001) Increased placental apoptosis in pregnancies complicated by preeclampsia. Am J Obstet Gynecol 184(6):1249–1250
72. Allaire AD, Ballenger KA et al (2000) Placental apoptosis in preeclampsia. Obstet Gynecol 96(2):271–276
73. Krebs C, Macara LM et al (1996) Intrauterine growth restriction with absent end-diastolic flow velocity in the umbilical artery is associated with maldevelopment of the placental terminal villous tree. Am J Obstet Gynecol 175(6):1534–1542
74. Macara L, Kingdom JC et al (1996) Structural analysis of placental terminal villi from growth-restricted pregnancies with abnormal umbilical artery Doppler waveforms. Placenta 17(1):37–48
75. Redline RW (2008) Placental pathology: a systematic approach with clinical correlations. Placenta 29(Suppl A):S86–S91
76. Myatt L (2010) Review: reactive oxygen and nitrogen species and functional adaptation of the placenta. Placenta 31(Suppl):S66–S69
77. Burton GJ (2009) Oxygen, the Janus gas; its effects on human placental development and function. J Anat 215(1):27–35
78. Many A, Hubel CA et al (2000) Invasive cytotrophoblasts manifest evidence of oxidative stress in preeclampsia. Am J Pathol 156(1):321–331
79. Myatt L, Eis AL et al (1997) Endothelial nitric oxide synthase in placental villous tissue from normal, pre-eclamptic and intrauterine growth restricted pregnancies. Hum Reprod 12(1):167–172
80. Bowen RS, Moodley J et al (2001) Oxidative stress in pre-eclampsia. Acta Obstet Gynecol Scand 80(8):719–725
81. Kharb S (2000) Vitamin E and C in preeclampsia. Eur J Obstet Gynecol Reprod Biol 93(1):37–39
82. Li H, Gu B et al (2005) Hypoxia-induced increase in soluble Flt-1 production correlates with enhanced oxidative stress in trophoblast cells from the human placenta. Placenta 26(2–3):210–217
83. Sankaralingam S, Arenas IA et al (2006) Preeclampsia: current understanding of the molecular basis of vascular dysfunction. Expert Rev Mol Med 8(3):1–20
84. Redman CW, Sargent IL (2000) Placental debris, oxidative stress and pre-eclampsia. Placenta 21(7):597–602
85. Hung TH, Skepper JN et al (2002) Hypoxia-reoxygenation: a potent inducer of apoptotic changes in the human placenta and possible etiological factor in preeclampsia. Circ Res 90(12):1274–1281
86. Levy R, Smith SD et al (2002) Trophoblast apoptosis from pregnancies complicated by fetal growth restriction is associated with enhanced p53 expression. Am J Obstet Gynecol 186(5):1056–1061
87. Endo H, Okamoto A et al (2005) Frequent apoptosis in placental villi from pregnancies complicated with intrauterine growth restriction and without maternal symptoms. Int J Mol Med 16(1):79–84
88. Kiserud T, Hellevik LR et al (1994) Estimation of the pressure gradient across the fetal ductus venosus based on Doppler velocimetry. Ultrasound Med Biol 20(3):225–232
89. Hitschold T, Mutefering H et al (1996) Does extremely low feto-placental impedence as estimated by umbilical artery Doppler velocimetry also indicate fetuses at risk? Ultrasound Gynecology 8:39A

90. Baker PN, Krasnow J et al (1995) Elevated serum levels of vascular endothelial growth factor in patients with preeclampsia. Obstet Gynecol 86(5):815–821
91. Sharkey AM, Cooper JC et al (1996) Maternal plasma levels of vascular endothelial growth factor in normotensive pregnancies and in pregnancies complicated by pre-eclampsia. Eur J Clin Invest 26(12):1182–1185
92. Brockelsby JC, Wheeler T et al (1998) Increased circulating levels of vascular endothelial growth factor in preeclampsia. Hypertens Pregnancy 17:283–290
93. Hunter A, Aitkenhead M et al (2000) Serum levels of vascular endothelial growth factor in preeclamptic and normotensive pregnancy. Hypertension 36(6):965–969
94. El-Salahy EM, Ahmed MI et al (2001) New scope in angiogenesis: role of vascular endothelial growth factor (VEGF), NO, lipid peroxidation, and vitamin E in the pathophysiology of pre-eclampsia among Egyptian females. Clin Biochem 34(4):323–329
95. Reuvekamp A, Velsing-Aarts FV et al (1999) Selective deficit of angiogenic growth factors characterises pregnancies complicated by pre-eclampsia. Br J Obstet Gynaecol 106(10):1019–1022
96. Livingston JC, Chin R et al (2000) Reductions of vascular endothelial growth factor and placental growth factor concentrations in severe preeclampsia. Am J Obstet Gynecol 183(6):1554–1557
97. Trollmann R, Amann K et al (2003) Hypoxia activates the human placental vascular endothelial growth factor system in vitro and in vivo: up-regulation of vascular endothelial growth factor in clinically relevant hypoxic ischemia in birth asphyxia. Am J Obstet Gynecol 188(2):517–523
98. Cooper JC, Sharkey AM et al (1996) VEGF mRNA levels in placentae from pregnancies complicated by pre-eclampsia. Br J Obstet Gynaecol 103(12):1191–1196
99. Lyall F, Young A et al (1997) Placental expression of vascular endothelial growth factor in placentae from pregnancies complicated by pre-eclampsia and intrauterine growth restriction does not support placental hypoxia at delivery. Placenta 18(4):269–276
100. Helske S, Vuorela P et al (2001) Expression of vascular endothelial growth factor receptors 1, 2 and 3 in placentas from normal and complicated pregnancies. Mol Hum Reprod 7(2):205–210
101. Ranheim T, Staff AC et al (2001) VEGF mRNA is unaltered in decidual and placental tissues in preeclampsia at delivery. Acta Obstet Gynecol Scand 80(2):93–98
102. Lash G, MacPherson A et al (2001) Abnormal fetal growth is not associated with altered chorionic villous expression of vascular endothelial growth factor mRNA. Mol Hum Reprod 7(11):1093–1098
103. Tse JY, Lao TT et al (2001) Expression of vascular endothelial growth factor in third-trimester placentas is not increased in growth-restricted fetuses. J Soc Gynecol Investig 8(2):77–82
104. Maynard SE, Karumanchi SA (2011) Angiogenic factors and preeclampsia. Semin Nephrol 31(1):33–46
105. Gu Y, Lewis DF et al (2008) Placental productions and expressions of soluble endoglin, soluble fms-like tyrosine kinase receptor-1, and placental growth factor in normal and preeclamptic pregnancies. J Clin Endocrinol Metab 93(1):260–266
106. Venkatesha S, Toporsian M et al (2006) Soluble endoglin contributes to the pathogenesis of preeclampsia. Nat Med 12(6):642–649
107. Farley AE, Graham CH et al (2004) Contractile properties of human placental anchoring villi. Am J Physiol Regul Integr Comp Physiol 287(3):R680–R685
108. Zenclussen AC, Lim E et al (2003) Heme oxygenases in pregnancy II: HO-2 is downregulated in human pathologic pregnancies. Am J Reprod Immunol 50(1):66–76
109. Cudmore M, Ahmad S et al (2007) Negative regulation of soluble Flt-1 and soluble endoglin release by heme oxygenase-1. Circulation 115(13):1789–1797
110. Al-Ani B, Hewett PW et al (2010) Activation of proteinase-activated receptor 2 stimulates soluble vascular endothelial growth factor receptor 1 release via epidermal growth factor receptor transactivation in endothelial cells. Hypertension 55(3):689–697

111. Zhao H, Wong RJ et al (2009) Effect of heme oxygenase-1 deficiency on placental development. Placenta 30(10):861–868
112. Rajakumar A, Whitelock KA et al (2001) Selective overexpression of the hypoxia-inducible transcription factor, HIF-2alpha, in placentas from women with preeclampsia. Biol Reprod 64(2):499–506
113. Rajakumar A, Brandon HM et al (2004) Evidence for the functional activity of hypoxia-inducible transcription factors overexpressed in preeclamptic placentae. Placenta 25(10):763–769
114. Khaliq A, Dunk C et al (1999) Hypoxia down-regulates placenta growth factor, whereas fetal growth restriction up-regulates placenta growth factor expression: molecular evidence for "placental hyperoxia" in intrauterine growth restriction. Lab Invest 79(2):151–170
115. Kingdom JC, Kaufmann P (1997) Oxygen and placental villous development: origins of fetal hypoxia. Placenta 18(8):613–621 discussion 623–626
116. Burton GJ (1997) On oxygen and placental villous development: origins of fetal hypoxia. Placenta 18:625–626
117. Ahmed A, Kilby MD (1997) Hypoxia or hyperoxia in placental insufficiency? Lancet 350(9081):826–827

Chapter 6
The Use of Antioxidants in Pre-eclampsia

Jean-François Bilodeau

Abstract Pre-eclampsia (PE) is a complex pathology diagnosed during the second part of the pregnancy. The clinical features of PE are hypertension and proteinuria. This syndrome occurs in 3–8 % of all pregnancies. The only cure for PE is delivery, and this syndrome is a leading cause of maternal and neonatal mortality and morbidity. The cause of PE is thought to originate from the placenta through the release of circulating factors that lead to a generalized systemic vascular endothelial dysfunction. The factors suspected include reactive oxygen species (ROS). It is strongly believed that ROS are major contributors of endothelial cell dysfunction leading to PE, since women affected by this syndrome show imbalanced ROS production, abnormal levels of antioxidant defenses, and increased blood and placental lipid peroxidation. Here, we discussed the rationale for the use of antioxidants during pregnancy to prevent PE. We will also review several approaches or trials involving antioxidants, such as vitamin C and E, coenzyme Q10, N-acetylcysteine, carotenoids, and selenium (Se), a precursor of antioxidant enzymes.

Keywords Use of antioxidants · Pre-eclampsia · ROS production · Clinical importance · Physiopathology of PE · Placental antioxidant imbalance · Carotenoids · Maternal antioxidant imbalance

---

J.-F. Bilodeau (✉)
Centre de Recherche du CHUQ, CHUL,
Obstetrics and Gynecology, Laval University,
Axe reproduction, santé périnatale et santé de l'enfant 2705,
boul. Laurier, local T-1-49, Québec, QC G1V 4G2, Canada
e-mail: jean-francois.bilodeau@crchul.ulaval.ca

6.1 Clinical Importance of PE

Pre-eclampsia (PE) usually occurs after 20 weeks of gestation with an incidence between 3 and 7 % of all pregnancies in industrialized countries [1]. This pregnancy-specific syndrome, unique to human, is characterized by hypertension (>90 mm Hg diastolic on two separate readings at least 4 h apart) and proteinuria ($\geq$0.3 g of protein in a 24 h urine sample) [1]. This pathological state can lead to complications, such as hemolysis, elevated liver enzymes and low platelet count (HELLP syndrome), and eclampsia, characterized by a convulsive condition associated or not with multiple organ dysfunctions [2]. PE is a leading cause of maternal and neonatal mortality and morbidity in developing countries [3, 4]. Actually, there is no effective treatment except inducing labor, which contributes to the higher incidence of premature births.

There are numerous risk factors associated with PE including primiparity, multiple pregnancy, psychosocial strain related to work during pregnancy, poor social background, the mother's low birth weight, young or old age of the mother (<20 or >40 years), pre-pregnancy obesity, family history of PE, and endometriosis [5–9]. The assisted reproductive technologies (ARTs) such as in vitro fertilization (IVF) are associated with increased risk of PE [10]. Abnormal implantation, excessive placental size, or microvascular disease from pre-existing hypertension and diabetes mellitus predispose also to PE [11]. Of note, women affected by PE are more likely to develop cardiovascular diseases later in life [12].

6.2 Physiopathology of PE

According to Redman and Roberts, PE is a two-stage disorder [13]. The pathology is believed to originate from the placenta (stage 1). PE is often associated with an abnormal cytotrophoblast differentiation and invasion [14]. Indeed, there is an abnormal vascular remodeling of spiral arteries in the placenta [15]. Consequently, the placental perfusion is reduced and this leads to placental ischemia and oxidative stress [16]. The exact cellular and molecular basis of this phenomenon is unknown and is thought to be multifactorial, involving genetic, immunological, and biochemical disorders. The consequence of stage 1 of PE is a general vasoconstrictive disorder resulting from abnormal maternal vascular endothelial function (stage 2) [17]. Factors released by the placenta that potentially cause PE include cytokines like TNF-α and IL-1-β, endothelin, neurokinin B, and various ROS [17, 18].

6.3 Oxidative Stress in PE

In pregnancy, increased levels of lipid peroxidation are normally found in comparison to the non-pregnant state in saliva, urine, and plasma [19, 20]. However, this state of oxidative stress in "normal" or normotensive pregnancy is further exacerbated in PE. One of the most sensitive marker of oxidative stress is the determination of the level of 8-isoprostane (8-iso-PGF$_{2\alpha}$) derived from the chemical oxidation of arachidonic acid in biological membranes. The 8-isoprostane is also a vasoconstrictor believed to contribute partly to the overall hypertension in PE. The levels of 8-iso-PGF$_{2\alpha}$ were measured in the fluids of normotensive and PE women. Higher levels of free, but not total 8-iso-PGF$_{2\alpha}$ were reported in plasma of PE women than normotensive controls [21]. Since only the free form of 8-iso PGF$_{2\alpha}$ exerts a biological effect (vasoconstrictor), this increase partly explains the higher blood pressure in PE. Indeed, the highest levels of free 8-iso PGF$_{2\alpha}$ are found in severe cases of PE ($\geq$160 mm Hg systolic, $\geq$110 mm Hg diastolic, proteinuria >300 mg/24 h). Elevated levels of free 8-IsoPs were also found in decidua basalis and homogenates of placenta of women with PE [22, 23]. The 8-iso-PGF$_{2\alpha}$ was reported higher in sera [24, 25], urine [26], and plasma [27, 28] of PE women.

There are now sufficient evidences to consider ROS as promoters of endothelial cell dysfunction leading to PE since women affected by this condition show imbalanced ROS production, mostly NO and O$_2^-$ [29], abnormal levels of antioxidant defenses [30], and increased blood and placental lipid peroxidation [23]. Recently, we have clearly shown that mRNA expression for a stress protein related to oxidative stress, the heme oxygenase 1 (HO-1), was significantly higher in blood and placenta of women affected by PE as well as in the fetal circulation [31, 32]. Women affected by PE have also clearly altered levels of endogenous antioxidants in blood, placenta, and decidua. The latter will be discussed in more details below.

6.3.1 Placental Antioxidant Imbalance

Placental oxidative stress has been strongly associated with the pathogenesis of PE. Most studies showed a decrease in placental GSH, vitamin C levels, glutathione peroxidase (GPx), glutathione-S-transferase (GST), and superoxide dismutase (SOD) activities in PE [33–36]. It has also been reported that specifically GST of class pi was reduced in placenta and decidua of PE women [37]. In contrast, it was shown that peroxiredoxin 3 was increased in mitochondria of placenta of PE pregnancies [38]. We recently showed that the reported decrease for SOD activity in PE placentas can be attributed to SOD1 mRNA and protein levels in absence of labor [39]. We also demonstrated that combination of labor and PE upregulate SOD1 in fetal membranes, and SOD2, and SOD3 in the whole placenta [39]. Moreover, we showed that GPX4 mRNA expression was deficient in the PE placentas in presence or absence of labor [32]. Similarly, Mistry et al. (2010) recently showed reduced GPx-1, GPx-3, and GPx-4 protein content in PE placental villi [40]. They attributed this decreased GPx level to a reduced selenium (Se) availability in PE placentas that was not observed in controls [41]. Of note, GPx are selenoproteins involved in the detoxification of lipid peroxides bound to biological membranes.

6.3.2 Maternal Antioxidant Imbalance

In PE, it was observed that the plasma concentration of nitric oxide and lipid peroxides were elevated with a concomitant decrease in catalase and SOD activities in comparison to normotensive pregnancies [35, 42, 43]. Other investigators found that the serum levels of lipid peroxides were significantly increased, and SOD activities and GSH level in erythrocytes were significantly decreased in women affected by PE and eclampsia [44]. Another study indicates that lipid peroxidation and total GPx activity increased in plasma and erythrocytes of patients with PE + HELLP syndrome, while GST, glutathione reductase (GR) and SOD activities remained unchanged [45]. In contrast to sera and plasma, we have shown that higher mRNA or protein expression for GPx-1 and GPx-4 in blood lymphocytes and monocytes occur in PE than in normotensive pregnancies [31].

6.4 Clinical Trials and Rational for the Use of Antioxidants in PE

In this section, we will discuss specifically for each antioxidant reviewed, the rational for its use in PE treatment or prevention. We will first cover the use of vitamin C and E in PE that caused several debates over the year for their efficacies to treat PE. Then, we will review less studied but promising antioxidants that might deserve to be tested extensively in PE treatment such as coenzyme Q_{10}, carotenoids, N-acetylcysteine, and Se.

6.4.1 Controversy in the Outcome of Vitamins C and E Trials and PE

Vitamin C, also known as ascorbate, is a water-soluble antioxidant and a scavenger of free radicals. One of the main advantage of ascorbate is when oxidized under the form of ascorbate radical (Fig. 6.1), the latter is relatively unreactive and can be easily regenerated back to its reduced form. Throughout normal pregnancy, plasma ascorbate reserves gradually decrease but, its level is further decreased by 20–50 % in PE pregnancies [46–49]. Some data suggest that vitamin C deficiency may explain the occurrence of PE since women who ingested less than 85 mg of vitamin C daily are at 2-fold increased risk of developing this adverse condition [49]. Of note, women who suffered from previous PE remained with a dysfunctional endothelium as measured by brachial artery flow-mediated dilation. The intravenous injection of 1 g of ascorbate can reverse this phenomenon in these women [50].

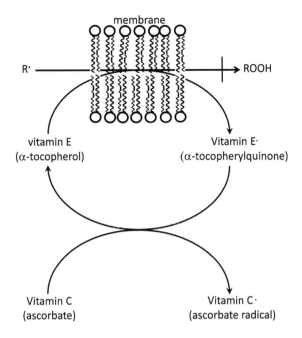

Fig. 6.1 Proposed protective mechanism of action of vitamin C and E against lipid oxidation. Highly reactive radical or oxyradical ($R^·$) will react with unsaturated lipids to yield a lipid hydroperoxide (ROOH) unless vitamin E intervenes in the reaction

The beneficial mode actions of ascorbate hypothesized to occur in PE include (1) the regeneration of oxidized vitamin E and glutathione; (2) to act as a scavenger of O_2^- before it react with NO to form $ONOO^-$, an aspect still debated to occur in vivo [51–53]; (3) possible enhancement of eNOS enzymatic activity and NO production favoring vasodilation [52]; and (4) promoting the decomposition of extracellular S-nitrosothiols and subsequent release of NO [54].

According to the Cochrane Data base review published in 2005, the use of vitamin C with dose of 500 or 1,000 mg may prevent PE using a fixed effect but not a random effect model [55]. More data would be required to make sure of the protective effect on PE incidence since a lot of the studies reviewed used vitamin C in combination with other antioxidants. It was also noted that vitamin C supplementation was associated with a moderate increase in the risk of preterm birth.

In contrast to the highly hydrophilic vitamin C, the vitamin E (tocopherol) is a major lipophilic and a potent chain-breaking antioxidant that limits the propagation of free radicals in polyunsaturated membrane lipids (Fig. 6.1). The oxidized form of vitamin E, the tocopherylquinone, has the ability to propagate free radicals [56]. When vitamin E intercepts a radical, a tocopheroxyl radical is formed that can be reduced back to tocopherol by vitamin C or other reducing agents (Fig. 6.1). Vitamins C and E are thus complementary in many respects and have been used together in numerous trials to prevent PE.

Plasma vitamin E concentration increases during normal pregnancy [57, 58], likely caused by the gestational increase in circulating lipoproteins, the transporters of vitamin E [59]. Plasma vitamin E concentrations are either unchanged or increased in PE per plasma volume [57, 60–66] but unaffected when adjusted for apolipoprotein B or cholesterol or cholesterol + triglycerides [47, 60, 62, 65–67].

At first, very encouraging results were obtained with antioxidant vitamin trials for the prevention of PE. The combination of two exogenous antioxidants, vitamin C and E, from 22 weeks gestation until birth was shown to reduce the incidence of this syndrome in high-risk women and was associated with a reduction in markers of endothelial dysfunction (plasminogen activator inhibitor ratio, PAI-1/PAI-2) [68, 69]. A follow-up study further demonstrated that vitamin C and E supplementation in women at risk of PE was associated with a reduction of vasoconstrictor and oxidative marker, 8-iso-PGF$_{2\alpha}$, in plasma. An inversed relationship between α-tocopherol and isoprostane levels strongly suggested that vitamin E prevents lipid peroxidation in PE [46].

Later, high doses of these antioxidant vitamins (1,000 mg vitamin C and 400 IU vitamin E) were shown to be detrimental to the newborn such as increasing the rate of low birth weight [70]. The efficacy of these treatments is still debated today [71, 72]. Recent results of two large North American clinical trials on vitamins C and E prevention of PE, with thousands of patients enrolled, showed no prevention of PE [73, 74]. However, these studies urge the scientific committee for more fundamental research because (1) Recent evidence indicates that vitamin E is generally not lacking in PE, but maybe in only a small subset of women [62, 74]. (2) Women were treated with specifically α-tocopherol, one of the several possible isomers of vitamin E. It is unknown whether α-tocopherol is the most efficient isomer against oxidative stress in PE. (3) The vitamin treatments were maybe started too late; after 9–16 weeks of pregnancy in the US trial and after 12–18 weeks in the Mexican/Canadian trial [73, 74]. This can be problematic since the initiation of intervillous blood flow in the placenta starts around 8–10 weeks of gestation, and is associated with oxidative stress [75]. (4) The absorption and the plasmatic values of antioxidant vitamins in these trials were not assessed and it is not possible to know if the supplementation really increased the levels of these vitamins in a subset or in all women.

6.4.2 Coenzyme Q_{10}

The coenzyme Q_{10} (CoQ$_{10}$) is an essential component of the mitochondrial respiratory chain and a very efficient antioxidant under its reduced form, the ubiquinol-10. Indeed, ubiquinol-10, a vitamin-like antioxidant, is more efficient than β-carotene, α-tocopherol, and lycopene in inhibiting low density (LDL) oxidation in vitro [76]. The CoQ$_{10}$ has also the ability to protect vitamin E from superoxide anion ($O_2^{-\cdot}$) attack [77]. The oxidized (ubiquinone-10)/reduced (ubiquinol-10) ratio of CoQ$_{10}$ can be used as an oxidative stress marker in human plasma [78, 79].

In PE, the total CoQ$_{10}$ level was shown to be decreased in serum and was associated with the severity of the syndrome [80]. The total CoQ$_{10}$ level was also reported to be decreased in PE plasma [81, 82]. However, CoQ_{10} level was shown to be increased in the PE cord blood and placentas when compared to normotensive pregnancies [82]. Interestingly, our group reported no increase of total CoQ_{10} level in plasma, but rather an increase in the ratio of oxidized to reduced form of CoQ_{10} in the blood of PE mothers [62]. Our study also indicated that CoQ_{10} is a more sensitive marker of oxidative stress than vitamin E in PE. Also, we showed that ubiquinol-10 was correlated with α-tocopherol in PE pregnancies only [62]. The positive correlation between ubiquinol-10 and α-tocopherol could thus represent a coordinated defense mechanism against oxidative stress in PE.

Beside its efficient antioxidant properties, CoQ_{10} may have unsuspected roles on immune functions and the control of blood pressure in PE. Indeed, we discovered that plasmatic IL-18 level, a cytokine known to induce tumor necrosis factor-α (TNF-α), was positively correlated with the ubiquinol-10 in PE pregnancies only [83]. Though, the exact impact of the latter on immune function remains to be determined. We also observed that CoQ_{10} (oxidized/reduced ratio) positively correlates with the prostacyclin (vasodilator) to thromboxane (vasoconstrictor) ratio [62]. This ratio may partly explain the hypertension in PE. More work still need to be done to understand the impact of the redox state on immune and vasoactive mediators.

A recent study indicated that CoQ10 supplementation at a dose of 200 mg/day from 20 weeks of pregnancy until birth can prevent PE [84]. In this randomized cohort, the placebo group was constituted of 74 women and the CoQ10 treated group of 80 women. The 25.6 % incidence of PE in the placebo group was significantly reduced to 14.4 % in the CoQ_{10}-treated group. This encouraging study was performed in Quito, Ecuador at 2,800 m of altitude, a factor know to increase the incidence of PE.

6.4.3 Carotenoids

It was observed that placental and serum levels of two carotenoids, β-carotene, and lycopene, were significantly lower during late gestation in PE than in normotensive pregnancies [85]. The red pigment, lycopene, derived from vegetables like tomatoes has antioxidant properties. Lycopene is a better scavenger of singlet oxygen than β-carotene [86]. Two trials using lycopene as an antioxidant were designed to prevent PE [87, 88]. The initial study of Sharma et al. [88] showed very promising results. In this randomized controlled study, 116 pregnant women were given 2 mg twice a day orally from 20–24 weeks of gestation until delivery. This group was compared to a control group of 135 pregnant women treated with a placebo. Interestingly, the incidence of PE was down to 8.6 % in the lycopene-treated group while the incidence was 17.7 % in the control group. In addition, the fetal weight was significantly higher in the lycopene-treated group by 3.5 %. The incidence of intrauterine growth restriction was also reduced from 23.7 % in controls to 12 % in the lycopene-treated group.

Later, Banerjee et al. [87] published rather discouraging results from their lycopene trial. No difference between lycopene-treated group of 77 women and the control group of 82 women was found for PE incidence [87]. Moreover, even adverse effects of the lycopene treatment were registered such as increased preterm labor and low birth weight.

Strong similarities between the two above studies on lycopene can be observed: (1) the same supplier of lycopene was used, (2) the two studies were performed in New Dehli, India; (3) the incidence of PE was rather high (~ 18 %) for control groups. However, the major differences were the dose of lycopene used and the timing of the initial treatment of pregnant women. In the study of Sharma et al. [88], the lycopene dose was 4 mg per day started at 20–24 weeks of gestation. In contrast, the Banerjee et al. [87] study was started earlier between 12 and 20 weeks with half the dose (2 mg/24 h). The discrepancies between experimental protocols may explain the different outcome of these two studies. More study would be required to ascertain the timings and dosage for the lycopene treatments.

6.4.4 N-acetylcysteine

The blood level of total cysteine decreases from the second to the third trimester in a normal pregnancy, while the concentration eliminated in urine remains unchanged [89]. The total plasma cysteine levels in PE were reported significantly higher compared to pregnant controls [90] but more oxidized to the total cysteine was observed in PE [91]. Despite of higher cysteine levels, precursor of GSH, the levels of the latter were often reported lower in the maternal circulation of PE than control pregnancies [92–95]. Also, the GSH levels were reported halved in PE placentas than controls [96], but another studies showed the opposite in the placenta and decidua of PE women [97].

It is why some suggested to replete the tripeptide GSH in PE using N-acetylcysteine (NAC) [98]. The NAC is rapidly deacetylated and imported into the cell under the form of cysteine. NAC have been widely used to treat hepatotoxic effect of an overdose of acetaminophen and is relatively well-tolerated at doses below 3 g/day [99]. The NAC protective actions are not only attributed to GSH repletion. Indeed, NAC may improve the bioavailability of NO, a known vasodilator by protecting the latter from other inactivating ROS. This was shown using an ex vivo perfusion cotyledon model [100]. Also, NAC may prevent immune cell adhesion to blood vessels. Indeed, NAC can decrease U937 monocyte adhesion to human endothelial vein cell HUVEC by 20 % in the presence of PE plasma in vitro, while normal pregnant plasma had no significant effect in presence of NAC [101].

So far, only one study is available for the use of NAC in PE [98]. Roes et al. (2006) hypothesized that oral NAC administration at the onset of PE would result in a 7 days prolongation of pregnancy. The dose used was 1.8 g every 8 h. The placebo ($n = 19$) and PE ($n = 19$) group were treated for an average of 6 and 5 days before birth respectively. The only findings between the two groups were decreased plasma GSH levels 24 h after delivery in the placebo groups, and increased plasma cysteine concentrations 6 weeks postpartum in women who received NAC. Also, there was no difference in maternal complications and neonatal outcomes between the two groups. As argued by the author, the power of their study was limited (75 %) and the intervention may be too late to alleviate complications associated with PE.

6.4.5 Selenium, the Precursor of Antioxidant Selenoproteins

Selenium is not an antioxidant by itself but, this mineral is essential for antioxidant selenoproteins synthesis and activity. There are few selenoproteins in mammals. Only 25 genes encode selenoproteins in human [102]. The biosynthesis of these selenoproteins relies on the presence of a special amino acid, the selenocystein, the 21st amino acid encoded by a stop codon [102]. Therefore, selenoprotein synthesis is affected by the lack of this trace element in the diet. Se deprivation causes thyroid and fertility problems. Moreover, low red cell Se level was associated with recurrent pregnancy loss [103]. Suboptimal or Se deficiency affects an important class of antioxidant, the SeGPx involved in the detoxification of various peroxides know to be increased in PE. Still today, the optimal concentration of Se for a healthy pregnancy is not precisely known.

A study indicates that Se levels in toe nail can be associated to higher risk of PE. Within the PE group, a lower Se status is significantly associated with the severity of PE [104]. In addition, there are indications that serum or plasma Se levels are lower in PE than in normal healthy pregnancies in various countries [40, 105, 106]. There is a positive association between serum or plasmatic maternal Se and maternal plasma GPx activity in normal and PE pregnancies [41]. In some case, this can also be associated to lower SeGPx activity in the placenta [105].

A recent human trial (2010) suggested that organic Se supplementation (100 μg of Se yeast/day) from the first trimester to the delivery may reduce the incidence of PE [107]. The design was randomized double-blind placebo-controlled trial. No side effects of the supplementation were noted. As expected, the serum Se levels increased by an average of 38 % in the Se-treated group only. Interestingly, PE was observed only in the non-treated group (3 cases on 83 women). The sample size did not allow to conclude statistically on the protective effect of Se against PE, but this approach looks promising.

6.5 Conclusion and Perspectives

A lot of work still need to be done to understand the roles and the regulation of antioxidants in PE through basic science studies. A better understanding of the pathophysiology of PE can lead to new therapeutic approaches to treat this pathology. Besides the previously discussed antioxidant therapies, other possible approaches are under consideration such as the role melatonin as a placental antioxidant response enhancer [108, 109]. Also, estrogen should be considered since intramuscular injection of estradiol (3×10 mg every 24 h) three days before labor induction was show to have antioxidative effect in erythrocytes and to lower blood pressure in PE women [108]. For the upcoming trials involving antioxidant, it is imperative to (1) better plan the begining the treatment considering the initiation of the intervillous blood flow in the placenta that starts around 8–10 weeks of gestation. (2) The clinical study must be accompanied by solid biochemical data that will reveal the actual level of the studied antioxidant before, during, and after the treatment. (3) To assess the impact of the antioxidant treatment on well-established markers of oxidative stress (e.g. isoprostane). (4) To carefully choose the population to be studied since malnutrition and poor social background increase the risk of PE adding more factors to be controlled for. Studies aiming toward personalized dose of antioxidants should be considered, but it would also require knowing the endogenous level of antioxidant before the treatment.

References

1. Report of the national high blood pressure education program working group on high blood pressure in pregnancy. Am J Obstet Gynecol. 200; 183: S1–S22
2. Weinstein L (1982) Syndrome of hemolysis, elevated liver enzymes, and low platelet count: a severe consequence of hypertension in pregnancy. Am J Obstet Gynecol 142:159–167
3. Canada S (1997) Causes of death 1993, 1994, 1995, 1996, 1997. Ottawa: Health Statistics Division
4. Roberts J (1998) Pregnancy related hypertension. In: Creasy R, Resnik R (eds) Maternal Fetal Medicine. WB Saunders, Philiadelphia
5. Frederick IO, Rudra CB, Miller RS, Foster JC, Williams MA (2006) Adult weight change, weight cycling, and prepregnancy obesity in relation to risk of preeclampsia. Epidemiology 17:428–434
6. Innes KE, Marshall JA, Byers TE, Calonge N (1999) A woman's own birth weight and gestational age predict her later risk of developing preeclampsia, a precursor of chronic disease. Epidemiology 10:153–160
7. Klonoff-Cohen HS, Cross JL, Pieper CF (1996) Job stress and pre-eclampsia. Epidemiology 7:245–249
8. Lie RT, Rasmussen S, Brunborg H, Gjessing HK, Lie-Nielsen E, Irgens LM (1998) Fetal and maternal contributions to risk of pre-eclampsia: population based study. BMJ 316:1343–1347
9. Stephansson O, Kieler H, Granath F, Falconer H (2009) Endometriosis, assisted reproduction technology, and risk of adverse pregnancy outcome. Hum Reprod 24:2341–2347
10. Shevell T, Malone FD, Vidaver J, Porter TF, Luthy DA, Comstock CH, Hankins GD, Eddleman K, Dolan S, Dugoff L, Craigo S, Timor IE, Carr SR, Wolfe HM, Bianchi DW, D'Alton ME (2005) Assisted reproductive technology and pregnancy outcome. Obstet Gynecol 106:1039–1045
11. Broughton Pipkin F (2001) Risk factors for preeclampsia. N Engl J Med 344:925–926

12. Roberts JM, Gammill H (2005) Pre-eclampsia and cardiovascular disease in later life. Lancet 366:961–962
13. Roberts JM (2000) Preeclampsia: what we know and what we do not know. Semin Perinatol 24:24–28
14. Lim KH, Zhou Y, Janatpour M, McMaster M, Bass K, Chun SH, Fisher SJ (1997) Human cytotrophoblast differentiation/invasion is abnormal in pre-eclampsia. Am J Pathol 151:1809–1818
15. Pijnenborg R, Bland JM, Robertson WB, Brosens I (1983) Uteroplacental arterial changes related to interstitial trophoblast migration in early human pregnancy. Placenta 4:397–413
16. Redman CW, Sargent IL (2000) Placental debris, oxidative stress and pre-eclampsia. Placenta 21:597–602
17. Roberts JM, Taylor RN, Musci TJ, Rodgers GM, Hubel CA, McLaughlin MK (1989) Preeclampsia: an endothelial cell disorder. Am J Obstet Gynecol 161:1200–1204
18. Page NM, Woods RJ, Gardiner SM, Lomthaisong K, Gladwell RT, Butlin DJ, Manyonda IT, Lowry PJ (2000) Excessive placental secretion of neurokinin B during the third trimester causes pre-eclampsia. Nature 405:797–800
19. Hung TH, Lo LM, Chiu TH, Li MJ, Yeh YL, Chen SF, Hsieh TT (2010) A longitudinal study of oxidative stress and antioxidant status in women with uncomplicated pregnancies throughout gestation. Reprod Sci 17:401–409
20. Little RE, Gladen BC (1999) Levels of lipid peroxides in uncomplicated pregnancy: a review of the literature. Reprod Toxicol 13:347–352
21. McKinney ET, Shouri R, Hunt RS, Ahokas RA, Sibai BM (2000) Plasma, urinary, and salivary 8-epi-prostaglandin f2alpha levels in normotensive and preeclamptic pregnancies. Am J Obstet Gynecol 183:874–877
22. Staff AC, Halvorsen B, Ranheim T, Henriksen T (1999) Elevated level of free 8-isoprostaglandin F2alpha in the decidua basalis of women with preeclampsia. Am J Obstet Gynecol 181:1211–1215
23. Walsh SW, Vaughan JE, Wang Y, Roberts LJ 2nd (2000) Placental isoprostane is significantly increased in preeclampsia. Faseb J 14:1289–1296
24. Harsem NK, Roald B, Braekke K, Staff AC (2007) Acute atherosis in decidual tissue: not associated with systemic oxidative stress in preeclampsia. Placenta 28:958–964
25. Mandang S, Manuelpillai U, Wallace EM (2007) Oxidative stress increases placental and endothelial cell activin A secretion. J Endocrinol 192:485–493
26. Peter Stein T, Scholl TO, Schluter MD, Leskiw MJ, Chen X, Spur BW, Rodriguez A (2008). Oxidative stress early in pregnancy and pregnancy outcome. Free Radic Res 42:841–848
27. Ouyang YQ, Li SJ, Zhang Q, Cai HB, Chen HP (2009) Interactions between inflammatory and oxidative stress in preeclampsia. Hypertens Pregnancy 28:56–62
28. Zhang J, Masciocchi M, Lewis D, Sun W, Liu A, Wang Y (2008) Placental anti-oxidant gene polymorphisms, enzyme activity, and oxidative stress in preeclampsia. Placenta 29:439–443
29. Norris LA, Higgins JR, Darling MR, Walshe JJ, Bonnar J (1999) Nitric oxide in the uteroplacental, fetoplacental, and peripheral circulations in preeclampsia. Obstet Gynecol 93:958–963
30. Walsh SW, Wang Y (1993) Deficient glutathione peroxidase activity in preeclampsia is associated with increased placental production of thromboxane and lipid peroxides. Am J Obstet Gynecol 169:1456–1461
31. Boutet M, Roland L, Thomas N, Bilodeau JF (2009) Specific systemic antioxidant response to preeclampsia in late pregnancy: the study of intracellular glutathione peroxidases in maternal and fetal blood. Am J Obstet Gynecol 200:530.e1–7
32. Roland-Zejly L, Moisan V, St-Pierre I, Bilodeau JF (2011) Altered placental glutathione peroxidase mRNA expression in preeclampsia according to the presence or absence of labor. Placenta 32:161–167
33. Mutlu-Turkoglu U, Ademoglu E, Ibrahimoglu L, Aykac-Toker G, Uysal M (1998) Imbalance between lipid peroxidation and antioxidant status in preeclampsia. Gynecol Obstet Invest 46:37–40

34. Padmini E, Lavany S, Uthra V (2009) Preeclamptic placental stress and over expression of mitochondrial HSP70. Clin Chem Lab Med 47:1073–1080
35. Patil SB, Kodliwadmath MV, Kodliwadmath M (2009) Lipid peroxidation and antioxidant activity in complicated pregnancies. Clin Exp Obstet Gynecol 36:110–112
36. Vanderlelie J, Gude N, Perkins AV (2008) Antioxidant gene expression in preeclamptic placentae: a preliminary investigation. Placenta 29:519–522
37. Zusterzeel PL, Peters WH, De Bruyn MA, Knapen MF, Merkus HM, Steegers EA (1999) Glutathione S-transferase isoenzymes in decidua and placenta of preeclamptic pregnancies. Obstet Gynecol 94:1033–1038
38. Shibata E, Nanri H, Ejima K, Araki M, Fukuda J, Yoshimura K, Toki N, Ikeda M, Kashimura M (2003) Enhancement of mitochondrial oxidative stress and up-regulation of antioxidant protein peroxiredoxin III/SP-22 in the mitochondria of human pre-eclamptic placentae. Placenta 24:698–705
39. Roland L, Beauchemin D, Acteau G, Fradette C, St-Pierre I, Bilodeau JF (2010) Effects of labor on placental expression of superoxide dismutases in preeclampsia. Placenta 31:392–400
40. Mistry HD, Kurlak LO, Williams PJ, Ramsay MM, Symonds ME, Pipkin FB (2010) Differential expression and distribution of placental glutathione peroxidases 1, 3 and 4 in normal and preeclamptic pregnancy. Placenta 31:401–408
41. Mistry HD, Wilson V, Ramsay MM, Symonds ME (2008) Broughton Pipkin F. Reduced selenium concentrations and glutathione peroxidase activity in preeclamptic pregnancies. Hypertension 52:881–888
42. Dordevic NZ, Babic GM, Markovic SD, Ognjanovic BI, Stajn AS, Zikic RV, Saicic ZS (2008) Oxidative stress and changes in antioxidative defense system in erythrocytes of preeclampsia in women. Reprod Toxicol 25:213–218
43. Kumar CA, Das UN (2000) Lipid peroxides, anti-oxidants and nitric oxide in patients with pre-eclampsia and essential hypertension. Med Sci Monit 6:901–907
44. Bayhan G, Atamer Y, Atamer A, Yokus B, Baylan Y (2000) Significance of changes in lipid peroxides and antioxidant enzyme activities in pregnant women with preeclampsia and eclampsia. Clin Exp Obstet Gynecol 27:142–146
45. Diedrich F, Renner A, Rath W, Kuhn W, Wieland E (2001) Lipid hydroperoxides and free radical scavenging enzyme activities in preeclampsia and HELLP (hemolysis, elevated liver enzymes, and low platelet count) syndrome: no evidence for circulating primary products of lipid peroxidation. Am J Obstet Gynecol 185:166–172
46. Chappell LC, Seed PT, Kelly FJ, Briley A, Hunt BJ, Charnock-Jones DS, Mallet A, Poston L (2002) Vitamin C and E supplementation in women at risk of preeclampsia is associated with changes in indices of oxidative stress and placental function. Am J Obstet Gynecol 187:777–784
47. Hubel CA (1997) Plasma vitamin E in preeclampsia. Am J Obstet Gynecol 177:484–485
48. Mikhail MS, Anyaegbunam A, Garfinkel D, Palan PR, Basu J, Romney SL (1994) Preeclampsia and antioxidant nutrients: decreased plasma levels of reduced ascorbic acid, alpha-tocopherol, and beta-carotene in women with preeclampsia. Am J Obstet Gynecol 171:150–157
49. Zhang C, Williams MA, King IB, Dashow EE, Sorensen TK, Frederick IO, Thompson ML, Luthy DA (2002) Vitamin C and the risk of preeclampsia–results from dietary questionnaire and plasma assay. Epidemiology 13:409–416
50. Chambers JC, Fusi L, Malik IS, Haskard DO, De Swiet M, Kooner JS (2001) Association of maternal endothelial dysfunction with preeclampsia. J Am Med Assoc 285:1607–1612
51. Dudgeon S, Benson DP, MacKenzie A, Paisley-Zyszkiewicz K, Martin W (1998) Recovery by ascorbate of impaired nitric oxide-dependent relaxation resulting from oxidant stress in rat aorta. Br J Pharmacol 125:782–786
52. Huang A, Vita JA, Venema RC, Keaney JF Jr (2000) Ascorbic acid enhances endothelial nitric-oxide synthase activity by increasing intracellular tetrahydrobiopterin. J Biol Chem 275:17399–17406

53. Jackson TS, Xu A, Vita JA, Keaney JF Jr (1998) Ascorbate prevents the interaction of superoxide and nitric oxide only at very high physiological concentrations. Circ Res 83:916–922
54. Xu A, Vita JA, Keaney JF Jr (2000) Ascorbic acid and glutathione modulate the biological activity of S-nitrosoglutathione. Hypertens 36:291–295
55. Rumbold A, Crowther CA (2005) Vitamin C supplementation in pregnancy. Cochrane Database Syst Rev CD004072
56. Banerjee S, Chambers AE, Campbell S (2006) Is vitamin E a safe prophylaxis for preeclampsia. Am J Obstet Gynecol 194:1228–1233
57. Morris JM, Gopaul NK, Endresen MJ, Knight M, Linton EA, Dhir S, Anggard EE, Redman CW (1998) Circulating markers of oxidative stress are raised in normal pregnancy and pre-eclampsia. Br J Obstet Gynaecol 105:1195–1199
58. Wang YP, Walsh SW, Guo JD, Zhang JY (1991) Maternal levels of prostacyclin, thromboxane, vitamin E, and lipid peroxides throughout normal pregnancy. Am J Obstet Gynecol 165:1690–1694
59. Traber MG (1994) Determinants of plasma vitamin E concentrations. Free Radical Biol Med 16:229–239
60. Hubel CA, Kagan VE, Kisin ER, McLaughlin MK, Roberts JM (1997) Increased ascorbate radical formation and ascorbate depletion in plasma from women with preeclampsia: implications for oxidative stress. Free Radical Biol Med 23:597–609
61. Jendryczko A, Drozdz M (1989) Plasma retinol, beta-carotene and vitamin E levels in relation to the future risk of pre-eclampsia. Zentralbl Gynakol 111:1121–1123
62. Roland L, Gagne A, Belanger MC, Boutet M, Berthiaume L, Fraser WD, Julien P, Bilodeau JF (2010) Existence of compensatory defense mechanisms against oxidative stress and hypertension in preeclampsia. Hypertens Pregnancy 29:21–37
63. Schiff E, Friedman SA, Stampfer M, Kao L, Barrett PH, Sibai BM (1996) Dietary consumption and plasma concentrations of vitamin E in pregnancies complicated by preeclampsia. Am J Obstet Gynecol 175:1024–1028
64. Uotila JT, Tuimala RJ, Aarnio TM, Pyykko KA, Ahotupa MO (1993) Findings on lipid peroxidation and antioxidant function in hypertensive complications of pregnancy. Br J Obstet Gynaecol 100:270–276
65. Williams MA, Woelk GB, King IB, Jenkins L, Mahomed K (2003) Plasma carotenoids, retinol, tocopherols, and lipoproteins in preeclamptic and normotensive pregnant Zimbabwean women. Am J Hypertens 16:665–672
66. Zhang C, Williams MA, Sanchez SE, King IB, Ware-Jauregui S, Larrabure G, Bazul V, Leisenring WM (2001) Plasma concentrations of carotenoids, retinol, and tocopherols in preeclamptic and normotensive pregnant women. Am J Epidemiol 153:572–580
67. Chappell LC, Seed PT, Briley A, Kelly FJ, Hunt BJ, Charnock-Jones DS, Mallet AI, Poston L (2002) A longitudinal study of biochemical variables in women at risk of preeclampsia. Am J Obstet Gynecol 187:127–136
68. Chappell LC, Seed PT, Briley AL, Kelly FJ, Lee R, Hunt BJ, Parmar K, Bewley SJ, Shennan AH, Steer PJ, Poston L (1999) Effect of antioxidants on the occurrence of pre-eclampsia in women at increased risk: a randomised trial. Lancet 354:810–816
69. Poston L, Raijmakers M, Kelly F (2004) Vitamin E in preeclampsia. Ann N Y Acad Sci 1031:242–248
70. Poston L, Briley AL, Seed PT, Kelly FJ, Shennan AH (2006) Vitamin C and vitamin E in pregnant women at risk for pre-eclampsia (VIP trial): randomised placebo-controlled trial. Lancet 367:1145–1154
71. Klemmensen A, Tabor A, Osterdal ML, Knudsen VK, Halldorsson TI, Mikkelsen TB, Olsen SF (2009) Intake of vitamin C and E in pregnancy and risk of pre-eclampsia: prospective study among 57 346 women. BJOG 116:964–974
72. Rahimi R, Nikfar S, Rezaie A, Abdollahi M (2009) A meta-analysis on the efficacy and safety of combined vitamin C and E supplementation in preeclamptic women. Hypertens Pregnancy 28:417–434

73. Roberts JM, Myatt L, Spong CY, Thom EA, Hauth JC, Leveno KJ, Pearson GD, Wapner RJ, Varner MW, Thorp JM Jr, Mercer BM, Peaceman AM, Ramin SM, Carpenter MW, Samuels P, Sciscione A, Harper M, Smith WJ, Saade G, Sorokin Y, Anderson GB (2010) Vitamins C and E to prevent complications of pregnancy-associated hypertension. N Engl J Med 362:1282–1291
74. Xu H, Perez-Cuevas R, Xiong X, Reyes H, Roy C, Julien P, Smith G, von Dadelszen P, Leduc L, Audibert F, Moutquin JM, Piedboeuf B, Shatenstein B, Parra-Cabrera S, Choquette P, Winsor S, Wood S, Benjamin A, Walker M, Helewa M, Dube J, Tawagi G, Seaward G, Ohlsson A, Magee LA, Olatunbosun F, Gratton R, Shear R, Demianczuk N, Collet JP, Wei S, Fraser WD (2010) An international trial of antioxidants in the prevention of preeclampsia (INTAPP). Am J Obstet Gynecol 202:239.e1–239.e10
75. Burton GJ, Jauniaux E (2004) Placental oxidative stress: from miscarriage to preeclampsia. J Soc Gynecol Investig 11:342–352
76. Stocker R, Bowry VW, Frei B (1991) Ubiquinol-10 protects human low density lipoprotein more efficiently against lipid peroxidation than does alpha-tocopherol. Proc Natl Acad Sci U S A 88:1646–1650
77. Stoyanovsky DA, Osipov AN, Quinn PJ, Kagan VE (1995) Ubiquinone-dependent recycling of vitamin E radicals by superoxide. Arch Biochem Biophys 323:343–351
78. Lagendijk J, Ubbink JB, Vermaak WJ (1996) Measurement of the ratio between the reduced and oxidized forms of coenzyme Q10 in human plasma as a possible marker of oxidative stress. J Lipid Res 37:67–75
79. Yamashita S, Yamamoto Y (1997) Simultaneous detection of ubiquinol and ubiquinone in human plasma as a marker of oxidative stress. Anal Biochem 250:66–73
80. Palan PR, Shaban DW, Martino T, Mikhail MS (2004) Lipid-soluble antioxidants and pregnancy: maternal serum levels of coenzyme Q10, alpha-tocopherol and gamma-tocopherol in preeclampsia and normal pregnancy. Gynecol Obstet Invest 58:8–13
81. Teran E, Racines-Orbe M, Vivero S, Escudero C, Molina G, Calle A (2003) Preeclampsia is associated with a decrease in plasma coenzyme Q10 levels. Free Radic Biol Med 35:1453–1456
82. Teran E, Vivero S, Racines-Orbe M, Castellanos A, Chuncha G, Enriquez G, Moya W (2005) Coenzyme Q10 is increased in placenta and cord blood during preeclampsia. BioFactors 25:153–158
83. Roland L, Gagne A, Belanger MC, Boutet M, Julien P, Bilodeau JF (2010) Plasma interleukin-18 (IL-18) levels are correlated with antioxidant vitamin coenzyme Q(10) in preeclampsia. Acta Obstet Gynecol Scand 89:360–366
84. Teran E, Hernandez I, Nieto B, Tavara R, Ocampo JE, Calle A (2009) Coenzyme Q10 supplementation during pregnancy reduces the risk of pre-eclampsia. Int J Gynaecol Obstet 105:43–45
85. Palan PR, Mikhail MS, Romney SL (2001) Placental and serum levels of carotenoids in preeclampsia. Obstet Gynecol 98:459–462
86. Tsen KT, Tsen SW, Kiang JG (2006) Lycopene is more potent than beta carotene in the neutralization of singlet oxygen: role of energy transfer probed by ultrafast Raman spectroscopy. J Biomed Opt 11:064025
87. Banerjee S, Jeyaseelan S, Guleria R (2009) Trial of lycopene to prevent pre-eclampsia in healthy primigravidas: results show some adverse effects. J Obstet Gynaecol Res 35:477–482
88. Sharma JB, Kumar A, Malhotra M, Arora R, Prasad S, Batra S (2003) Effect of lycopene on pre-eclampsia and intra-uterine growth retardation in primigravidas. Int J Gynaecol Obstet 81:257–262
89. Viskova H, Vesela K, Janosikova B, Krijt J, Visek JA, Calda P (2007) Plasma cysteine concentrations in uncomplicated pregnancies. Fetal Diagn Ther 22:254–258
90. Raijmakers MT, Zusterzeel PL, Steegers EA, Hectors MP, Demacker PN, Peters WH (2000) Plasma thiol status in preeclampsia. Obstet Gynecol 95:180–184

91. Raijmakers MT, Zusterzeel PL, Roes EM, Steegers EA, Mulder TP, Peters WH (2001) Oxidized and free whole blood thiols in preeclampsia. Obstet Gynecol 97:272–276
92. Kaur G, Mishra S, Sehgal A, Prasad R (2008) Alterations in lipid peroxidation and antioxidant status in pregnancy with preeclampsia. Mol Cell Biochem 313:37–44
93. Kharb S (2000) Low whole blood glutathione levels in pregnancies complicated by preeclampsia and diabetes. Clin Chim Acta 294:179–183
94. Kharb S, Gulati N, Singh V, Singh GP (2000) Superoxide anion formation and glutathione levels in patients with preeclampsia. Gynecol Obstet Invest 49:28–30
95. Knapen MF, Mulder TP, Van Rooij IA, Peters WH, Steegers EA (1998) Low whole blood glutathione levels in pregnancies complicated by preeclampsia or the hemolysis, elevated liver enzymes, low platelets syndrome. Obstet Gynecol 92:1012–1015
96. Rani N, Dhingra R, Arya DS, Kalaivani M, Bhatla N, Kumar R (2010) Role of oxidative stress markers and antioxidants in the placenta of preeclamptic patients. J obstetrics gynaecology res 36:1189–1194
97. Knapen MF, Peters WH, Mulder TP, Merkus HM, Jansen JB, Steegers EA (1999) Glutathione and glutathione-related enzymes in decidua and placenta of controls and women with pre-eclampsia. Placenta 20:541–546
98. Roes EM, Raijmakers MT, Boo TM, Zusterzeel PL, Merkus HM, Peters WH, Steegers EA (2006) Oral N-acetylcysteine administration does not stabilise the process of established severe preeclampsia. Eur J Obstet Gynecol Reprod Biol 127:61–67
99. Dodd S, Dean O, Copolov DL, Malhi GS, Berk M (2008) N-acetylcysteine for antioxidant therapy: pharmacology and clinical utility. Expert Opin Biol Ther 8:1955–1962
100. Bisseling TM, Maria Roes E, Raijmakers MT, Steegers EA, Peters WH, Smits P (2004) N-acetylcysteine restores nitric oxide-mediated effects in the fetoplacental circulation of preeclamptic patients. Am J Obstet Gynecol 191:328–333
101. Ryu S, Huppmann AR, Sambangi N, Takacs P, Kauma SW (2007) Increased leukocyte adhesion to vascular endothelium in preeclampsia is inhibited by antioxidants. Am J Obstet Gynecol 196:400.e1–7, discussion 400.e7–8
102. Kryukov GV, Castellano S, Novoselov SV, Lobanov AV, Zehtab O, Guigo R, Gladyshev VN (2003) Characterization of mammalian selenoproteomes. Science 300:1439–1443
103. Kumar KS, Kumar A, Prakash S, Swamy K, Jagadeesan V, Jyothy A (2002) Role of red cell selenium in recurrent pregnancy loss. J obstet gynaecol: j Inst Obstet Gynaecol 22:181–183
104. Rayman MP, Bode P, Redman CW (2003) Low selenium status is associated with the occurrence of the pregnancy disease preeclampsia in women from the United Kingdom. Am J Obstet Gynecol 189:1343–1349
105. Atamer Y, Kocyigit Y, Yokus B, Atamer A, Erden AC (2005) Lipid peroxidation, antioxidant defense, status of trace metals and leptin levels in preeclampsia. Eur J Obstet Gynecol Reprod Biol 119:60–66
106. Maleki A, Fard MK, Zadeh DH, Mamegani MA, Abasaizadeh S (2010) Mazloomzadeh S. The Relationship between Plasma Level of Se and Preeclampsia, Hypertens Pregnancy 30:180–187
107. Tara F, Maamouri G, Rayman MP, Ghayour-Mobarhan M, Sahebkar A, Yazarlu O, Ouladan S, Tavallaie S, Azimi-Nezhad M, Shakeri MT, Boskabadi H, Oladi M, Sangani MT, Razavi BS, Ferns G (2010) Selenium supplementation and the incidence of preeclampsia in pregnant Iranian women: a randomized, double-blind, placebo-controlled pilot trial. Taiwan J Obstet Gynecol 49:181–187
108. Milczarek R, Hallmann A, Sokolowska E, Kaletha K, Klimek J (2010) Melatonin enhances antioxidant action of alpha-tocopherol and ascorbate against NADPH- and iron-dependent lipid peroxidation in human placental mitochondria. J Pineal Res 49:149–155
109. Okatani Y, Wakatsuki A, Shinohara K, Kaneda C, Fukaya T (2001) Melatonin stimulates glutathione peroxidase activity in human chorion. J Pineal Res 30:199–205

Chapter 7
Recurrent Pregnancy Loss and Oxidative Stress

Nabil Aziz

Abstract Physiologic levels of Reactive oxygen species (ROS) are important in embryo development and organogenesis. On the other hand, high ROS levels have adverse developmental outcome, including embryopathies, embryonic mortality, and miscarriage. First-trimester miscarriage is associated with significantly increased markers of oxidative stress in the maternal blood. The negative impact of maternal obesity on pregnancy outcome may be mediated through excess circulating fatty acids and oxidative stress. The increased risk of miscarriage in diabetic pregnancies is attributed to increased lipid peroxidation in the embryo and low prostaglandin E2 levels. The male contribution to oxidative stress related adverse pregnancy outcomes is transmitted through the oxidative damage of the sperm DNA. Taking any vitamin supplements prior to pregnancy or in early pregnancy does not appear to prevent women experiencing early or late miscarriage.

Keywords Oxidative stress · Miscarriage · Redox in embryonic implantation · Placentation abnormalities · Obesity · Maternal nutrition

7.1 Introduction

Reactive oxygen species (ROS) and their control by antioxidants are important in embryo development and organogenesis through cellular signaling pathways involved in proliferation, differentiation, or apoptosis. Pregnancy as such is

N. Aziz (✉)
Gynecology and Reproductive Medicine,
Liverpool Women's Hospital, Crown Street,
Liverpool, L8 7SS, UK
e-mail: naziz@liv.ac.uk

characterized by increased generation of ROS. The generation of ROS is enhanced by increased placental mitochondrial respiratory chain activity [1] and the greatly increased placental production of the radical superoxide [2, 3]. In either pregnant or non-pregnant states ROS can also be formed as a result of exogenous exposures to alcohol, tobacco smoke, and environmental pollutants. Enzymatic and non-enzymatic Antioxidants (including vitamins C and E) and antioxidant cofactors (such as selenium, zinc, and copper) are capable of disposing, scavenging, or suppressing the formation of ROS. Full accounts of the different types and the sources of free radicals and of antioxidants are given in Chap. 1.

Oxidative stress takes place when the generation of ROS exceeds the scavenging capacity by antioxidants due to excessive production of ROS and/or inadequate intake or increased utilization of antioxidants. When ROS are present in excessive amounts they induce cellular toxic effects including DNA damage and altered cell development [4].

During their early development emberyos have a high energy demand which is met by the Mitochondrial Oxidative Phosphorylation (OXPHOS) metabolic pathway that uses energy released by the oxidation of nutrients to produce ATP [5, 6]. The mitochondria also provide ATP for the regeneration of NADPH and reduced glutathione involved in ROS scavenging [7]. Excessive oxidative stress can have deleterious effects on the cellular milieu and can result in impaired cellular growth in the embryo or apoptosis resulting in what is clinically described as embryo fragmentation.

Initially, the preimplantation embryo develops and grows under uterine hypoxic environment. As a result the embryonic tissues cannot produce large amount of ATP. However, this physiological hypoxia of the early gestational sac protects the developing embryo against the deleterious and teratogenic effects of O_2 free radicals. A stable O_2 gradient between the maternal uterine decidua and the feto-placental tissue is also an important factor in trophoblast differentiation and migration, normal villous development and angiogenesis [8].

With the initiation of conceptus vascularization the switch from preimplantation embryonic anaerobic metabolism to postimplantation aerobic metabolism takes place. The establishment of utero-placental blood flow exposes embryonic and extraembryonic tissues to ROS produced as normal by-products of OXPHOS. These metabolic changes start to take place during the period of organogenesis (first-trimester), when the embryo is most susceptible to environmental factors. Therefore, implanted embryo may be particularly vulnerable to oxidative stress caused by mitochondrial dysfunction due to pathological and/or environmental insults, thus triggering apoptosis in the embryo and compromises early developmental processes and birth outcome [1, 4]. Electrons leak from the electron transport chain at the inner mitochondrial membranes. These electrons are transferred to the oxygen molecule, resulting in an unpaired electron in the orbit leading to the generation of the superoxide molecule. On the other hand, it has been shown that abnormal or reduced mitochondrial activity leading to altered ROS production reduces implantation rates in women [9].

The antioxidant status of the embryo and the surrounding uterine environment play vital roles in protecting embryonic and extraembryonic tissues from the deleterious effects of non-physiological levels of endogenous ROS and the exposure to exogenous environmental factors during the critical period of organogenesis.

7.2 The Role of Redox in Embryonic Implantation

A network of signaling molecules that mediate cell-to-cell communications between the receptive endometrium and embryonic trophectoderm promotes embryonic development and implantation. In rodents, ROS production from both mitochondrial and non-mitochondrial sources functions as a physiological component of the early embryonic development. Glutathione-dependent antioxidant mechanisms are developmentally regulated in the inner blastocyst cell mass and H_2O_2 is a potential mediator of apoptosis in the blastocyst [10]. On the endometrial level, studies on the rat demonstrated a uterine O_2^- burst at proestrus suggesting its involvement in regulating uterine edema and cell proliferation [11]. A peak of O_2^- in the uterus at day five of mice pregnancy suggested a contribution of this radical in vascular permeability at the initiation of implantation [12]. Moreover, an NADPH-dependent O_2^- production pathway associated with the uterus of pregnant mice was demonstrated to increases across the preimplantation stages [13]. Antioxidant enzyme systems have been shown as components of the developing embryo and its receptive uterine endometrium [14]. These antioxidant enzyme systems may have vital role in regulating fetal development and survival through the control of placental ROS production and propagation [15].

7.3 Placentation Abnormalities and Miscarriage

Maternal blood directly bathes the fetal trophoblast. Extravillous trophoblastic invades the tips of the spiral arteries transforming them from small caliber high resistance arteries into large caliber, low resistance, and high capacitance uteroplacental arteries. Abnormal placentation has been implicated in the pathogenesis of pre-eclampsia and miscarriage [16]. Unlike humans, pre-eclampsia is not known and miscarriage is a rare event in other animals [17]. A sharp peak in the expression of the markers of oxidative stress in the trophoblast is detected in normal pregnancies at around 9–10 weeks gestation, which is evidenced by an increase in HSP70 activity mainly in the periphery of the primitive placenta [18]. The villous changes observed in the periphery of the placenta during the formation of the fetal membranes are identical to those found in the missed-miscarriage indicating a common mechanism mediated by oxidative stress [18]. Moreover, threatened miscarriage (diagnosed when normally grown live fetus is found on ultrasound in the presence of vaginal bleeding) is associated with focal oxidative stress in the definitive placenta and increased risk of adverse pregnancy outcomes such as miscarriage, preterm delivery, and premature rupture of the membranes [19]. In early pregnancy loss, the development of the placento-decidual interface is severely impaired leading to early and widespread onset of maternal blood flowing continuously inside the placenta, which is associated with major oxidative stress induced tissue degeneration [16, 18, 20]. In more than two-thirds of the cases of missed miscarriage, there is anatomical evidence of defective placentation with reduced cytotrophoblast invasion of the endometrium, and reduced transformation and incomplete plugging of the spiral arteries [21].

Because the first-trimester human placenta has limited antioxidant enzyme capacity, it was speculated that if this oxidative burst was excessive may have a detrimental effects on the syncytiotrophoblast leading to early pregnancy loss [33]. Jauniaux et al. [17] studied the distribution and transfer pathways of antioxidant molecules inside the first-trimester gestational sac. They reported that the coelomic fluid of the exocoelomic cavity, which borders the inside of the first-trimester placenta, contained a very low level of reduced glutathione. This suggested that the role of glutathione-related detoxification system is limited in fetal fluid compartments. The levels of alpha- and gamma-tocopherol were lower whereas the concentrations of ascorbic and uric acid were similar in coelomic fluid compared with maternal plasma. It was suggested that the presence of these molecules inside the early gestational sac might play an essential role in the fetal tissues' antioxidant capacity at a time when the fetus is most vulnerable to oxidative stress. The same study indicated that the uterine glands and the secondary yolk sac play key roles in supplying alpha-tocopherol to the developing fetus before the placental circulations are established.

7.4 Markers of Oxidative Stress and Miscarriage

Miscarriage and pregnancy appear to be associated with increased oxidative stress. In a successful pregnancy, however, changes occurred within the peripheral blood that offered protection from oxidant attack. The roles of the peripheral blood-plasma antioxidants thiol and ceruloplasmin—and two extracellular parameters—superoxide dismutase (SOD) and red cell lysate thiol have been examined in healthy pregnancy and in women suffering first-trimester miscarriage. It was found that pregnancies that went successfully to term were associated with increased levels of ceruloplasmin and SOD early in the first-trimester [22]. These changes were thought to offer the cell protection from the damage caused by the increased oxidative stress associated with pregnancy. First-trimester miscarriage was associated with significantly reduced levels of SOD. A subgroup of patients who miscarried in their first pregnancy, but whose second pregnancies were successful, had higher levels of plasma thiol and significantly reduced levels of red cell lysate thiol in the on-going pregnancy compared to levels at the time of miscarriage.

The antioxidants Glutathione and glutathione transferase family of enzymes have been investigated in patients who experience recurrent pregnancy loss (RPL) [23, 24]. In a large case controlled study the relationship between RPL and polymorphisms in two genes, glutathione S-transferases (GST) M1 and T1, which are involved in the metabolism of a wide range of environmental toxins and carcinogens, was examined [24]. The results suggested that women with GSTM1 null polymorphism may have an increased risk of RPL. In another study, elevated glutathione levels in pregnant patients with history of recurrent pregnancy loss were associated with pregnancy loss [23].

Early pregnancy physiologic oxidative stress is associated with changes in placental protein production including angiogenic and anti-angiogenic placental proteins [25]. Maternal serum soluble vascular endothelial growth factor receptor 1 and maternal serum placental growth factor have been found to be markedly decreased in threatened miscarriage patients who subsequently have a miscarriage suggesting these proteins are sensitive predictive markers of subsequent pregnancy loss [25].

The primary recognized health risk from common deficiencies in glucose-6-phosphate dehydrogenase (G6PD), a cytoprotective enzyme against oxidative stress, is non-immune red blood cell hemolysis. The protective role of the G6PD enzyme against oxidative stress has been demonstrated in animal studies [26]. G6PD-deficient mice had higher embryonic DNA oxidation and more fetal death and birth defects. Embryopathies were prevented by protecting the embryos against oxidative stress and resulted in improved pregnancy outcome. G6PD enzyme deficiencies accordingly may have a broader biological relevance as important determinants of infertility, miscarriage, and teratogenesis.

7.5 Obesity and Miscarriage

Although lean and obese pregnant women gain similar amount of fat during pregnancy lean women deposit the excess fat in the lower part (thighs) of the body, where obese women preferentially deposit their excess fat around the trunk and as visceral fat (centrally obesity) [27]. It has been suggested that these differences in fat deposition may be linked to higher adverse pregnancy outcomes such as pregnancy loss and pre-eclampsia [28]. The underlying hypothesis is that the preferential storage of fat in central rather than 'safer' lower body depots in obese pregnancy leads to lipotoxicity. Lipotoxicity is associated with two unfavorable features namely excess fatty acids and oxidative stress [27]. In addition, it has been demonstrated that lipid excess and oxidative stress provokes endothelial dysfunction [29]. Oxidized lipids can inhibit trophoblast invasion and influence placental development, lipid metabolism and transport, and fetal developmental pathways. As lipotoxicity has the capability of influencing both maternal endothelial function and placental function, it may link maternal obesity and placental-related adverse pregnancy outcomes such as miscarriage and pre-eclampsia [27].

7.6 Maternal Nutrition

Maternal under- or poor nutrition are associated with multiple micronutrient deficiencies, a major cause of IUGR and miscarriage [30]. Folate deficiency in the mother result in elevated homocysteine levels. Homocysteine induced oxidative stress has been suggested as a potential factor for causing apoptosis that disrupts palate development, which ultimately leads to cleft palate [31]. Similarly, it has been suggested that thalidomide-induced embryopathy and other embryopathies are caused as a result of oxidative stress-mediated damage of the macromolecules [32, 33].

Pregnancy outcomes in type 1 diabetes mellitus remain worse than in the general population. A recent study from Dublin, Ireland, demonstrated that the first trimester spontaneous miscarriage rate was 15 % in diabetic pregnant women versus 8 % in the control non-diabetic pregnancy group ($p < 0.0001$) [34]. In the rat model hyperglycemia/diabetes both in vitro and in vivo induced down-regulation of cycloxygenase-2 gene expression, increased lipid peroxidation in the embryo, and low prostaglandin E2 levels. This resulted in diabetic embryopathy and dysmorphogenesis [35]. This may explain the increased risk of miscarriage in diabetic pregnancies.

7.7 Maternal Age and Miscarriage

The incidence of early pregnancy loss in older women is higher than their young counterparts. This may at least in part relate to mitochondrial dysfunction [9]. In oocytes, mitochondria provide sufficient energy for cell division by supporting spindle formation during meiosis II. Mitochondrial OXPHOS process is also crucial for embryo development, playing a role in blastocyst differentiation, expansion, and hatching [36]. Since mitochondria are inherited from mother to child, it is important that oocyte mitochondria should be intact. Older women seem to have more mitochondrial DNA mutations leading to inadequate capacity to generate ATP at levels sufficient to support normal chromosomal segregation or normal biosynthetic activities within blastomeres [37]. Also the meiotic spindles, that are crucial for normal chromosomes segregation, may not be formed normally because of the poor ATP levels and altered spindles may result in aneuploidy embryos. These factors combined can be responsible for poor implantation and early pregnancy loss, two conditions that occur more often in older women.

7.8 Role of the Sperm

The male gamete contributes 50 % of the genomic material to the embryo and contributes to placental and embryonic development [38]. Paternal genes involved in early embryonic development are located close to the surface of the male pronucleus in the fertilized egg. Therefore, it is conceivable that genetic and epigenetic alterations of the sperm may have important consequences on early pregnancy. The possible epigenetic alterations in the sperm include altered chromatin packing, imprinting errors, telomeric shortening, absence or alteration of the centrosome, and absence of sperm RNA. These alterations have the potential to affect some of the functional characteristics of paternal genes leading to early embryo loss [39]. The study of potential causes of unexplained recurrent pregnancy loss have indicated that a proportion of the male partners of recurrent pregnancy loss demonstrated significant increase in sperm chromosome aneuploidy, abnormal chromatin condensation, DNA fragmentation, increased apoptosis, and abnormal sperm morphology compared with fertile men [40–47]. These genetic abnormalities may be induced by oxidative stress.

The detrimental impact of oxidative stress on the sperm cell membrane and its DNA integrity is well-documented. The main sources of ROS in semen are seminal leukocytes and morphologically abnormal spermatozoa, although ROS can also be produced by precursor germ cells [48–50]. Spermatozoa are particularly susceptible to oxidative stress damage, because their plasma membrane is rich in polyunsaturated fatty acids susceptible to fatty acid peroxidation, its sparse cytoplasm contains low concentrations of scavenging enzymes and posses limited capacity for DNA repair [51]. Excessive amounts of ROS cause DNA damage, leading to formation of 7-hydro-8-oxo-2-deoxy- guanosine (8-oxodG), the major oxidative product of sperm DNA, which causes DNA fragmentation [52, 53]. Oxidative damage of sperm membrane and DNA have a detrimental effect on sperm-egg interaction, implantation, and early embryo development compromising pregnancy outcome [54–57].

Seminal plasma contains high amounts of antioxidants that protect the spermatozoa from DNA damage and lipid peroxidation [58]. In one study Men who fathered normal pregnancies had higher antioxidant capacity in the semen and lower sperm cell membrane lipid peroxidation compared with men from the recurrent pregnancy loss group higher [59]. In a follow on study [60] Male partners of recurrent early pregnancy loss (<12 weeks gestation) couples and with increased sperm DNA fragmentation index or high serum thiobarbituric acid reactive substances (oxidative stress marker) were instructed to consume a diet rich in antioxidants or commercial multivitamins containing b-carotene, vitamin C, vitamin E, and zinc for at least 3 months. All couple that subsequently became pregnant had successful pregnancy outcome. Further larger studies are needed to substantial this study finding.

7.9 Oxidative Stress in Miscarriage and Clinical Intervention

Pharmacological interventions to overcome oxidative stress in spontaneous miscarriage and unexplained recurrent pregnancy loss have been extensively reviewed [61]. Since oxidative stress has been linked to spontaneous and recurrent miscarriage, intake of antioxidant vitamins such as vitamin C and vitamin E may be an important factor associated with the risk of miscarriage. One observational study has demonstrated an association between the risk of spontaneous early miscarriage and dietary factors, with a high risk associated with poor intake of green vegetables, fruits, and dairy products coupled with a high intake of fat [62]. However, the vast majority of women who experienced spontaneous miscarriage have a successful outcome in subsequent pregnancies without any specific interventions. A Cochran systematic review of the use of vitamin supplementation to prevent miscarriage has been updated in 2011 [63]. The author concluded that taking any vitamin supplements prior to pregnancy or in early pregnancy does not prevent women experiencing early or late miscarriage or stillbirth. However, women taking vitamin supplements may be more likely to have a multiple pregnancy. There was insufficient evidence to examine the effects of different combinations of vitamins on miscarriage or stillbirth.

7.10 Summary

Physiological levels ROS play important role in the regulation of reproductive processes, including embryonic development, uterine receptivity, embryonic implantation, placental development and endocrine functions, and fetal development. On the other hand, high ROS levels have adverse developmental outcome, including embryopathies, embryonic mortality, fetal growth restriction, low birth weight, and miscarriage. The male contribution to such adverse pregnancy outcomes may be attributed to oxidative damage of the sperm.

References

1. Shibata E, Nanri H, Ejima K, Araki M, Fukuda J, Yoshimura K, Toki N, Ikeda M, Kashimura M (2003) Enhancement of mitochondrial oxidative stress and up-regulation of antioxidant protein peroxiredoxin III/SP-22 in the mitochondria of human pre-eclamptic placentae. Placenta 24:698–705
2. Sikkema JM, van Rijn BB, Franx A, Bruinse HW, de Roos R, Stroes ES, van Faassen EE (2001) Placental superoxide is increased in preeclampsia. Placenta 22:304–308
3. Wang Y, Walsh SW (2001) Increased superoxide generation is associated with decreased superoxide dismutase activity and mRNA expression in placental trophoblast cells in pre-eclampsia. Placenta 22:206–212
4. Dennery PA (2007) Effects of oxidative stress on embryonic development. Birth Defects Res Part C Embryo Today. 81:155–162
5. Wilding M, Dale B, Marino M et al (2011) Mitochondrial aggregation patterns and activity in human oocytes and preimplantation embryos. Hum Reprod 16:909–917
6. Smith LC, Thundathil J, Filion F (2005) Role of mitochondrial genome in preimplantation development and assisted reproductive techniques. Reprod Fert Develop 17:15–22
7. Dumollard R, Carroll J, Duchen MR, Campbell K, Swann K (2009) Mitochondrial function and redox state in mammalian embryos. Semin Cell Dev Biol 20:346–353
8. Burton GJ (2009) Oxygen, the Janus gas; its effects on human placental development and function. J Anat 215:27–35

9. Bartmann AK, Romão GS, Ramos Eda S, Ferriani RA (2004) Why do older women have poor implantation rates? a possible role of the mitochondria. J Assist Reprod Genet 21:79–83
10. Pierce GB, Parchment RE, Lewellyn AL (1991) Hydrogen peroxide as a mediator of programmed cell death in the blastocyst. Differentiation 46:181–186
11. Laloraya M, Kumar GP, Laloraya MM (1991) Changes in the superoxide radical and superoxide dismutase levels in the uterus of Rattus norvegicus during the estrous cycle and a possible role for superoxide radical in uterine oedema and cell proliferation at proestrus. Biochem Cell Biol 69:313–316
12. Laloraya M, Kumar GP, Laloraya MM (1989) A possible role of superoxide anion radical in the process of blastocyst implantation in mus musculus. Biochem Biophys Res Commun 161:762–770
13. Jain S, Saxena D, Kumar GP, Laloraya M (2000) NADPH dependent superoxide generation in the ovary and uterus of mice during estrous cycle and early pregnancy. Life Sci 66:1139–1146
14. Guérin P, El Mouatassim S, Ménézo Y (2001) Oxidative stress and protection against reactive oxygen species in the pre-implantation embryo and its surroundings. Hum Reprod 7:175–189
15. Al-Gubory KH, Garrel C, Delatouche L, Heyman Y, Chavatte-Palmer P (2010) Antioxidant adaptive responses of extraembryonic tissues from cloned and noncloned bovine conceptuses to oxidative stress during early pregnancy. Reproduction 140:175–181
16. Jauniaux E, Hempstock J, Greenwold N, Burton GJ (2003) Trophoblastic oxidative stress in relation to temporal and regional differences in maternal placental blood flow in normal and abnor- mal early pregnancies. Am J Pathol 162:115–125
17. Jauniaux E, Cindrova-Davies T, Johns J, Dunster C, Hempstock J, Kelly FJ, Burton GJ (2004) Distribution and transfer pathways of antioxidant molecules inside the first trimester human gestational sac. J Clin Endocrinol Metab 89:1452–1458
18. Burton GJ, Woods AW, Jauniaux E, Kingdom JC (2009) Rheological and physiological consequences of conversion of the maternal spiral arteries for uteroplacental blood flow during human pregnancy. Placenta 30:473–482
19. Johns J, Hyett J, Jauniaux E (2003) Obstetric outcome after threatened miscarriage with and without a hematoma on ultrasound. Obstet Gynecol 102:483–487
20. Muttukrishna S, Suri S, Groome NP, Jauniaux E (2008) Relationships between TGFβ proteins and oxygen concentrations inside the first trimester human gestational sac. PLoS ONE 31: e2302 1–7
21. Jauniaux E, Burton GJ (2005) Pathophysiology of histological changes in early pregnancy loss. Placenta 26:114–123
22. Jenkins C, Wilson R, Roberts J, Miller H, McKillop JH, Walker JJ (2000) Antioxid Redox Signal 2:623–628
23. Miller H, Wilson R, Jenkins C, MacLean MA, Roberts J, Walker JJ (2000) Glutathione levels and miscarriage. Fertil Steril 74:1257–1258
24. Sata F, Yamada H, Kondo T, Gong Y, Tozaki S, Kobashi G, Kato EH, Fujimoto S, Kishi R (2003) Glutathione S-transferase M1 and T1 poly- morphisms and the risk of recurrent pregnancy loss. Mol Hum Reprod 9:165–169
25. Muttukrishna S, Swer M, Suri S, Jamil A, Calleja-Agius J (2011) Soluble Flt-1 and PlGF: new markers of early pregnancy loss? PLoS ONE 6:e18041. doi:10.1371/journal.pone.0018041
26. Nicol CJ, Zielenski J, Tsui LC, Wells PG (2000) An embryoprotective role for glucose-6-phosphate dehydrogenase in developmental oxidative stress and chemical teratogenesis. Faseb J 14:111–127
27. Jarvie E, Hauguel-de-Mouzon S, Nelson SM, Sattar N, Catalano PM, Freeman DJ (2010) Lipotoxicity in obese pregnancy and its potential role in adverse pregnancy outcome and obesity in the offspring. Clin Sci (Lond) 119:123–129
28. Jarvie E, Ramsay JE (2010) Obstetric management of obesity in pregnancy. Semin Fetal Neonatal Med 15:83–88
29. Paradisi G, Biaggi A, Ferrazzani S, De Carolis S, Caruso A (2002) Abnormal carbohydrate metabolism during pregnancy: association with endothelial dysfunction. Diabetes Care 25:560–564

30. Fall CH, Yajnik CS, Rao S, Davies AA, Brown N, Farrant HJ (2003) Micronutrients and fetal growth. J Nutr 133(Suppl 2):1747S–1756S
31. Knott L, Hartridge T, Brown NL, Mansell JP, Sandy JR (2003) Homo-cysteine oxidation and apoptosis: a potential cause of cleft palate. In Vitro Cell Dev Biol Anim 39:98–105
32. Parman T, Wiley MJ, Wells PG (1999) Free radical-mediated oxidative DNA damage in the mechanism of thalidomide teratogenicity. Nat Med 5:582–585
33. Burton GJ, Hempstock J, Jauniaux E (2003) Oxygen, early embryonic metabolism and free radical-mediated embryopathies. Reprod Biomed Online 6:84–96
34. Al-Agha R, Firth RG, Byrne M, Murray S, Daly S, Foley M, Smith SC (2011) Kinsley BTOutcome of pregnancy in type 1 diabetes mellitus (T1DMP): results from combined diabetes-obstetrical clinics in Dublin in three university teaching hospitals (1995–2006). Ir J Med Sci Nov 5. [Epub ahead of print] PMID: 22057636
35. Wentzel P, Welsh N, Eriksson UJ (1999) Developmental damage, increased lipid peroxidation, diminished cyclooxygenase-2 gene expression, and lowered prostaglandin E2 levels in rat embryos exposed to a diabetic environment. Diabetes 48:813–820
36. Sathananthan AH, Trounson AO (2000) Mitochondrial morphology during preimplantation human embryogenesis. Hum Reprod 15:148–159
37. Bartmann AK, Romão GS, Ferriani RA, Ramos Eda S (2004) Inadequate capacity to generate ATP at levels sufficient to support normal chromosomal segregation or normal biosynthetic activities within blastomeres. J Assist Reprod Genet 21:79–83
38. Sutovsky P, Schatten G (2000) Paternal contributions to the mammalian zygote: fertilization after sperm–egg fusion. Int Rev Cytol 195:1–65
39. Gil Villa AM, Cardona-Maya WD, Cadavid Jaramillo AP (2007) Early embryo death: does the male factor play a role? Arch Esp Urol 60:1057–1068
40. Evenson DP, Jost LK, Marshall D, Zinaman MJ, Clegg E, Purvis K (1999) Utility of the sperm chromatin structure assay as a diagnostic and prognostic tool in the human fertility clinic. Hum Reprod 14:1039–1049
41. Rubio C, Simon C, Blanco J, Vidal F, Minguez Y, Egozcue J (1999) Implications of sperm chromosome abnormalities in recurrent miscarriage. J Assist Reprod Genet 16:253–258
42. Gopalkrishnan K, Padwal V, Meherji PK, Gokral JS, Shah R, Juneja HS (2000) Poor quality of sperm as it affects repeated early pregnancy loss. Arch Androl 45:111–117
43. Larson KL, DeJonge CJ, Barnes AM, Jost LK, Evenson DP (2000) Sperm chromatin structure ssay parameters as predictors of failed pregnancy following assisted reproductive techniques. Hum Reprod 15:1717–1722
44. Spano M, Bonde JP, Hjollund HI, Kolstad HA, Cordelli E, Leter G (2000) Sperm chromatin damage impairs human fertility. The danish first pregnancy planner study team. Fertil Steril 73:43–50
45. Carrell DT, Wilcox AL, Lowy L, Peterson CM, Jones KP, Erickson L et al (2003) Elevated sperm chromosome aneuploidy and apoptosis in patients with unexplained recurrent pregnancy loss. Obstet Gynecol 101:1229–1235
46. Carrell DT, Liu L, Peterson CM, Jones KP, Hatasaka HH, Erickson L et al (2003) Sperm DNA fragmentation is increased in couples with unexplained recurrent pregnancy loss. Arch Androl 49:49–55
47. Bernardini LM, Costa M, Bottazzi C, Gianaroli L, Magli MC, Venturini PL et al (2004) Sperm aneuploidy and recurrent pregnancy loss. Reprod Biomed Online 9:312–320
48. Aziz N, Novotny J, Oborna I, Fingerova H, Brezinova J, Svobodova M (2010) Comparison of chemiluminescence and flowcytometry in the estimation of reactive oxygen and nitrogen species in human semen. Fertil Steril 94:2604–2608
49. Aziz N, Saleh RA, Sharma RK, Lewis-Jones I, Esfandiari N, Thomas AJ Jr, Agarwal A (2004) Novel association between sperm reactive oxygen species production, sperm morphological defects, and the sperm deformity index. Fertil Steril 81:349–354
50. Aziz N, Saleh RA, Sharma RK, Lewis-Jones I, Esfandiari N, Thomas AJ Jr, Agarwal A (2004) Novel association between sperm reactive oxygen species production, sperm morphological defects, and the sperm deformity index. Fertil Steril 81:349–354

51. Saleh RA, Agarwal A (2002) Oxidative stress and male infertility: from research bench to clinical practice. J Androl 23:737–752
52. Lopes S, Jurisicova A, Sun JG, Casper RF (1998) Reactive oxygen species: potential cause for DNA fragmentation in human spermatozoa. Hum Reprod 13:896–900
53. Loft S, Kold-Jensen T, Hjollund NH, Giwercman A, Gyllemborg J, Ernst E et al (2003) Oxidative DNA damage in human sperm influences time to pregnancy. Hum Reprod 18:1265–1272
54. Aitken RJ, Clarkson JS, Fishel S (1989) Generation of reactive oxygen species, lipid peroxidation, and human sperm function. Biol Reprod 41:183–197
55. Sharma RK, Agarwal A (1996) Role of reactive oxygen species in male infertility. Urology 48:835–850
56. Sikka SC (2001) Relative impact of oxidative stress on male reproductive function. Curr Med Chem 8:851–862
57. Agarwal A, Said TM (2005) Oxidative stress, DNA damage and apoptosis in male infertility: a clinical approach. BJU Int 95:503–507
58. Potts RJ, Notarianni LJ, Jefferies TM (2000) Seminalplasmareducesexogenous oxidative damage to human sperm, determined by the measurement of DNA strand breaks and lipid peroxidation. Mutat Res 447:249–256
59. Gil-Villa AM, Cardona-Maya W, Agarwal A, Sharma R, Cadavid A (2009) Assessment of sperm factors possibly involved in early recurrent pregnancy loss.2010. Fertil Steril 94: 1465–1472 Epub Jun 21
60. Gil-Villa AM, Cardona-Maya W, Agarwal A, Sharma R, Cadavid A (2009) Role of male factor in early recurrent embryo loss: do antioxidants have any effect. Fertil Steril 92:565–571
61. Gupta S, Agarwal A, Banerjee J, Alvarez JG (2007) The role of oxidative stress in spontaneous abortion and recurrent pregnancy loss: a systematic review. Obstet Gynecol Surv 62:335–347
62. Parazzini F, Chatenoud L, Bettoni G, Tozzi L, Turco S, Surace M, Di Cintio E, Benzi G (2001) Selected food intake and risk of multiple pregnancies. Hum Reprod 16(2):370–373
63. Rumbold A, Middleton P, Pan N, Crowther CA (2011) Vitamin supplementation for preventing miscarriage. Cochrane Database Syst Rev 1:CD004073

Chapter 8
Premature Rupture of Membranes and Oxidative Stress

Anamar Aponte and Ashok Agarwal

Abstract Premature rupture of membranes (PROM) is rupture of the chorioamniotic membranes before the onset of labor. The chorioamniotic sac requires a balance between collagen formation and enzymatic collagenolytic activity expressed in the fetal membrane. The amniotic membrane is a complex tissue in which collagen is fundamental for mechanical integrity and stress tolerance. Membrane rupture is associated with biochemical disturbances between collagen produced by fibroblasts and fetal membranes. An association exists between PROM and oxidative stress (OS) in pregnancy. Reactive oxygen species (ROS), which are unstable molecules produced in the body, may be the reason for the collagen damage. OS produced by augmented ROS production debilitates the collagen's elasticity and strength, leading to PROM. Although antioxidants act as defense agents against oxidation, they have not been found to be protective against ROS and subsequent OS damage in PROM.

Keywords Premature rupture of membranes · Chorioamniotic membranes · Oxidative stress · Enzymatic collagenolytic activity expressed · Fetal membrane

---

A. Aponte · A. Agarwal (✉)
Center for Reproductive Medicine, Lerner College of Medicine, Cleveland Clinic, 9500 Euclid Avenue, Cleveland, OH 44195, USA
e-mail: agarwaa@ccf.org

A. Aponte
e-mail: anamar06@hotmail.com

8.1 Introduction

Parturition is determined by ripening of the cervix, augmented contractility of the myometrium, and positive signaling of the maternal decidua and the chorioamniotic membranes. Activation of genes involved in inflammatory activities promotes the chorioamniotic membrane to undergo complex biochemical changes [1]. Premature rupture of membranes (PROM) is defined as rupture of membranes before the onset of labor and membrane rupture that takes place before 37 weeks of gestation is known as preterm PROM [2]. At term, weakening of the membranes occurs from physiologic changes in association with shearing forces created by uterine contractions [3, 4]. Normal rupture of membranes at term is thought to begin at a specific weakened para-cervical region which forms late in gestation due to collagen disruption [5]. Intraamniotic infection is commonly associated with preterm PROM, especially if it occurs at an earlier gestational age[2]. Low socioecononomic status, second- and third-trimester bleeding, low body mass index less than 19.8, nutritional deficiencies of copper and ascorbic acid, connective tissue disorders, maternal cigarette smoking, pulmonary disease in pregnancy, uterine overdistention, and amniocentesis are some of the risk factors associated with PROM occurrence [6, 7].

The amnion is an avascular entity that depends on the amniotic fluid for its stability and may be involved in the pathogenesis of PROM [8]. The chorioamniotic sac requires a balance between collagen formation by fibroblasts and the collagenolytic activity by enzymes expressed in the fetal membrane [9]. Membrane rupture is associated with biochemical disturbances between collagen produced by fibroblasts and fetal membranes [9]. The elasticity and strength of fetal membranes is primary determined by collagen. Reduced concentrations of it or altered cross-linked organization may contribute to PROM [10].

MnSOD formation. MnSOD is a member of an iron/manganese superoxide dismutase family which is located in the mitochondria and inhibits reactive oxygen species (ROS) produced by oxidative phosphorylation [1]. A moderate increase in ROS can stimulate cell growth and proliferation, while an excessive ROS accumulation will lead to cellular injury, such as damage to DNA, protein, and lipid membrane.ROS enhance NF-κB and promote the expression of cyclooxygenase-2 (COX-2), which upregulates the inflammatory pathway and the propagation of labor [1]. These inflammatory processes may lead to increased formation of ROS and subsequent oxidative stress (OS). OS occurs when the production of ROS exceeds the endogenous antioxidant defense. On the contrary, MnSOD may be activated by inflammatory processes, which in turn downregulates OS and inflammation by inhibiting NF-κB and mitogen activated protein pathways (MAPK) [1].

8.2 Chorioamniotic Membrane: What is it Made of?

The amnion is a single layer of non-ciliated cuboidal cells with its outermost part near the amniotic fluid. Beneath the amnion there is a layer that consists of loosely packed fibrils and where the chorion, which composed of reticular fibers, a basement membrane, and the trophoblast cells, is found. Under the chorion is the maternal decidual layer, which consists of epithelial cells that allow the separation of the chorioamnion from the endometrial and myometrial layers of the uterus [11].

The amnion originates its strength from collagen. At least five (types I, III, IV, V, and VI) types of collagen are encountered in the chorioamnion which are organized in triple helices [10]. Collagen strength, which is primarily attributed to collagen I, is derived from hydroxyproline and hydroxylysine bridges around the helix. The reticular layers contain collagen I, III, IV, V, and VI. The chorionic basement is composed of collagen IV and the chorion has types IV and V [12].

The chorioamnion is a biologically active membrane whose collagenolytic enzyme is susceptible to ROS proliferation [13].

8.3 Premature Rupture of Membranes, Oxidative Stress, and Antioxidants

In the normal physiological capacity, a delicate balance exists at the molecular level between oxidants and antioxidants. ROS may play a role in the collagen damage in the chorioamniotic sac leading to tearing [9]. OS occurs when the production of reactive oxygen species exceeds the antioxidant defense. OS may altercate the elasticity and strength of collagen and promote PROM [10, 14]. Isoprostanes (F2-IP) can be used to a marker of OS [15], especially of lipid peroxidation [9]. They are prostaglandin-like products produced by free radical catalyzed nonenzymatic peroxidation of arachidonic acid [9]. Peroxidation decreases, disrupts membrane barrier function and lowers its fluidity [16]. F2-IP levels in the amniotic fluid at 15–18 weeks gestations are a predictor of PROM in preterm deliveries [9].

The antioxidants play a role in the protection of the chorioamniotic sac from oxidant damage [17–20]. Vitamin C (ascorbic acid) is a known redox catalyst, which has the ability to reduce and neutralize ROS. It plays a primordial role in the formation of collagen triple helix and the fortification of collagen cross-links[5]. Vitamin E is a lipid soluble vitamin with antioxidant activity. It constitutes of eight tocopherols and tocotrienols. α-Tocopherol is considered to play a central role due to its capability to react with lipid radicals produced during lipid peroxidation reaction. In vitro, Vitamins C and E have been successful in the protection of the amnion and chorion from damage induced by ROS [8]. Mechanisms have proposed that Vitamin C may salvage oxygen species in the amniotic fluid and Vitamin E may prevent lipid peroxidation [8]. In contrast to the reports implying a protective action, Bolisetty et al. [20] found that antioxidant vitamins, especially vitamin E, diminish markers of OS at birth. Mathews et al. [8] found no evidence to support the hypothesis that antioxidants against PROM. Finally, Mercer et al. [5] postulated that antioxidant treatment with Vitamin C and E may inhibit ROS formation and resultant fetal membrane weakening. Two studies were conducted, one in vitro and the other in vivo, and concluded that neither antioxidants were effective in the prevention of PROM.

Superoxide dismutases are antioxidant enzymes that protect cells against ROS and damage [21–23]. There are four classes that differ by, their protein configuration, the intra- or extracellular localization and the active metal ions (Cu/Zn, Mn, Fe, or Ni) in their catalytic centers [21]. In humans, MnSOD is encoded by SOD2, which is synthesized in the cytosol and imported into the mitochondrial matrix [1]. Than et al. [1] showed MnSOD mRNA expression in the fetal membranes and in chorioamnionitis of PROM suggesting an antioxidant mechanism to counteract the inflammatory process in the chorioamniotic membranes.

8.4 Conclusion

PROM is characterized by reduced collagen concentrations, altered collagen cross-link profiles, and increased concentrations of biomarkers of oxidative damage. In the amniotic membrane, collagen is primordial for mechanical integrity and stress tolerance. The OS that occurs during pregnancy increases the risk for PROM that is caused by changes in collagen integrity. In the future, trials should focus on determining if supplementation with antioxidants could protect the fetal membranes from premature rupture.

References

1. Than NG, Romero R et al (2009) Mitochondrial manganese superoxide dismutase mRNA expression in human chorioamniotic membranes and its association with labor, inflammation, and infection. J Matern Fetal Neonatal Med 22(11):1000–1013
2. American College of Obstetricians and Gynecologists. (2007) Premature rupture of membranes. Obstet Gynecol 80:975–985
3. Lavery JP, Miller CE et al (1982) The effect of labor on the rheologic response of chorioamniotic membranes. Obstet Gynecol 60(1):87–92
4. McLaren J, Taylor DJ et al (1999) Increased incidence of apoptosis in non-labour-affected cytotrophoblast cells in term fetal membranes overlying the cervix. Hum Reprod 14(11):2895–2900
5. Mercer BM, Abdelrahim A et al (2010) The impact of vitamin C supplementation in pregnancy and in vitro upon fetal membrane strength and remodeling. Reprod Sci 17(7):685–695
6. Taylor J, Garite TJ (1984) Premature rupture of membranes before fetal viability. Obstet Gynecol 64(5):615–620
7. Naeye RL, Peters EC (1980) Causes and consequences of premature rupture of fetal membranes. Lancet 1(8161):192–194

8. Mathews F, Neil A (2005) Antioxidants and preterm prelabour rupture of the membranes. BJOG 112(5):588–594
9. Longini M, Perrone S et al (2007) Association between oxidative stress in pregnancy and preterm premature rupture of membranes. Clin Biochem 40(11):793–797
10. Woods JR, Jr (2001) Reactive oxygen species and preterm premature rupture of membranes-a review. Placenta 22 (Suppl A): S38–S44
11. Malak TM, Ockleford CD et al (1993) Confocal immunofluorescence localization of collagen types I, III, IV, V and VI and their ultrastructural organization in term human fetal membranes. Placenta 14(4):385–406
12. Fortunato SJ, Menon R et al (1999) Stromelysins in placental membranes and amniotic fluid with premature rupture of membranes. Obstet Gynecol 94(3):435–440
13. Buhimschi IA, Kramer WB et al (2000) Reduction-oxidation (redox) state regulation of matrix metalloproteinase activity in human fetal membranes. Am J Obstet Gynecol 182(2):458–464
14. Wall PD, Pressman EK et al (2002) Preterm premature rupture of the membranes and antioxidants: the free radical connection. J Perinat Med 30(6):447–457
15. Winterbourn CC, Chan T, Buss IH, Inder TE, Mogridge N, Darlow BA (2000) Protein carbonyls and lipid peroxidation products as oxidation markers in preterm infant plasma: associations with chronic lung disease and retinopathy and effects of selenium supplementation. Pediatr Res 48:84–90
16. Ogino M, Hiyamuta S et al (2005) Establishment of a prediction method for premature rupture of membranes in term pregnancy using active ceruloplasmin in cervicovaginal secretion as a clinical marker. J Obstet Gynaecol Res 5:1421–1426
17. Aplin JD, Campbell S et al (1986) Importance of vitamin C in maintenance of the normal amnion: an experimental study. Placenta 7(5):377–389
18. Casanueva E, Ripoll C et al (2005) Vitamin C supplementation to prevent premature rupture of the chorioamniotic membranes: a randomized trial. Am J Clin Nutr 81(4):859–863
19. Borna S, Borna H et al (2005) Vitamins C and E in the latency period in women with preterm premature rupture of membranes. Int J Gynaecol Obstet 90(1):16–20
20. Bolisetty S, Naidoo D et al (2002) Antenatal supplementation of antioxidant vitamins to reduce the oxidative stress at delivery-a pilot study. Early Hum Dev 67(1–2):47–53
21. Whittaker JW (2003) The irony of manganese superoxide dismutase. Biochem Soc Trans 31(Pt 6):1318–1321
22. Landis GN, Tower J (2005) Superoxide dismutase evolution and life span regulation. Mech Ageing Dev 126(3):365–379
23. Culotta VC, Yang M et al (2006) Activation of superoxide dismutases: putting the metal to the pedal. Biochim Biophys Acta 1763(7):747–758

Chapter 9
Endometriosis and Oxidative Stress

Lucky H. Sekhon and Ashok Agarwal

Abstract Endometriosis is a chronic gynecologic disease process with multifactorial etiology. Increased oxidative stress, a result of increased production of free radicals or depletion of the body's endogenous antioxidant defense, has been implicated in its pathogenesis. Oxidative stress is thought to promote angiogenesis and the growth and proliferation of endometriotic implants. Oxidative stress in the reproductive tract microenvironment is known to negatively affect sperm count and quality and may also arrest fertilized egg division leading to embryo death. Increased DNA damage in sperm, oocytes, and resultant embryos may account for the increase in miscarriages and fertilization and implantation failures seen in patients with endometriosis. The evidence linking endometriosis and infertility to endogenous pro-oxidant imbalance provides a rationale for the empiric use of antioxidant therapy. Vitamin C and E deficiency has been demonstrated in women with endometriosis. Observational and randomized controlled studies have shown vitamin C and E combination therapy to decrease markers of oxidative stress.

Keywords Endometriosis · Oxidative stress · OS induced infertility · Antioxidant treatment · Curcumin · Melatonin · Pentoxifylline

---

L. H. Sekhon
Department of Obstetrics and Gynecology, Mount Sinai School of Medicine,
1176 5th Ave Klingenstein Pavilion, 9th Floor, New York, NY 10029, USA
e-mail: sekhon@hotmail.com; lucky.sekhon@mountsinai.org

A. Agarwal (✉)
Center for Reproductive Medicine, Lerner College of Medicine, Cleveland Clinic,
9500 Euclid Avenue, Cleveland, OH 44195, USA
e-mail: agarwaa@ccf.org

9.1 Introduction

Endometriosis is defined by the development of endometrial tissue, including both glandular epithelium and stroma, outside the uterine cavity, in the pelvic peritoneum, ovaries and the recto-vaginal septum and rarely in remote locations such as the pericardium, pleura, and brain [1–3]. It is a benign, chronic gynecologic disease and is clinically associated with dysmenorrhea, dyspareunia, pelvic pain, and subfertility. It affects 10–15 % of all women of reproductive age and 30 % of infertile women [4, 5].

Despite a large number of studies on endometriosis, its etiology has yet to be clearly defined as the disease is known to have multifactorial characteristics. There is a growing body of evidence suggests that a combination of genetic, hormonal, environmental, immunological, and anatomical factors play a role in the pathogenesis of this disorder [6–8].

The widely accepted Sampson's theory asserts that endometriosis originates from the implantation and invasion of cells from retrograde menstruation to particularly the pelvic peritoneal cavity (Fig. 9.1). This reflux of menstrual endometrial tissue through the fallopian tubes into the peritoneal cavity is a common physiologic event which results in red blood cells being present in the peritoneal fluid of most women [9]. Recent findings indicate that the influence of the local environment is crucial in the development of endometriosis [10]. Iron overload, from lysis of pelvic red blood cells has been identified in different components of the peritoneal cavity of endometriosis patients, including peritoneal fluid, ectopic endometrial tissue and peritoneum adjacent to lesions, and macrophages. It is hypothesized that the peritoneal protective mechanisms of patients with endometriosis might be overwhelmed by menstrual reflux, either because of the abundance of reflux or because of defective scavenging systems [11]. Bleeding from endometriotic lesions may further contribute to the accumulation of iron in peritoneal fluid. Iron can act as a catalyst which generates free radicals. Peritoneal iron overload encountered in lesions, peritoneal fluid and peritoneal macrophages of endometriosis patients may contribute to oxidative stress (OS) which impairs the functionality of protective immune cells, thereby contributing to the development of the disease. In a study by Yamaguchi et al. abundant free iron in the contents of endometriotic cysts was found to strongly associated with OS and frequent DNA mutations [12]. Therefore, the iron-rich environment within endometriotic cysts during may also play a crucial role in carcinogenesis through the iron-induced persistent oxidative stress.

Metaplasia of celomic epithelial cells lining the pelvic peritoneum is one of several theories regarding the pathogenesis of endometriosis. This may explain the mechanism by which endometriosis occurs in the ovary. Endometriotic implants proliferate on the ovarian surface epithelium, as a single cell layer on the surface of ovaries, which invaginates to form cortical inclusion cysts [13]. Both theories of implantation and celomic metaplasia are possible mechanisms of endometriotic lesion initiation. Both estrogen production and progesterone dysregulation may

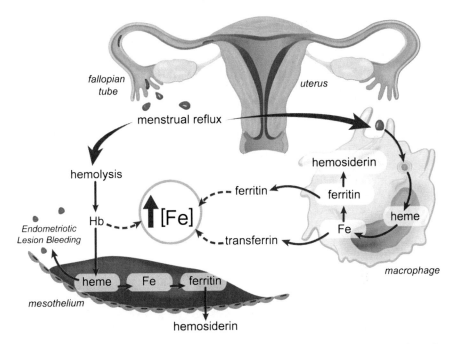

Fig. 9.1 The reflux of menstrual endometrial tissue through the fallopian tubes into the peritoneal cavity results in red blood cells being present in the peritoneal fluid. Iron overload, from lysis of pelvic red blood cells has been identified in different components of the peritoneal cavity of endometriosis patients, including peritoneal fluid, ectopic endometrial tissue and peritoneum adjacent to lesions, and macrophages. (Reprinted with permission, Cleveland Clinic Center for Medical Art & Photography © 2004–2011. All Rights Reserved)

also play a major role in the initiation and promotion of endometriosis [13]. The initial development of endometriosis occurs as a result of induction of attachment, invasion angiogenesis, cell growth, and survival. Additional factors contributing to the establishment and persistence of these endometriotic lesions involve hormonal imbalance, genetic predisposition, and altered immune surveillance.

Mediators of fibrosis and inflammatory changes in the follicular fluid and peritoneal fluid environments are likely involved in the development of the symptoms associated with endometriosis. An increased percentage of B lymphocytes, natural killer cells, and monocyte-macrophages in the follicular fluid have been noted in a case-controlled study of patients with endometriosis compared to patients with other causes of infertility, providing evidence of altered immunologic function in the follicular fluid of patients with endometriosis [14]. Impaired natural killer cell activity may result in inadequate removal of refluxed menstrual debris leading to the development of endometriotic implants. Although the peritoneal fluid of women with endometriosis contains increased numbers of immune cells, these seem to facilitate rather than inhibit the development of endometriosis [15]. Macrophages, that would be expected to clear the peritoneal cavity from endometrial cells, appear to enhance their proliferation by secreting growth factors and cytokines.

Increased concentrations of interleukins IL-6, IL-1b, IL-10, and tumor necrosis factor-a (TNF-a), as well as decreased vascular endothelial growth factor (VEGF) have been documented in the follicular fluid of endometriosis patients [16–18].

As current evidence suggests that endometriosis induces local inflammatory processes, many studies have focused on markers of inflammation and OS in an effort to find less invasive methods of diagnosis [19–21]. OS has been implicated in the pathogenesis of endometriosis. Moreover, evidence is emerging that women with endometriosis experience a greater degree of OS than healthy fertile women.

9.2 Oxidative Stress

Oxidative stress has been implicated in endometriosis and develops when there is an imbalance between the generation of free radicals and the scavenging capacity of antioxidants in the reproductive tract. Free radicals are defined as any species with one or more unpaired electrons in the outer orbit [22]. There are two types of free radicals: reactive oxygen species (ROS) and reactive nitrogen species (RNS). The main free radicals are the superoxide radical, hydrogen peroxide, hydroxyl, and singlet oxygen radicals. ROS are intermediate products of normal oxygen metabolism. Oxygen is required to support life, but its metabolites can modify cell functions, endanger cell survival, or both [23]. Almost all major classes of biomolecules, including lipids, proteins, and nucleic acids, are potential targets for ROS. Hydroxyl radicals are the most reactive free radical species known and have the ability to react with a wide range of cellular constituents, including amino-acid residues, and purine and pyrimidine bases of DNA, as well as attacking membrane lipids to initiate a free radical chain reaction known as lipid peroxidation. Therefore, ROS must be continuously inactivated to keep only a small amount necessary to maintain normal cell function. Both enzymatic and non-enzymatic antioxidant systems scavenge and deactivate excessive free radicals, helping to prevent cell damage. The body's complex antioxidant system is influenced by dietary intake of nonenzymatic antioxidants such as manganese, copper, selenium and zinc, beta-carotenes, vitamin C, vitamin E, taurine, hypotaurine, and B vitamins [24]. On the other hand, the body produces several antioxidant enzymes such as catalase, superoxide dismutase, glutathione reductase, glutathione peroxidase, and molecules like glutathione and NADH. Glutathione is produced by the cell and plays a crucial role in maintaining the normal balance between oxidation and antioxidation. NADH is considered as an antioxidant in biological systems due to its high reactivity with some free radicals, its high intracellular concentrations and the fact that it has the highest reduction power of all biologically active

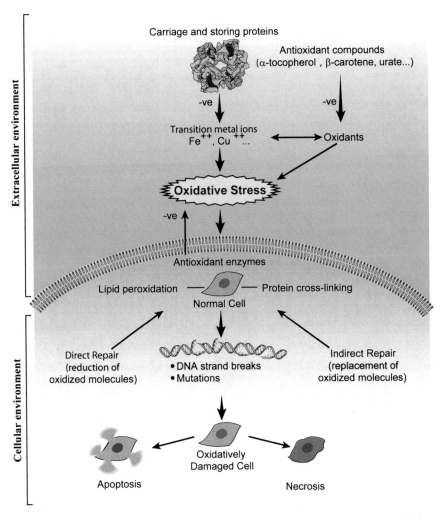

Fig. 9.2 Peritoneal fluid containing ROS-generating iron, macrophages, and environmental contaminants such as polychlorinated biphenyls may disrupt the balance between ROS and antioxidants, resulting in increased proliferation of tissue and adhesions, direct cytotoxic actions, and higher rates of apoptosis. (Reprinted with permission, Cleveland Clinic Center for Medical Art & Photography © 2004–2011. All Rights Reserved)

compounds [25]. When the balance between ROS production and antioxidant defense is disrupted, higher levels of ROS are generated and OS may occur, leading to harmful effects (Fig. 9.2). OS is implicated as a major factor involved in the pathophysiology of endometriosis.

9.3 Oxidative Stress and Endometriosis

Peritoneal fluid containing ROS-generating iron, macrophages, and environmental contaminants such as polychlorinated biphenyls may disrupt the balance between ROS and antioxidants, resulting in increased proliferation of tissue and adhesions [26–29]. OS is thought to have a biphasic dose-response, where only moderate doses of ROS induce endometriotic growth and proliferation, whereas higher doses do not, due to its direct cytotoxic actions and higher rates of apoptosis [30]. OS may have a role in promoting angiogenesis in ectopic endometrial implants by increasing VEGF production [31]. This effect is partly mediated by glycodelin, a glycoprotein whose expression is stimulated by OS. Glycodelin may act as an autocrine factor within ectopic endometrial tissue by augmenting VEGF expression [31].

Altered molecular genetic pathways may also contribute to the effects of OS in the pathogenesis of endometriosis and endometriosis-associated infertility. Differential gene expression of ectopic and normal endometrial tissue has been identified, including differential gene expression of glutathione-S-transferase, an enzyme in the metabolism of the potent antioxidant glutathione [32]. Thioredoxin (TRX), an endogenous redox regulator that protects cells against OS, and TRX-binding protein-2 play a crucial role in the homeostasis of eutopic endometrium. A study by Seo et al. showed that altered TRX and TRX-binding protein-2 mRNA expression in the endometrium is associated with the endometriosis. Therefore, altered molecular genetic pathways may determine the development of OS and its ability to induce cellular proliferation and angiogenesis in women with endometriosis [33].

The peritoneal fluid of women with endometriosis has been reported to exhibit increased ROS generation by activated peritoneal macrophages [34]. Increased macrophage activity is accompanied by the release of cytokines and other immune mediator. After adjusting for confounding factors such as age, BMI, gravidity, serum vitamin E, and serum lipid levels, Jackson et al. found a weak trend involving elevated levels of thiobarbituric acid reactive substances (TBARS), an overall measure of OS, in women with endometriosis [29]. Ota et al. demonstrated that there was a consistently high expression of xanthine oxidase, an enzyme producing ROS, in the endometrium of women with endometriosis, in contrast to the cyclic variations seen in normal subjects. Similarly, they showed that enzymes associated with free radicals are present in the glandular epithelium of endometrium, at levels which are pronounced in endometriosis [35]. These findings suggest that free radical metabolism is abnormal, overall, in endometriosis. Levels of the OS marker, 8-hydroxy 1-deoxyguanosine, were seen to be higher in patients with endometriosis than in patients with tubal, male factor, or idiopathic infertility. A 6-fold increase in the levels of 8-hydroxy 1- deoxyguanosine and lipid peroxide was demonstrated in ovarian endometriomas compared with normal endometrial tissue [36]. Increased NItric oxide (NO) production and lipid peroxidation have been reported in the endometrium of women with endometriosis [37, 38]. NO is a pro-inflammatory free radical that decreases fertility by increasing the amount of OS in the peritoneal fluid, an environment that hosts processes such as ovulation, gamete transportation, sperm–oocyte interaction, fertilization, and early embryonic development [37, 39, 40]. However, several studies failed to find significant differences in the peritoneal fluid levels of NO, lipid peroxide and ROS in women with, and without endometriosis associated infertility. The failure of some studies to confirm alterations in peritoneal fluid NO, lipid peroxide and antioxidant status in women with endometriosis may be explained by the fact that OS may occur locally, without affecting total peritoneal fluid ROS concentration. Also, markers of OS may be transient and not detected at the time endometriosis is diagnosed. In a study by Lambinoudaki et al. stable stress-induced heat shock proteins were used as serum markers of systemic oxidative stress. Women with endometriosis demonstrated increased systemic OS expressed by higher levels of heat shock protein 70bo, which indicated that OS may not be confined to the peritoneal cavity in women with endometriosis [41]. Many of the studies which failed to show increased OS in the peritoneal fluid or systemically [39, 42, 43] often measured ROS or total antioxidant capacity (TAC), parameters that can be affected by handling, whereas heat shock proteins are considered more stable and easily detected.

Increased OS may be due to increased production of ROS or due to depletion of antioxidant reserve. Many recent studies in women with endometriosis have shown altered expression of enzymes involved in defence against OS [35, 44]. Murphy et al. showed that vitamin E levels are significantly lower in the peritoneal fluid of women with endometriosis, possibly due to a local decrease of antioxidants caused by excessive OS [45]. Enzymes associated with free radicals are present in the glandular epithelium of the endometrium and these levels vary dynamically throughout the menstrual cycle. In healthy women, levels of superoxide dismutase (SOD) and nitric oxide synthase (NOS) in the endometrium are low during the proliferative phase and increase during the early and midsecretory phase. However, in women with endometriosis, this variation is lost and the levels of SOD and NOS are seen to remain constant throughout the menstrual cycle [46]. Furthermore, expression of glutathione peroxidase also ceases to vary during the menstural cycle in endometriosis [47]. Women with endometriosis have been shown to have significantly lower levels of antioxidants than women without endometriosis and significantly higher levels of lipid peroxides [48]. Szczepanska et al. reported that women with endometriosis have significantly lower levels of SOD and glutathione peroxidase in peritoneal fluid compared with fertile control women [37]. Excessive OS is thought to contribute to formation of endometriosis-related adhesions. Portz et al. found that injection of antioxidant enzymes, such as SOD and catalase, into the peritoneal cavity prevented the formation of intraperitoneal adhesions at endometriosis sites in rabbits [49]. Therefore, there is evidence to suggest that antioxidants oppose the processes involved in the pathogenesis of endometriosis by controlling OS in the female reproductive tract and peritoneal cavity.

9.4 OS Induced Infertility in Endometriosis

An association between endometriosis and infertility has been often reported in the literature, but a direct causal relationship has yet to be confirmed [50–52]. Severe cases of endometriosis are thought to render a woman infertile by mechanical hindrance of the sperm–egg encounter due to adhesions, endometriomata, and pelvic anatomy disruption. However, in less severe cases where there is no pelvic anatomical distortion, the mechanism by which their fertility is reduced is poorly understood. Numerous mechanisms have been proposed to account for fertility impairment. Endometriosis can cause ovulatory dysfunction, poor oocyte quality [53], luteal phase defects [54], and abnormal embryogenesis [55] which may lead to poor fertilization [56]. Peritoneal macrophages from women with endometriosis associated infertility expressed higher levels of NOS2, had higher NOS enzyme activity and produced more NO in response to immune stimulation in vitro [57]. Peritoneal fluid from women with endometriosis, with high concentrations of cytokines, growth factors, and activated macrophages, has been shown to have levels of ROS that are toxic to the sperm plasma and acrosomal membranes, resulting in a loss of motility and decreased spermatozoal ability to bind, and penetrate the oocyte [58, 59]. Increased iron in the peritoneal fluid results in OS in the reproductive tract microenvironment which negatively affects sperm motility and may also arrest division of the fertilized egg leading to embryo death. Spermatozoa have been shown to exhibit increased DNA fragmentation when incubated with peritoneal fluid from endometriosis patients, with the extent of fragmentation correlating with the stage of endometriosis and duration of infertility [60]. Oocytes exhibited increased DNA damage as they were incubated in peritoneal fluid of endometriosis patients, and the extent of the damage was dependent on the duration of peritoneal fluid exposure [61]. Embryos incubated in the peritoneal fluid of endometriosis patients also exhibited DNA fragmentation as indicated by increased apoptosis [62]. It has been proposed that IVF may improve the conception rate in women with endometriosis as it avoids contact between the gametes and embryos with potentially toxic peritoneal and oviductal factors [57].

Several studies on assisted reproduction have suggested lower than normal rates of pregnancy among women with endometriosis. A meta-analysis of most of these studies showed that the pregnancy rate in women with endometriosis was about half of that in women with tubal-factor infertility, after controlling for confounding factors. Excessive ROS can also interfere with IVF by decreasing the likelihood of fertilization, inducing embryonic fragmentation when intracytoplasmic sperm injection is used, and hampering the in vitro development of blastocysts [63]. The results of studies focusing on IVF treatment suggest poor ovarian reserve in more advanced endometriosis, low oocyte and embryo quality, and poor implantation [64]. Changes in granulosa cell cycle kinetics may be responsible for impaired follicle growth and oocyte maturation in endometriosis patients [36]. Flow cytometric analysis was used to determine the cell cycle of granulosa cells in endometriosis and nonendometriosis patients. A decreased number of granulosa cells in the G2/M phase and an increase in both the S phase and apoptotic cells were documented in women with endometriosis [36, 65]. Oocyte quality may be influenced by granulosa cell apoptosis as well. Granulosa cell apoptosis increased proportionally with the severity of disease and resulted in poor oocyte quality and a reduction in fertilization and pregnancy rates [66]. A higher percentage of granulosa cell apoptosis was associated with significantly reduced pregnancy rates in patients with endometriosis or tubal-factor infertility undergoing IVF [67]. In an observational IVF study using natural cycles, the follicular phase was significantly longer and the fertilization rate was lower in patients with minimal to mild endometriosis compared with women with tubal factor and unexplained infertility [68]. Women with endometriosis were noted to have a slower follicular growth rate [53] and reduced dominant follicle size compared with women with unexplained infertility [69]. Trinder and Cahill also concluded that endometriosis patients have abnormal follicle development, ovulation, and luteal function [70]. Conversely, Mahmood et al. found that women with endometriosis did not experience significant differences in the duration of their follicular phase, and that dominant follicle development was not effected by the disease [71].

The increased DNA damage in sperm, oocytes, and the resultant embryos is proposed to be accountable for increased miscarriages and fertilization and implantation failures among endometriosis patients [60].

Amelioration of infertility associated with endometriosis has been investigated with medical and surgical therapeutic modalities, individually, and in combination. Medical treatments have uniformly been unsuccessful and the outcomes of surgical trials have been inconsistent. Two randomized controlled trials investigating the effects of surgical treatment of mild endometriosis yielded conflicting results [72, 73]. Studies on the surgical management of ovarian endometriomas before assisted reproduction also produced contradictory outcomes [74, 75]. The excessive ROS implicated in the pathogenesis of endometriosis-induced infertility may be a potential target for medical treatment of these patients.

9.5 Antioxidant Treatment of Endometriosis

Several studies have shown that the peritoneal fluid of women with endometriosis-associated infertility have insufficient antioxidant defense, with lower TAC and significantly reduced SOD levels [37, 39]. An early study used a simple rabbit model to demonstrate the beneficial effect of antioxidant therapy in halting progression of the disease [49]. SOD and catalase were instilled in the rabbit peritoneal cavity and were shown to significantly reduce the formation of intraperitoneal adhesions at endometriosis sites by blocking the toxic effects of the superoxide anion and hydrogen peroxide radicals [49]. An in vitro study by Foyouzi et al. was conducted to compare the effects of culturing endometrial stromal cells with antioxidants or with agents inducing oxidative stress. OS induced by hypoxanthine and xanthine oxidase was seen to stimulate endometrial stromal proliferation and DNA synthesis. However, culture with antioxidants such as vitamin E, ebselen, and N-acetylcysteine was shown to inhibit proliferation of endometrial stromal cells in a dose-dependent manner [30].

Lifestyle factors such as inadequate dietary intake of antioxidants may contribute to the OS seen in women with endometriosis. Parazzini et al. reported a significant reduction in risk of endometriosis in women with a greater intake of green vegetables and fresh fruit [76]. Mier-Cabrera et al. conducted a study which reported that women with endometriosis have lower vitamin A, C, E, zinc, and copper intake compared to women without endometriosis. The application of a high antioxidant diet in women with endometriosis increased the peripheral concentration of vitamins A, C, and E after 3 months of intervention in comparison to the control diet group. The antioxidant diet also increased the peripheral enzymatic SOD and glutathione peroxidase activity after 3 months of intervention while decreasing the peripheral concentration of malondialdehyde and lipid hydroperoxides in women with endometriosis [77]. Westphal et al. studied the impact of a nutritional supplementation formula called FertilityBlend on the reproductive health of women who had unsuccessfully attempted to become pregnant for 6–36 months. After 5 months, 33 % of the women in the supplementation group were pregnant compared to 0 % in the placebo group. Therefore, dietary supplementation with antioxidants to alleviate OS may be an effective alternative to conventional fertility therapy [78].

Women with endometriosis are likely to be prescribed a number of empirical therapies. There is a rationale to support the use of antioxidants in these patients. The low cost and relatively low risk of toxicity of these compounds is appealing to both patients and clinicians. Several studies have examined the potential use of antioxidant supplementation to treat OS associated symptoms and complications in endometriosis.

9.5.1 Vitamin E and Vitamin C

The daily requirement of vitamin E varies from 50 to 800 mg, depending on the intake of fruits, vegetables, tea, or wine [79]. Vitamin E (a-tocopherol) is an important lipid-soluble antioxidant molecule in the cell membrane. It is thought to interrupt lipid peroxidation and enhance the activity of various antioxidants that scavenge free radicals generated during the univalent reduction of molecular oxygen and during the normal activity of oxidative enzymes [80, 81]. Vitamin E works synergistically with selenium as an antiperoxidant [82]. Compared with healthy fertile women, women with endometriosis have been shown to have a lower overall intake of vitamin E [83] as well as lower levels of vitamin E within their peritoneal fluid [45]. Afamin, a specific carrier protein of vitamin E in extravascular fluids, was found to be lower in women with endometriosis [84, 85]. A possible explanation of vitamin E deficient intake observed in women with endometriosis could be attributed to nutritional customs and behavioral habits, such as decreased dietary consumption of nuts, wheat germ, sunflower seeds, and extra virgin olive oil [86]. Previous studies done in the US population have shown that only 8–11 % of men and 2–8 % of women meet the new estimated average requirement for vitamin E [87]. Given the strong association between vitamin E deficiency and endometriosis, its use as a supplement may be beneficial to patients with uncontrolled levels of oxidative stress. It is important to mention, however, that vitamin E should be used cautiously in women who are on anticoagulants, because it can have antiplatelet properties and daily intake should be limited to 400 IU or less [88, 89].

Vitamin C (ascorbic acid) is a water-soluble ROS scavenger with high potency. It protects the reproductive microenvironment against endogenous oxidative damage by neutralizing hydroxyl, alkolyl, peroxyl and superoxide anions, hydroperoxyl radicals, and reactive nitrogen radicals such as NO and peroxinitrite. Vitamin C and vitamin E are often prescribed in combination as they act synergistically, with vitamin C exerting its antioxidant function in the aqueous phase, scavenging radicals and regenerating the tocopheroxyl radical [24], whereas Vitamin E scavenges peroxide radicals in the hydrophobic phase of cellular lipid membranes and lipoproteins, protecting them from lipoperoxidation. In addition, Bruno et al. found that supplementation with vitamin C decreased plasma α-tocopherol disappearance rates in smokers [90].

However, this concept must be verified by further prospective controlled clinical studies in selected patients with endometriosis with identified raised markers of oxidative stress.

9.5.2 Pentoxifylline

Another drug being investigated for its potential use in the treatment of endometriosis-associated infertility is pentoxifylline, a 3′,5′-nucleotide phosphodiesterase inhibitor that raises intracellular cAMP and reduces inflammation by inhibiting TNF-a and leukotriene synthesis. Pentoxifylline has potent immunomodulatory properties and has been shown to significantly reduce the embryotoxic effects of hydrogen peroxide [91]. Zhang et al., conducted a recent randomized control trial in which pentoxifylline treatment failed to demonstrate significant reduction in endometriosis-associated symptoms such as pain. Furthermore, there was no evidence of an increase in the clinical pregnancy rates in the pentoxifylline group compared with placebo [92]. Currently, there is not enough evidence to warrant the use of pentoxifylline in the management of premenopausal women with endometriosis-associated pain and infertility.

9.5.3 Curcumin

Curcumin is a polyphenol derived from turmeric (Curcuma longa) with antioxidant, anti-inflammatory and antiproliferative properties. This compound has been shown to have an anti-endometriotic effect by targeting aberrant matrix remodeling in a mouse model. Matrix metalloproteinase-9 (MMP-9) levels are thought to positively correlate with the severity of endometriosis. In randomized controlled trials, curcumin treatment was seen to reverse MMP-9 activity in endometriotic implants near to control values. Furthermore, the anti-inflammatory property of curcumin was demonstrated by the fact that the attenuation of MMP-9 was accompanied by a reduction in cytokine release. Decreased expression of TNF-a was demonstrated during regression and healing of endometriotic lesions within the mouse model. Pretreatment of endometriotic lesions with curcumin has been shown to prevent lipid peroxidation and protein oxidation within the experimental tissue, attesting to its therapeutic potential to provide antioxidant defense against OS-mediated infertility in endometriosis [93].

9.5.4 Melatonin

MMP-9 also was identified as a therapeutic target for melatonin in the treatment of OS-mediated endometriosis in another study evaluating the effectiveness of melatonin in treating experimental endometriosis in a mouse model [94]. Melatonin is a major secretory product of the pineal gland with anti-oxidant properties that has been shown to arrest lipid peroxidation and protein oxidation, while downregulating MMP-9 activity and expression in a time and dose dependent manner. Tissue inhibitors of metalloproteinase (TIMP)-1 were found to be elevated in response to melatonin treatment. Regression of peritoneal endometriotic lesions was seen to accompany the alteration in metalloproteinase expression [94]. Guney et al. confirmed these findings in that treatment with melatonin was seen to cause regression and atrophy of endometriotic lesions in an experimental rat model [95]. Endometrial lesions treated with melatonin demonstrated lower MDA levels and significantly increased SOD and catalase activity [95], further substantiating the usefulness of this hormone to neutralizing endometriosis associated OS.

9.5.5 Green Tea

As previously mentioned, OS stimulates factors that increase VEGF expression and promote angiogenesis of endometriotic lesions. The green tea-containing compound, epigallocatechin gallate (EGCG) has been evaluated as a treatment for endometriosis due to its powerful antioxidant and anti-angiogenic properties. Xu et al. conducted a study in which eutopic endometrium transplanted subcutaneously into a mouse model was used to compare the effects of EGCG treatment on endometriotic implants to the effects seen with vitamin E treatment or untreated controls [96]. Lesions treated with EGCG were seen to have significantly downregulated levels of VEGF-A mRNA. While the control endometrial implants exhibited newly developed blood vessels with proliferating glandular epithelium, the EGCG group demonstrated significantly smaller endometriotic lesions and smaller and more eccentrically distributed glandular epithelium. Despite its widely studied benefits as a potent antioxidant in the treatment of female infertility, vitamin E was not shown to control or decrease angiogenesis compared with baseline controls [96]. As EGCG was shown to significantly inhibit the development of experimental endometriotic lesions in a mouse model, its effectiveness as an oral therapy in female patients to limit progression and induce remission of their endometriosis should be further investigated.

9.5.6 Other

Treatment with an iron chelator could be beneficial in the case of endometriosis to prevent iron overload in the pelvic cavity [97], thereby diminishing the deleterious effects of the resulting OS. However, in women suffering from endometriosis, menstrual periods are often longer and heavier [98]. Sanfilippo et al. and cycles tend to be shorter [99]. Therefore, iron overload observed in these patients is generally localized in the pelvic cavity, whereas body iron content may actually be decreased due to abundant menstruation. For this reason, iron chelator treatment may only be helpful if applied locally, within the peritoneal cavity, by means of intrapelvic implants that release deferoxamine over several months or years.

Guney et al. evaluated caffeic acid phenethyl ester (CAPE), an active component of propolis from honeybee hives that is known to have antimitogenic, anticarcinogenic, antiinflammatory, and immunomodulatory properties. The effect of this compound on experimental endometriosis in a rat model, and the levels of peritoneal SOD and catalase activity, and MDA [100]. Treatment with CAPE was seen to decrease peritoneal MDA levels and antioxidant enzyme activity in rats. Endometriotic lesions treated with CAPE were histologically demonstrated to undergo atrophy and regression, compared with untreated controls [100].

Medical treatments which modulate the hormonal imbalances associated with endometriosis may also have an antioxidant mechanism of action. More recently, mifepristone (RU486)- a potent antiprogestational agent with antioxidant activity, was shown to effectively decrease the proliferation of epithelial and stromal cells in endometriosis [101].

9.6 Conclusion

ROS have been shown to have an important role in the normal functioning of the reproductive system and in the pathogenesis of infertility in females and is thought to play a role in the pathogenesis of endometriosis. Although many studies have investigated the factors that might be involved in the development of different stages of endometriosis, the precise mechanism by which this disease is established remains unclear. Decreased antioxidant protection within the peritoneal fluid of patients with endometriosis may render the reproductive tract more susceptible to damage by OS. The identification of highly sensitive and specific markers of oxidative stress in peritoneal fluid, serum, and tissue biopsies may facilitate the development of reliable non-invasive techniques for endometriosis diagnosis and prognosis. At present, there are many medical or surgical interventions for treating endometriosis. A multidisciplinary and integrative approach may offer expanded therapeutic solutions for this disorder. Endometriosis is associated with hormonal, chemical, and immunologic that may affect ovulation and oocyte quality, tubal function, sperm function, fertilization, and implantation. A greater understanding of these mechanisms is necessary to develop noninvasive methods of detection and diagnosis and to shift from surgical management of disease to medical treatment options.

Further studies to evaluate the effects of ROS and antioxidants on endometrial implants and on endometrial epithelial cells both in vitro and in vivo may provide a basis for clinical use of antioxidants in the treatment of endometriosis. However, the current data evaluating antioxidant supplements is derived from randomised controlled trials that often differ in terms of the selection of the control population, eligibility criteria, markers of OS and antioxidant status and the biological medium in which OS markers were measured, making it difficult to come to a definitive conclusion. Dietary supplements with antioxidants may be a potential strategy in the long-term treatment of endometriosis that is better accepted by patients due to increased cost-effectiveness and lower risk of toxicity. Future research should be directed towards implementing robust, large scale, randomized controlled trials in order to determine the efficacy, safety profiles, and effective doses of specific therapeutic regimens.

References

1. Snesky TE, Liu DT (1980) Endometriosis: associations with menorrhagia, infertility, and oral contraceptives. Int J Gynaecol Obstet 17:573–576
2. Cramer DW (1987) Epidemiology of endometriosis in adolescents. In: Wilson EA (ed) Endometriosis, vol 1. Alan Liss, New York, pp 5–8
3. Giudice L, Kao L (2004) Endometriosis. Lancet 364:1789–1799
4. Koninckx PR (1999) The physiopathology of endometriosis: pollution and dioxin. Gynecol Obstet Invest 47:47–49

5. Donnez J, Chantraine F, Nisolle M (2002) The efficacy of medical and surgical treatment of endometriosis-associated infertility: arguments in favour of a medico-surgical approach. Hum Reprod Update 8:89–94
6. Nisolle M, Donnez J (1997) Peritoneal endometriosis, ovarian endometriosis, and adenomyotic nodules of the rectovaginal septum are three different entities. Fertil Steril 68:585–596
7. Van Langendonckt A, Casanas-Roux F, Dolmans MM, Donnez J (2002) Potential involvement of hemoglobin and heme in the pathogenesis of peritoneal endometriosis. Fertil Steril 77:561–570
8. Heilier JF, Donnez J, Lison D (2008) Organochlorines and endometriosis: a mini-review. Chemosphere 71:203–210
9. Halme J, Hammond MG, Hulka JF, Raj SG, Talbert LM (1984) Retrograde menstruation in healthy women and in patients with endometriosis. Obstet Gynecol 64:151–154
10. Nap AW, Groothuis PG, Demir AY, Evers JL, Dunselman GA (2004) Pathogenesis of endometriosis. Best Pract Res Clin Obstet Gynaecol 18:233–244
11. Van Langendonckt A, Casanas-Roux F, Donnez J (2002) Iron overload in the peritoneal cavity of women with pelvic endometriosis. Fertil Steril 78:712–718
12. Yamaguchi K, Mandai M, Toyokuni S, Hamanishi J, Higuchi T, Takakura K, Fujii S (2008) Contents of endometriotic cysts, especially the high concentration of free iron, are a possible cause of carcinogenesis in the cysts through the iron-induced persistent oxidative stress. Hum Cancer Biol 14(1):32–40
13. Templeton DM, Liu Y (2003) Genetic regulation of cell function in response to iron overload or chelation. Biochim Biophys Acta 1619:113–124
14. Lachapelle MH, Hemmings R, Roy DC, Falcone T, Miron P (1996) Flow cytometric evaluation of leukocyte subpopulations in the follicular fluids of infertile patients. Fertil Steril 65:1135–1140
15. Seli E, Arici A (2003) Endometriosis: interaction of immune and endocrine systems. Semin Reprod Med 21:135–144
16. Pellicer A, Albert C, Mercader A, Bonilla-Musoles F, Remohi J, Simon C (1998) The follicular and endocrine environment in women with endometriosis: local and systemic cytokine production. Fertil Steril 70:425–431
17. Garrido N, Navarro J, Remohi J, Simon C, Pellicer A (2000) Follicular hormonal environment and embryo quality in women with endometriosis. Hum Reprod Update 6:67–74
18. Wunder DM, Mueller MD, Birkhauser MH, Bersinger NA (2006) Increased ENA-78 in the follicular fluid of patients with endometriosis. Acta Obstet Gynecol Scand 85:336–342
19. Bedaiwy MA, Falcone T, Sharma RK, Goldberg JM, Attaran M, Nelson DR, Agarwal A (2002) Prediction of endometriosis with serum and peritoneal fluid markers: a prospective controlled trial. Hum Reprod 17:426–431
20. Darai E, Detchev R, Hugol D, Quang NT (2003) Serum and cyst fluid levels of interleukin (IL) -6, IL-8 and tumour necrosis factor-alpha in women with endometriomas and benign and malignant cystic ovarian tumours. Hum Reprod 18:1681–1685
21. Ulukus M, Ulukus EC, Seval Y, Zheng W, Arici A (2005) Expression of interleukin-8 receptors in endometriosis. Hum Reprod 20:794–801
22. Agarwal A, Gupta S (2005) Role of reactive oxygen species in female reproduction. Part I. Oxidative stress: a general overview. Women Health 1:21–25
23. de Lamirande E, Gagnon C (1995) Impact of reactive oxygen species on spermatozoa: a balancing act between beneficial and detrimental effects. Hum Reprod 10:15–21
24. Alul RH, Wood M, Longo J, Marcotte AL, Campione AL, Moore MK, Lynch SM (2003) Vitamin C protects low-density lipoproteins from homocysteine-mediated oxidation. Free Radic Biol Med 34(7):881–891
25. Olek RA, Ziolkowski W, Kaczor JJ, Greci L, Popinigis J, Antosiewicz J (2004) Antioxidant activity of NADH and its analogue—an in vivo study. J Biochem Mol Biol 37:416–421

26. Reubinoff BE, Har-El R, Kitrossky N, Friedler S, Levi R, Lewin A, Chevion M (1996) Increased levels of redox-active iron in follicular fluid: a possible cause of free radical mediated infertility in beta-thalassemia major. Am J Obstet Gynecol 174(3):914–918
27. Arumugam K, Yip YC (1995) De novo formation of adhesions in endometriosis: the role of iron and free radical reactions. Fertil Steril 64(1):62–64
28. Murphy AA, Palinski W, Rankin S, Morales AJ, Parthasarathy S (1998) Evidence for oxidatively modified lipid-protein complexes in endometrium and endometriosis. Fertil Steril 69(6):1092–1094
29. Donnez J, Van Langendonckt A, Casanas-Roux F et al (2002) Current thinking on the pathogenesis of endometriosis. Gynecol Obstet Invest 54(Suppl 1):52–58. Discussion 9–62
30. Foyouzi N, Berkkanoglu M, Arici A, Kwintkiewicz J, Izquierdo D, Duleba AJ (2004) Effects of oxidants and antioxidants on proliferation of endometrial stromal cells. Fertil Steril 82(3):1019–1022
31. Park JK, Song M, Dominguez CE, Walter MF, Santanam N, Parthasarathy S, Murphy AA (2006) Glycodelin mediates the increase in vascular endothelial growth factor in response to oxidative stress in the endometrium. Am J Obstet Gynecol 195(6):1772–1777
32. Wu Y, Kajdacsy-Balla A, Strawn E, Basir Z, Halverson G, Jailwala P, Wang Y, Wang X, Ghosh S, Guo SW (2006) Transcriptional characterizations of differences between eutopic and ectopic endometrium. Endocrinology 147(1):232–246
33. Seo SK, Yang HI, Lee KE, Kim HY, Cho S, Choi YS, Lee BS (2010) The roles of thioredoxin and thioredoxin-binding protein-2 in endometriosis. Hum Reprod 25(5):1251–1258
34. Zeller JM, Henig I, Radwanska E, Dmowski WP (1987) Enhancement of human monocyte and peritoneal macrophage chemiluminescence activities in women with endometriosis. Am J Reprod Immunol Microbiol 13:78–82
35. Ota H, Igarashi S, Tanaka T (2001) Xanthine oxidase in eutopic and ectopic endometrium in endometriosis and adenomyosis. Fertil Steril 75:785–790
36. Saito H, Seino T, Kaneko T, Nakahara K, Toya M, Kurachi H (2002) Endometriosis and oocyte quality. Gynecol Obstet Invest 53(1):46–51
37. Szczepanska M, Kozlik J, Skrzypczak J, Mikolajczyk M (2003) Oxidative stress may be a piece in the endometriosis puzzle. Fertil Steril 79:1288–1293
38. Gupta S, Agarwal A, Krajcir N, Alvarez JG (2006) Role of oxidative stress in endometriosis. Reprod Biomed Online 13(1):126–134
39. Polak G, Koziol-Montewka M, Gogacz M, Blaszkowska I, Kotarski J (2001) Total antioxidant status of peritoneal fluid in infertile women. Eur J Obstet Gynecol Reprod Biol 94(2):261–263
40. Dong M, Shi Y, Cheng Q, Hao M (2001) Increased nitric oxide in peritoneal fluid from women with idiopathic infertility and endometriosis. J Reprod Med 46:887–891
41. Lambrinoudaki IV, Augoulea A, Christodoulakos GE, Economou EV, Kaparos G, Kontoravdis A, Papadias C, Creatsas G (2009) Measurable serum markers of oxidative stress response in women with endometriosis. Fertil Steril 91(1):46–50
42. Wang Y, Sharma RK, Falcone T, Goldberg J, Agarwal A (1997) Importance of reactive oxygen species in the peritoneal fluid of women with endometriosis or idiopathic infertility. Fertil Steril 68:826–830
43. Ho HN, Wu MY, Chen SU, Chao KH, Chen CD, Yang YS (1997) Total antioxidant status and nitric oxide do not increase in peritoneal fluids from women with endometriosis. Hum Reprod 12:2810–2815
44. Ota H, Igarashi S, Hatazawa J, Tanaka T (1999) Endometriosis and free radicals. Gynecol Obstet Invest 48:29–35
45. Murphy AA, Santanam N, Morales AJ, Parthasarathy S (1998) Lysophosphatidyl choline, a chemotactic factor for monocytes/T-lymphocytes is elevated in endometriosis. J Clin Endocrinol Metabol 83:2110–2113

46. Ota H, Igarashi S, Hatazawa J, Tanaka T (1999) Immunohistochemical assessment of superoxide dismutase expression in the endometrium in endometriosis and adenomyosis. Fertil Steril 72:129–134
47. Ota H, Igarashi S, Kato N, Tanaka T (2000) Aberrant expression of glutathione peroxidase in eutopic and ectopic endometrium in endometriosis and adenomyosis. Fertil Steril 74:313–318
48. Jackson L, Schisterman E, Day-Rao R, Browne R, Armstrong D (2005) Oxidative stress and endometriosis. Hum Reprod 20:2014–2020
49. Portz DM, Elkins TE, White R, Warren J, Adadevoh S, Randolph J (1991) Oxygen free radicals and pelvic adhesion formation: I. Blocking oxygen free radical toxicity to prevent adhesion formation in an endometriosis model. Int J Fertil 36:39–42
50. ASRM (2004) Endometriosis and infertility. Fertil Steril 82(1):S40–S45
51. Mahutte NG, Arici A (2002) New advances in the understanding of endometriosis related infertility. J Reprod Immunol 55:73–83
52. Alpay Z, Saed GM, Diamond MP (2006) Female infertility and free radicals: potential role in adhesions and endometriosis. J Soc Gynecol Investig 13(6):390–398
53. Doody MC, Gibbons WE, Buttram VC Jr (1988) Linear regression analysis of ultrasound follicular growth series: evidence for an abnormality of follicular growth in endometriosis patients. Fertil Steril 49:47–51
54. Grant A (1966) Additional sterility factors in endometriosis. Fertil Steril 17:514–519
55. Wardle PG, Mitchell JD, McLaughlin EA, Ray BD, McDermott A, Hull MG (1985) Endometriosis and ovulatory disorder: reduced fertilisation in vitro compared with tubal and unexplained infertility. Lancet 2:236–239
56. Garrido N, Navarro J, Garcia-Velasco J, Remoh J, Pellice A, Simon C (2002) The endometrium versus embryonic quality in endometriosis-related infertility. Hum Reprod Update 8:95–103
57. Osborn B, Haney AF, Misukonis M, Weinberg JB (2002) Inducible nitric oxide synthase expression by peritoneal macrophages in endometriosis-associated infertility. Fertil Steril 77:46–51
58. Curtis P, Lindsay P, Jackson AE, Shaw RW (1993) Adverse effects on sperm movement characteristic in women with minimal and mild endometriosis. Br J Obstet Gynaecol 100:165–169
59. Oak MK, Chantler EN, Wiliams CA, Elstein M (1985) Sperm survival studies in peritoneal fluid from infertile women with endometriosis and unexplained infertility. Clin Reprod Fertil 3:297–303
60. Mansour G, Aziz N, Sharma R, Falcone T, Goldberg J, Agarwal A (2009) The impact of peritoneal fluid from healthy women and from women with endometriosis on sperm DNA and its relationship to the sperm deformity index. Fertil Steril 92:61–67
61. Mansour G, Agarwal A, Radwan E, Sharma R, Goldberg J, Falcone T (2007) DNA damage in metaphase II oocytes is induced by peritoneal fluid from endometriosis patients. ASRM 63rd annual meeting
62. Mansour G, Radwan E, Sharma R, Agarwal A, Falcone T, Goldberg J (2007) DNA damage to embryos incubated in the peritoneal fluid of patients with endometriosis: role in infertility. ASRM 63rd annual meeting
63. Agarwal A, Gupta S, Sikka S (2006) The role of free radicals and antioxidants in reproduction. Curr Opin Obstet Gynecol 18:325–332
64. Polak G, Rola R, Gogacz M, Koziol-Montewka M, Kotarski J (1999) Malonyldialdehyde and total antioxidant status in the peritoneal fluid of infertile women. Ginecol PII 70:135–140
65. Toya M, Saito H, Ohta N, Saito T, Kaneko T, Hiroi M (2000) Moderate and severe endometriosis is associated with alterations in the cell cycle of granulosa cells in patients undergoing in vitro fertilization and embryo transfer. Fertil Steril 73:344–350
66. Nakahara K, Saito H, Saito T, Ito M, Ohta N, Sakai N et al (1997) Incidence of apoptotic bodies in membrana granulosa of the patients participating in an in vitro fertilization program. Fertil Steril 67:302–308

67. Sifer C, Benifla JL, Bringuier AF, Porcher R, Blanc-Layrac G, Madelenat P, Feldman G (2002) Could induced apoptosis of human granulosa cells predict in vitro fertilization-embryo transfer outcome? A preliminary study of 25 women. Eur J Obstet Gynecol Reprod Biol 103:150–153
68. Cahill DJ, Wardle PG, Maile LA, Harlow CR, Hull MG (1997) Ovarian dysfunction in endometriosis-associated and unexplained infertility. J Assist Reprod Genet 14:554–557
69. Tummon IS, Maclin VM, Radwanska E, Binor Z, Dmowski WP (1988) Occult ovulatory dysfunction in women with minimal endometriosis or unexplained infertility. Fertil Steril 50:716–720
70. Trinder J, Cahill DJ (2002) Endometriosis and infertility: the debate continues. Hum Fertil 5:S21–S27
71. Mahmood TA, Templeton A (1991) Folliculogenesis and ovulation in infertile women with mild endometriosis. Hum Reprod 6:227–231
72. Marcoux S, Maheux R, Berube S (1997) Laparoscopic surgery in infertile women with minimal or mild endometriosis. Canadian collaborative group on endometriosis. N Engl J Med 337:217–222
73. Parazzini F (1999) Ablation of lesions or no treatment in minimal–mild endometriosis in infertile women: a randomized trial. Gruppo Italiano per lo Studio dell'Endometriosi. Hum Reprod 14:1332–1334
74. Garcia-Velasco JA, Arici A (2004) Surgery for the removal of endometriomas before in vitro fertilization does not increase implantation and pregnancy rates. Fertil Steril 81:1206
75. Suzuki T, Izumi S, Matsubayashi H, Awaji H, Yoshikata K, Makino T (2005) Impact of ovarian endometrioma on oocytes and pregnancy outcome in in vitro fertilization. Fertil Steril 83:908–913
76. Parazzini F, Chiaffarino F, Surace M, Chatenoud L, Cipriani S, Chiantera V et al (2004) Selected food intake and risk of endometriosis. Hum Reprod 19:1755–1759
77. Mier-Cabrera J, Genera-Garcia M, De la Jara-Diaz J, Perichart-Perera O, Vadillo-Ortega F, Hernandez-Guerrero C (2008) Effect of vitamins C and E supplementation on peripherals oxidative stress markers and pregnancy rate in women with endometriosis. Int J Gynecol Obstet 100:252–256
78. Westphal LM, Polan ML, Trant AS, Mooney SB (2004) A nutritional supplement for improving fertility in women: a pilot study. J Reprod Med 49:289–293
79. National Academy of Sciences (1989) Recommended dietary allowances. 10th edn. National Academy Press, Washington
80. Ehrenkranz R (1980) Vitamin E and the neonate. Am J Dis Child 134:1157–1168
81. Palamanda JR, Kehrer JR (1993) Involvement of vitamin E and protein thiols in the inhibition of microsomal lipid peroxidation by glutathione. Lipids 28:427–431
82. Burton GW, Traber MG (1990) Vitamin E: antioxidant activity, biokinetics, and bioavailability. Annu Rev Nutr 10:357–382
83. Hernandez-Guerrero CA, Bujalil-Montenegro L, De la Jara-Diaz J, Mier-Cabrera J, Bouchan-Valencia P (2006) Endometriosis and deficient intake of antioxidant molecules related to peripherals and peritoneal oxidative stress. Ginecol Obstet Mex 74:20–28
84. Jackson D, Craven RA, Hutson RC, Graze I, Lueth P, Tonge RP, Hartley JL, Nickson JA, Rayner SJ, Johnston C, Dieplinger B, Hubalek M, Wilkinson N, Perren TJ, Kehoe S, Hall GD, Daxenbichler G, Dieplinger H, Selby PJ, Banks RE (2007) Proteomic profiling identifies afamin as a potential biomarker for ovarian cancer. Clin Cancer Res 13:7370–7379
85. Dieplinger H, Ankerst DP, Burges A, Lenhard M, Lingenhel A, Fineder L, Buchner H, Stieber P (2009) Afamin and apolipoprotein A-IV: novel protein markers for ovarian cancer. Cancer Epidemiol Biomarkers Prev 18:1127–1133
86. de Muñoz CM, Antonio RJ, Angel LJ, Eduardo M, Adolfo C, Fernando P-G, Sonia H, Alejandra C (1996) Tablas de valor nutritivo de los alimentos. Editorial Pax México, México D.F.

87. Gao X, Martin A, Lin H, Bermudez OI, Tucker KL (2006) α-tocopherol intake and plasma concentrations of hispanic and non-hispanic white elders is associated with dietary intake patters. J Nutr 136:2574–2579
88. Dennehy CE (2006) The use of herbs and dietary supplements in gynecology: an evidence-based review. J Midwifery Womens Health 51:402–409
89. Ziaei S, Faghihzadeh S, Sohrabvand F, Lamyian M, Emamgholy T (2001) A randomised placebo-controlled trial to determine the effect of vitamin E in treatment of primary dysmenorrhoea. Br J Obstet Gynecol 108:1181–1183
90. Bruno RS, Leonard SW, Atkinson J, Montine TJ, Ramakrishnan R, Bray TM, Traber MG (2006) Faster plasma vitamin E disappearance in smokers is normalized by vitamin C supplementation. Free Radic Biol Med 40:689–697
91. Zhang X, Sharma RK, Agarwal A, Falcone T (2005) Effect of pentoxifylline in reducing oxidative stress-induced embryotoxicity. J Assist Reprod Genet 22(11–12):415–417
92. Lv D, Song H, Clarke J, Shi G (2009) Pentoxifylline versus medical therapies for subfertile women with endometriosis. Cochrane Database Syst Rev 8(3):CD007677
93. Swarnaker S, Paul S (2009) Curcumin arrests endometriosis by downregulation of matrix metalloproteinase-9 activity. Indian J Biochem Biophys 46(1):59–65
94. Paul S, Sharma AV, Mahapatra PS, Bhattacharya P, Reiter RJ, Swarnakar S (2008) Role of melatonin in regulating matrix metalloproteinase-9 via tissue inhibitors of metalloproteinase-1 during protection against endometriosis. J Pineal Res 44(4):439–449
95. Guney M, Oral B, Karahan N, Mungan T (2008) Regression of endometrial explants in a rat model of endometriosis treated with melatonin. Fertil Steril 89(4):934–942
96. Xu H, Liu WT, Chu CY, Ng PS, Wang CC, Rogers MS (2009) Antiangiogenic effects of green tea catechin on an experimental endometriosis mouse model. Hum Reprod 24(3):608–618
97. Defre're S, Van Langendonckt A, Vaesen S, Jouret M, Gonza'lez Ramos R, Gonzalez D, Donnez J (2006) Iron overload enhances epithelial cell proliferation in endometriotic lesions induced in a murine model. Hum Reprod 21:2810–2816
98. Sanfilippo JS, Wakim NG, Schikler KN, Yussman MA (1986) Endometriosis in association with uterine anomaly. Am J Obstet Gynecol 154:39–43
99. Arumugam K, Lim JM (1997) Menstrual characteristics associated with endometriosis. Br J Obstet Gynaecol 104:948–950
100. Guney M, Nasir S, Oral B, Karahan N, Mungan T (2007) Effect of caffeic acid phenethyl ester on the regression of endometrial explants in an experimental rat model. Reprod Sci 14(3):270–279
101. Murphy AA, Zhou MH, Malkapuram S, Santanam N, Parthasarathy S, Sidell N (2000) RU486-induced growth inhibition of human endometrial cells. Fertil Steril 71:1014–1019

Chapter 10
Oxidative Stress Impact on the Fertility of Women with Polycystic Ovary Syndrome

Anamar Aponte and Ashok Agarwal

Abstract Polycystic ovary syndrome is a common endocrine abnormality in reproductive-age women. The pathophysiology of this condition remains unclear. Women with polycystic ovary syndrome present a diverse combination of clinical complications including, psychological problems, reproductive alterations, and metabolic sequelae. In affected women, hyperglycemia, independent of obesity, promotes reactive oxygen species. The resultant oxidative stress causes extensive cellular injury, demonstrated by protein oxidation, lipid peroxidation, and DNA damage. This oxidative stress may directly stimulate hyperandrogenism. Additionally, serum total antioxidant status, is diminished in women with polycystic ovary syndrome, decreasing the body's defense against an oxidative environment. Treatment through lifestyle intervention and medical/surgical therapy may improve metabolic consequences of polycystic ovary syndrome, including insulin resistance and reproductive status.

Keywords Oxidative stress impact · Fertility of women · Polycystic ovary syndrome · Metabolic syndrome · Psychological problems · Reproductive alterations · Metabolic sequelae

A. Aponte
Center for Reproductive Medicine, Cleveland Clinic Desk A19.1,
9500 Euclid Avenue, Cleveland, OH 44195, USA
e-mail: anamar06@hotmail.com

A. Agarwal (✉)
Center for Reproductive Medicine, Lerner College of Medicine Cleveland Clinic,
9500 Euclid Avenue, Cleveland, OH 44195, USA
e-mail: agarwaa@ccf.org

10.1 Background

Polycystic ovary syndrome (PCOS) is a condition characterized by ovulatory dysfunction, hyperandrogenism, and polycystic ovaries. It is the most common endocrine abnormality in reproductive-aged women, with the Rotterdam Criteria [1] (See Fig. 10.1) showing the prevalence of PCOS at approximately 18 %. Studies show that insulin resistance (IR) may be central to the etiology of the syndrome [2]. In IR, adipose, muscle, and liver cells do not respond appropriately to insulin, causing circulating glucose levels to remain high and leading to glucose intolerance and hyperinsulinemia [3].

Compensatory hyperinsulinemia causes decreased levels of sex hormone binding globulin (SHBG), a glycoprotein that binds testosterone and estradiol. This process increases the bioavailability of circulating androgens, which further stimulates increased androgen production in the adrenal gland and ovary [1]. Overweight or obesity affects 50–80 % of women with this disorder and can aggravate IR and anovulation [4].

The current recommendation by Androgen Excess Society (AES) is to measure free testosterone for measuring circulating androgens. Both the adrenal glands and ovaries contribute to the concentration of circulating androgen in women. Still, the ovary is the primary source of testosterone, accounting for 75 % of its circulating levels. The ovary produces androstenedione, which is converted in the liver, fat, and skin into testosterone. The adrenal glands secrete dehydroepiandrosterone (DHEAS), which may be elevated in PCOS. These hormones may act as precursors for more potent androgens such as testosterone and dihydrotestosterone. Their measurement may be useful in cases of rapid virilization [1].

The approach to diagnosis and treatment of PCOS is a challenge based on the complexity and diversity of its pathophysiology, including the association between metabolic syndrome and oxidative stress (OS) and their impacts on female fertility.

10.2 Clinical Manifestations

Menstrual disorders are among the most common signs of women with PCOS, ranging from amenorrhea to menorrhagia. In women with PCOS, abnormal menstruation is usually attributed to chronic anovulation, a steady state in which monthly rhythms associated with ovulation are not functional. This lack of ovulation creates fertility problems in these women.

Peripheral androgen excess manifests externally as an increase in acne and hirsutism. In hirsutism, hair is commonly seen on the upper lip, chin, around the nipples, and along the linea alba of the lower abdomen. Other signs of hyperandrogenism such as clitoromegaly, increased muscle mass, and voice deepening are more characteristic of an extreme form of PCOS termed hyperthecosis. Patients with PCOS may have dark, pigmented skin on the nape of their neck, skin folds,

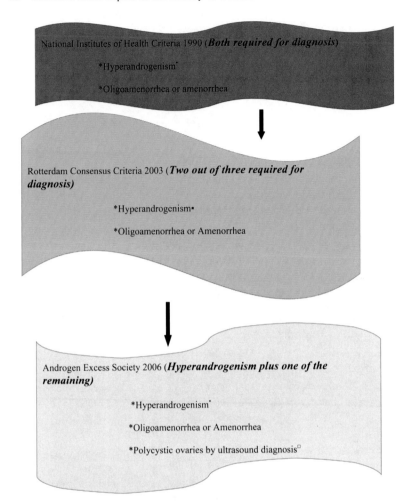

•May be either hirsutism or biochemical hyperandrogenemia
▫In one or both ovaries, either 12 or more follicles measuring 2-9mm in diameter, or increased ovarian volume (>10cm^3)

Fig. 10.1 Recommended diagnostic criteria for polycystic ovary syndrome

knuckles, and/or elbows, called acanthosis nigricans, and is attributed to insulin resistance, or in some cases, to visceral malignancies such as stomach cancer [1]. Signs of IR such as hypertension, obesity, centripetal fat distribution, and acanthosis nigricans may predispose the patient to metabolic syndrome, nonalcoholic fatty liver [5], and obesity-related disorders. In turn, these conditions are risk factors for long-term metabolic sequelae like cardiovascular disease (CVD) and diabetes mellitus (DM) type 2. Approximately 10 % of women with PCOS have type 2 DM, and 30–40 % of women with PCOS have impaired glucose tolerance by the age of 40 [1].

Obesity is a comorbidity that may exacerbate the effects of PCOS. However, it is not a diagnostic criterion for PCOS, and nearly 20 % of women with this condition are not obese [1]. Obesity itself increases hyperandrogenism, infertility, and pregnancy complications by aggravating PCOS. Along with IR, obesity may further augment the risks for developing DM type 2 and CVD.

10.3 Metabolic Syndrome

Numerous patients with PCOS have characteristics of metabolic syndrome. One study showed 43 % prevalence of metabolic syndrome in women with PCOS [6].

Metabolic syndrome involves a clustering of hyperglycemia/IR, visceral obesity, and dyslipidemia. In 2001, the National Cholesterol Education Program Adult Treatment Panel III (2001) established a definition for the metabolic syndrome [3], which was renewed in 2005 by the American Heart Association (AHA) and the National Heart Lung and Blood Institute in 2005 [7]. According to this revised definition, metabolic syndrome requires the presence of three or more of the five criteria below [3] (See Table 10.1).

In 2005, The International Diabetes Foundation (IDF) published new criteria for metabolic syndrome [8]. It includes the same criteria as the other definitions, but requires obesity, as opposed to IR. The IDF definition has been challenged for its emphasis on obesity rather than IR, although visceral obesity is a known important factor [9]. In addition, the concurrence of obesity, oxidative stress, and platelet aggregation exponentially increases the risk of CVD morbidity and mortality [10, 11].

Mechanisms of metabolic syndrome are based on four clinical features:

A. Insulin Resistance

The pancreas produces insulin in response to increased levels of glucose. Subsequently, different tissues of the body such as the liver, adipose tissue, and skeletal muscle use this glucose as fuel. In adipose tissue, insulin prevents fat breakdown and promotes glucose uptake, while in skeletal muscle and the liver, it promotes glycogen synthesis and prevents glycogenolysis. In IR, these tissues do not respond adequately to insulin, which exacerbates hyperglycemia [12].

Physiological insulin signaling takes place after the binding of insulin to its receptor, a ligand-activated tyrosine kinase, resulting in tyrosine phosphorylation and activating two parallel pathways: the phosphoinositide 3-kinase (PI3 K) pathway and the mitogen-activated protein (MAP) kinase pathway [3]. The PI3 K is responsible for many of the metabolic effects of insulin including the activation of endothelial nitric oxide synthetase (eNOS). The MAP kinase pathway leads to vasoconstriction; growth effects on vascular smooth muscle cells; and expression of the vascular adhesion molecules VCAM-1 and E-selectin [3]. In IR the PI3 K pathway is affected and the MAP pathway is not. This misbalance results in a reduced nitric oxide (NO) formation, leading to endothelial dysfunction. A reduction also occurs in GLUT4, an insulin-responsive glucose transporter, translocation leading to decreased skeletal and fat glucose uptake [3].

Table 10.1 Metabolic Syndrome Criteria

Blood Pressure >130/85
Waist circumference (>35 in, 88 cm)
HDL ≤50 mg/dL
Fasting glucose levels ≥100 mg/dL
Triglycerides ≥150 mg/dL
3/5 needed for diagnosis

B. Visceral adiposity

Visceral adiposity is closely associated with IR. Tumor necrosis factor α (TNF α) and interlukin6 (IL-6), inflammatory molecules produced by the adipose tissue, are thought to be closely involved in the pathogenesis of IR and vascular disturbance [13].

C. Dyslipidemia

Dyslipidemia comprises high levels of triglycerides, low-density lipoprotein (LDL), and low levels of high-density lipoprotein (HDL). IR leads to dyslipidemia mainly by preventing lipolysis, thereby increasing free fatty acid (FFA) concentration. In the liver, FFAs participate in the synthesis of triglycerides and in the formation of very low-density lipoprotein (VLDL) particles [14].

D. Endothelial dysfunction

The endothelium regulates physiological and pathological stimuli by maintaining homeostasis and preventing the development of atherosclerosis [12, 15, 16]. Endothelial dysfunction may result from the influence of FFAs, cytokines, hyperglycemia, or OS. IR and visceral adiposity are intimately related in promoting the damaging activity of the endothelium by increasing inflammatory reactants such as, diminishing blood flow to skeletal muscle, and further increasing reactive oxygen species (ROS). This creates a vicious cycle of endothelial dysfunction and formation of OS [3].

10.4 Polycystic Ovary Syndrome, Metabolic Syndrome, and Oxidative Stress

Oxidative Stress occurs (OS) when the production of ROS exceeds the endogenous anti-oxidant defense. The anti-oxidant status of plasma suggests the extent of both OS and anti-oxidant levels. A moderate increase in ROS can promote cell growth and proliferation. On the contrary, excessive ROS accumulation will result in cellular injury, such as damage to DNA, protein, and lipid membranes. This is why PCOS is considered an oxidative state. In women with PCOS, diminished total antioxidant capacity (TAC) is the sum of concentration of individual antioxidants such as thiol, carotene, Vitamin C and E, which leads to a decrease in antioxidants defense.

Metabolic IR may occur due to endothelial nitric oxide (NO) impairment. NO is a regulator of important nervous, immune and cardiovascular processes such as arterial vasodilation; it increases blood flow and promotes vascular smooth muscle relaxation. In both obesity and PCOS, stimulation of muscle by insulin is inhibited by a defect in endothelial synthesis [17, 18].

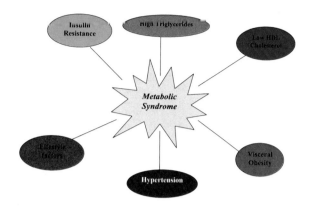

Fig. 10.2 Risk factors for metabolic syndrome

Markers of OS have been proposed to increase exponentially between partial and full metabolic syndrome during childhood and puberty. 15-F2t-Isoprostane is a marker for OS in humans [19, 20] and is synthesized non-enzymatically. Within the cell membrane, this process takes place through the impact of free radicals on the arachnoid acids of phospholipids and plasma lipoproteins, culminating in lipid peroxidation [21].

Protein carbonyl content, a marker for protein oxidation and another important indicator for OS, was found to increase gradually during the progression from partial to full metabolic syndrome in childhood. Attack by hydrogen peroxide (H_2O_2) or oxygen (O_2) can facilitate redox cycling cations such as Fe^+ and Cu^{2+} to attach to binding sites on proteins. Some amino acid residues are converted to carbonyl derivatives and as a result of oxidative modification, these proteins become highly susceptible to degradation. This oxidative modification is an early marker of oxygen radical-mediated tissue damage [22]. In addition to these OS markers, the concentration of tromboxane (TXB2), a potent vasoconstrictor produced by lipid peroxidation, substantially increases the risks for the development of atherosclerosis and other precursors of CVD [10, 11] (See Fig. 10.2).

Acute hyperglycemia causes an increase in the generation of ROS from mononuclear cells (MNC). MNC's promote the release of tumor necrosis factor (TNF) α, a known mediator of IR, and activate the proinflammatory transcription factor nuclear factor (NFkB), which further increases TNFα concentration [23]. MNCs of women with PCOS are in a proinflammatory state as evidenced by their increased sensitivity to physiologic hyperglycemia and elevated C-reactive protein, a marker of inflammation. ROS generation is directly related to androgen levels, which can explain why OS in response to hyperglycemia may be capable of directly stimulating hyperandrogenism [23].

The mitochondria are one of the main sites for ROS production by peripheral blood leukocytes [24]. IR in PCOS patients disrupts mitochondrial function, leading to increased ROS production, a reduction in glutathione levels and decreased oxygen consumption by the mitochondria [25].

Homocysteine is an intermediate formed during disruption of the amino acid methionine, or during trans-sulfuration to cystathione and cysteine. Homocysteine is closely implicated in the risk of developing CVD and other complications, by promoting OS in vascular endothelium, activation of platelets, blockage of blood flow, and stimulation of vascular smooth muscle proliferation. Homocysteine affects endometrial blood flow and vascular structure, thereby impairing adequate implantation. Both impaired implantation and increased rates of miscarriage might be due in part to elevated homocysteine and are more frequent in PCOS, even after controlling for ovulatory abnormalities, increased LH, and hyperandrogenism [26]. Increased homocysteine exacerbates some of the long-term complications of metabolic syndrome such as DM, hypertension, nephropathy, dyslipidemia, and CVD; IR alone is a risk factor for all of these complications. In metabolic disorders, PCOS may conceivably be considered an early indicator of IR syndrome (IRS).

The preovulatory follicle, which is a metabolically active environment, is likely to have multiple sources of ROS production, making it susceptible to OS. The follicular fluid (FF) environment is composed of the oocyte, granulosa cells, and surrounding cells, such as endothelial and theca cells. FF contains cytokines, neutrophils, and macrophages, all of which can participate in the production of oxygen free radicals. Monooxygenase reaction, required for the steroidogenic process and mediated by cytochrome P450, invariably results in ROS production [27]. For better reproductive results in in vitro fertilization (IVF) and intracytoplasmic sperm injection (ICSI) cycles, certain cutoff levels of ROS in FF are needed for pregnancy. On the other hand, excessive ROS levels in the FF lead to toxic effects.

Superoxide radical is a free radical that is produced by the activity of NADPH oxidase. Lean women with PCOS showed elevated ROS generation and $p47^{phox}$, a key protein component of NADPH oxidase. ROS production parallels androgen levels, while $p47^{phox}$ is inversely correlated to insulin sensitivity. The association among $p47^{phox}$ protein expression and percent visceral fat implies that elevated abdominal adiposity may be a key factor of the ROS-induced OS and a promoter of insulin action as seen in obese women with PCOS [23]. This suggests that ROS-induced OS may play a role in the associated development of IR and hyperandrogenism.

Another important consideration is the presence or absence of meiotic spindle (MS) in oocytes. The presence of MS is associated with lower OS and has improved intracytoplasmic sperm injection (ICSI) outcome both in women with PCOS and tubal factor infertility. However, in the absence of MS, there is significant reduction in the formation of good quality embryo and fertilization rate, though statistically not significant, is decreased in PCOS women [27]. Glutathione, an enzymatic antioxidant, plays an important role in maintaining MS morphology and in early blastocyst stage development [28].

A larger number of preantral follicles occur in PCOS ovaries. This disorderly growth takes place due to impaired apoptosis at an early stage of follicular growth, promoting many follicles to synthesize androgens in their thecal component. ROS initiates the apoptotic cascade in granulosa cells, while antioxidants antagonize these effects on cultured preovulatory follicles [29–31]. Free radicals can activate

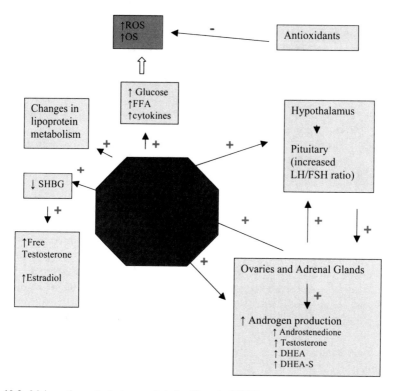

Fig. 10.3 Main pathogenic factors and their effects in PCOS

meiosis in immature oocytes, which is inhibited by antioxidants. Antioxidants block gonadotropin-induced oocyte maturation in follicles and oxygen radicals promote the activations of meiosis. This process may explain the persistence of immature follicles in PCOS [32]. Therefore, oxygen radicals may be categorized as potential mediators for the resumption of meiosis in the oocyte, which is activated by the ovulatory surge of LH.

TAC has been known to promote cell proliferation on ovaries. Development of basement membrane during follicular growth is possible due to adequate collagen synthesis, which is attained from antioxidant functions [31]. Antioxidants are encountered in tissues where steroid hormones are produced and maintain adequate control to ensure that steroidogenesis takes place despite the occurrence of lipid peroxidation [33]. Antioxidants, particularly vitamin E, have been shown to promote LH release from the pituitary gland upon activation by gonadotropin secretion [34].

OS and anti-oxidant status is universally in all PCOS patients, including those who are lean and without metabolic disturbances [35] (See Fig. 10.3).

10.5 Treatment

Women with PCOS who are not attempting to conceive:
A. Lifestyle modifications
Exercise of 30 minutes 3 times a week and a calorie-restricted diet are key factors shown to reduce DM and cardiovascular risk [36]. Even a 5–10 % weight loss has been found to lower androgen levels, decrease hirsutism, and reduced cardiovascular morbidity and mortality. In addition, this weight loss has been shown to increase SHBG, improve glucose and lipid levels, and promote spontaneous resumption of menses [37–40].
B. Combined oral contraceptives
Combination low-dose hormonal contraceptives are most commonly used for long-term management and are considered the primary treatment of menstrual disorders. They offer benefit by suppressing LH levels, ovarian androgen secretion, and increasing SHBG [1].

Oral contraceptives (OCPs) improve circulating markers of endothelial function. Estrogen contained in OCPs improved markers of endothelial function which counteracts the negative effects of impaired IR on endothelial activity [41]. There is no evidence that proves that women with PCOS who use OCPs suffer more cardiovascular events than the general population does [42, 43].
C. Insulin sensitizing agents
Insulin sensitizing agents improve peripheral insulin sensitivity by decreasing circulating androgen levels as well as improving ovulation rate and glucose tolerance. Studies have focused on agents that alleviate insulin sensitivity, including biguanides (e.g., metformin) and thiazolidinediones (e.g., rosiglitazone) [44, 45]. Thiazolidinediones have been unsatisfactory among PCOS patients due to increased weight gain.

Metformin reduces a significant number of risk factors for type 2 DM and CVD, including IR, inflammatory markers, and circulating markers of endothelial function in women with PCOS. This suggests an indirect association between IR and endothelial function and provides new understanding into the relationships between them [41].

A long-term study suggested that metformin continued to improve metabolic complications in women with PCOS over a 36-month treatment course, with particular improvement in circulating HDL (HDL-C), diastolic blood pressure, and BMI [46]. However, current data is insufficient to safely recommend metformin to all women with PCOS.

10.6 Conclusion and What's Next

OS and anti-oxidant status is omnipresent in all PCOS patients, including those who are lean and without metabolic disturbances. ROS generation from MNCs in response to hyperglycemia may serve as an inflammatory trigger for the induction of IR in PCOS. The resultant OS induces a proinflammatory state that may further contribute to IR and hyperandrogenism in PCOS. This IR increases risk for DM2 and several metabolic abnormalities that predispose patients to CVD.

Further study on OS defenses in PCOS will provide a better understanding into the mechanisms that result in decline of fertility. The in vitro maturation (IVM) technique has been used as an alternative treatment for PCOS-related infertility, however the clinical results of IVM have not been encouraging [47, 48]. As IVM does not require priming with FSH or hCG, it possesses several advantages for women with PCOS such as avoiding ovarian hyperstimulation [49]. Future research on PCOS needs to focus on improving the IVM technique to achieve high embryo implantation rates. Currently, new research emphasizes a familial basis for PCOS and the related pattern of diabetes, which may unveil the true nature of insulin resistance.

References

1. American College of Obstetricians and Gynecologists (2009) Polycystic ovary syndrome. Obstet Gynecol 114:936–949
2. Dunaif A (1997) Insulin resistance and the polycystic ovary syndrome: mechanism and implications for pathogenesis. Endocr Rev 18(6):774–800
3. Huang PL (2009) A comprehensive definition for metabolic syndrome. Dis Model Mech 2(5–6):231–237
4. Teede HJ, Meyer C et al (2010) Endothelial function and insulin resistance in polycystic ovary syndrome: the effects of medical therapy. Fertil Steril 93(1):184–191
5. Setji TL, Holland ND et al (2006) Nonalcoholic steatohepatitis and nonalcoholic Fatty liver disease in young women with polycystic ovary syndrome. J Clin Endocrinol Metab 91(5):1741–1747
6. Azziz R, Woods KS et al (2004) The prevalence and features of the polycystic ovary syndrome in an unselected population. J Clin Endocrinol Metab 89(6):2745–2749
7. Grundy SM, Cleeman JI et al (2005) Diagnosis and management of the metabolic syndrome: an American heart association/national heart, lung, and blood institute scientific statement. Circulation 112(17):2735–2752
8. Zimmet P, Magliano D et al (2005) The metabolic syndrome: a global public health problem and a new definition. J Atheroscler Thromb. 12(6):295–300
9. Reaven GM (2006) The metabolic syndrome: is this diagnosis necessary? Am J Clin Nutr 83:1237–1247
10. Urakawa H, Katsuki A et al (2003) Oxidative stress is associated with adiposity and insulin resistance in men. J Clin Endocrinol Metab 88(10):4673–4676
11. Davi G, Guagnano MT et al (2002) Platelet activation in obese women : role of inflammation and oxidative stress. JAMA 288:2008–2014
12. Kim JA, Montagnano M et al (2006) Reciprocal relationships between insulin resistance and endothelial dysfunction: molecular and pathophysiological mechanisms. Circulation 113:1888–1904
13. Kershaw EE, Flier JS (2004) Adipose tissue as an endocrine organ. J Clin Endocrinol Metab 89:2548–2556
14. Semenkovich CF (2006) Insulin resistance and atherosclerosis. J Clin Invest 116:1813–1822
15. Gimbrone MA Jr, Topper JN et al (2000) Endothelial dysfunction, hemodynamic forces, and atherogenesis. Ann NY Acad Sci 902:230–239
16. Huang PL (2005) Unraveling the links between diabetes, obesity, and cardiovascular disease. Circ Res 96:1129–1131
17. Fleming R (2006) The use of insulin sensitising agents in ovulation induction in women with polycystic ovary syndrome. Hormones (Athens) 5(3):171–178

18. Pasquali R, Gambineri A et al (2006) The impact of obesity on reproduction in women with polycystic ovary syndrome. BJOG 113(10):1148–1159
19. Roberts LJ, Morrow JD (2000) Measurement of F(2)-isoprostanes as an index of oxidative stress in vivo. Free Radic Biol Med 28(4):505–513
20. Morrow JD (2005) Quantification of isoprostanes as indices of oxidative stress and the risk of atherosclerosis in humans. Arterioscler Thromb Vasc Biol 25:279–286
21. Cracowski JL, Durand T et al (2002) Isoprostanes as a biomarker of lipid peroxidation in humans: physiology, pharmacology and clinical implications. Trends Pharmacol Sci 23:360–366
22. Stadtman ER, Oliver CN (1999) Metal-catalyzed oxidation of protein. Physiological consequences. J Biol Chem 266:2005–2008
23. Gonzalez F, Rote NS et al (2006) Reactive oxygen species-induced oxidative stress in the development of insulin resistance and hyperandrogenism in polycystic ovary syndrome. J Clin Endocrinol Metab 91(1):336–340
24. Victor VM, Rocha M et al (2009) Mitochondrial complex I impairment in leukocytes from polycystic ovary syndrome patients with insulin resistance. J Clin Endocrinol Metab 94(9): 3505–3512
25. Victor VM, Rocha M et al (2011) Induction of oxidative stress and human leukocyte/ endothelial cell interactions in polycystic ovary syndrome patients with insulin resistance. J Clin Endocrinol Metab 96(10):3115–3122
26. Schachter M, Raziel A et al (2003) Insulin resistance in patients with polycystic ovary syndrome is associated with elevated plasma homocysteine. Hum Reprod 18(4):721–727
27. Chattopadhayay R, Ganesh A et al (2009) Effect of follicular fluid oxidative stress on meiotic spindle formation in infertile women with polycystic ovarian syndrome. Gynecol Obstet Invest 69(3):197–202
28. Luberda Z (2005) The role of glutathione in mammalian gametes. Reprod Biol 5(1):5–17
29. Eppig JJ, Hosoe M et al (2000) Conditions that affect acquisition of developmental competence by mouse oocytes in vitro: FSH, insulin, glucose and ascorbic acid. Mol Cell Endocrinol 163:109–116
30. Murray AA, Molinek MD et al (2001) Role of ascorbic acid in promoting follicle integrity and survival in intact mouse ovarian follicles in vitro. Reproduction 121:89–96
31. Thomas FH, Leask R et al (2001) Effect of ascorbic acid on health and morphology of bovine preantral follicles during long-term culture. Reproduction 122(3):487–495
32. Behrman HR, Kodaman PH et al (2001) Oxidative stress and the ovary. J Soc Gynecol Investig 8(1 Suppl Proceedings):S40–S42
33. Verit FF, Erel E (2008) Oxidative stress in nonobese women with polycystic ovary syndrome: correlations with endocrine and screening parameters. Gynecol Obstet Invest 65(4):233–239
34. Karanth S, Yu WH et al (2003) Vitamin E stimulates luteinizing hormone-releasing hormone and ascorbic acid release from medial basal hypothalami of adult male rats. Exp Biol Med (Maywood) 228(7):779–785
35. Kuscu NK, Var A (2009) Oxidative stress but not endothelial dysfunction exists in non-obese, young group of patients with polycystic ovary syndrome. Acta Obstet Gynecol Scand 88(5):612–617
36. Knowler WC, Barrett-Connor E et al (2002) Reduction in the incidence of type 2 diabetes with lifestyle intervention or metformin. N Engl J Med 346(6):393–403
37. Pasquali R, Antenucci D et al (1989) Clinical and hormonal characteristics of obese amenorrheic hyperandrogenic women before and after weight loss. J Clin Endocrinol Metab 68(1):173–179
38. Guzick DS, Wing R et al (1994) Endocrine consequences of weight loss in obese, hyperandrogenic, anovulatory women. Fertil Steril 61(4):598–604
39. Clark AM, Thornley B et al (1998) Weight loss in obese infertile women results in improvement in reproductive outcome for all forms of fertility treatment. Hum Reprod 13(6):1502–1505
40. Huber-Buchholz MM, Carey DG et al (1999) Restoration of reproductive potential by lifestyle modification in obese polycystic ovary syndrome: role of insulin sensitivity and luteinizing hormone. J Clin Endocrinol Metab 84(4):1470–1474

41. Teede HJ, Meyer C et al (2010) Endothelial function and insulin resistance in polycystic ovary syndrome: the effects of medical therapy. Fertil Steril 93(1):184–191
42. Korytkowski MT, Mokan M et al (1995) Metabolic effects of oral contraceptives in women with polycystic ovary syndrome. J Clin Endocrinol Metab 80:3327–3334
43. Meyer C, McGrath BP et al (2007) Effects of medical therapy on insulin resistance and the cardiovascular system in polycystic ovary syndrome. Diabetes Care 30:471–478
44. Nestler JE, Jakubowicz DJ (1997) Lean women with polycystic ovary syndrome respond to insulin reduction with decreases in ovarian p450c17 alpha activity and serum androgens. J Clin Endocrinol Metab 82:4075–4079
45. Nestler JE, Jakubowicz DJ et al (1998) Effects of metformin on spontaneous and clomiphene induced ovulation in the polycystic ovary syndrome. N Engl J Med 338:1876–1880
46. Cheang KI, Huszar JM et al (2009) Long-term effect of metformin on metabolic parameters in the polycystic ovary syndrome. Diab Vasc Dis Res. 6(2):110–119
47. Mikkelsen AL, Lindenberg S (2001) Benefit of FSH priming of women with PCOS to the in vitro maturation procedure and the outcome: a randomized prospective study. Reproduction 122(4):587–592
48. Mikkelsen AL, Smith SD et al (1999) In vitro maturation of human oocytes from regularly menstruating women may be successful without follicle stimulating hormone priming. Hum Reprod 14(7):1847
49. Zhao JZ, Zhou W et al (2009) In vitro maturation and fertilization of oocytes from unstimulated ovaries in infertile women with polycystic ovary syndrome. Fertil Steril 91(6):2568–2571

Chapter 11
The Menopause and Oxidative Stress

Lucky H. Sekhon and Ashok Agarwal

Abstract Reproductive aging resulting in menopause is characterized by the permanent cessation of ovarian follicular activity. The signs and symptoms resulting from estrogen withdrawal can significantly disrupt a woman's activities of daily living and sense of well being, while predisposing them to osteoporosis and heart disease. Current medical therapies are targeted at symptomatic relief or alleviating the hormonal deficiency itself to prevent its harmful sequelae. The progressive loss of estrogen and its protective effects, combined with deficient endogenous antioxidant, results in oxidative stress—which is implicated in the pathogenesis of vasomotor disturbances, loss of bone mass and heart disease in menopause. The link between oxidative stress and estrogen deficiency has been demonstrated by numerous studies. Based on this, hormonal replacement therapy, antioxidant supplementation, and lifestyle modification have been investigated for their efficacy and safety in the treatment and prevention of menopause-related symptoms and chronic disease processes.

Keywords Reproductive aging · Menopause · Antioxidant vitamins · Deficient endogenous antioxidant · Loss of estrogen · Herbal extracts · Vitamin C · Vitamin E · Vitamin A · Phytoestrogens · Curcuma longa · Lycopene · Grape polyphenols · Melatonin

L. H. Sekhon
Mount Sinai School of Medicine, OB/GYN, New York, NY, USA
e-mail: lucky.sekhon@mountsinai.org

A. Agarwal (✉)
Lerner College of Medicine, Cleveland Clinic, Center for Reproductive Medicine, Cleveland, OH, USA
e-mail: agarwaa@ccf.org

11.1 Introduction

Reproductive aging involves the permanent cessation of the primary female reproductive functions—the ripening and release of ova and the release of hormones that modulate the endometrial proliferation and shedding. This loss of ovarian follicular activity can be a natural process or a result of an iatrogenic insult such as surgery, chemotherapy, or radiotherapy. In the US, menopause is typically reached at an average of 51 years and affects approximately 40 million women. Premature menopause occurs when a women experiences menopause before 40 years of age, and can result from gynecologic disorders such as polycystic ovaries and endometriosis. In certain women, the changes that can occur during the menopause transition years can significantly disrupt their daily activities and their sense of well being. These may include irregular menses, vasomotor instability (hot flashes and night sweats), genitourinary tissue atrophy, increased stress, breast tenderness, vaginal dryness, forgetfulness, mood changes and sometimes osteoporosis and heart disease. These effects are a direct result of estrogen decline and may affect each woman to a different extent. Currently, established medical treatment targets the altered hormonal milieu of women experiencing menopause. Therapy may also include lifestyle modifications, such as exercise and dietary measures. Free radicals and oxidative stress have been implicated in the pathogenesis of various menopause-related symptoms and complications. As such, vitamins and foods rich in antioxidant compounds might be an effective strategy to alleviate oxidative stress and the associated symptoms and complications affecting women experiencing menopause.

11.2 The Pathophysiology of Hormonal Changes in Menopause

The transition from reproductive to non-reproductive is the result of a major reduction hormone production by the ovaries. This transition is normally not sudden or abrupt, tends to occur over a period of years, and is a natural consequence of aging. The early phase of postmenopause consists of the first 5 years. The late phase of postmenopause is the time from 5 years after the onset of menopause until death [1].

The terminal phase of reproductive aging is preceded by many hormonal changes. These hormonal changes result in age-related fertility decline and a gradual decrease in the number of ovarian follicles and have physical manifestations which often negatively impact the quality of life of perimenopausal and postmenopausal women. The earliest hormonal alteration noted in the perimenopause is the rise in follicle stimulating hormone (FSH) levels, followed several years later by a rise in luteinizing hormone (LH) levels [2, 3]. Inhibin, a dimeric glycoprotein known to suppress FSH, shows a marked decline at and before menopause. Therefore, the decrease in inhibin B is a hormonal change that is an early indicator of reproductive aging [4]. Inhibin B exhibits greater potency than estradiol in exerting negative feedback on pituitary FSH secretion [4]. Thus, increased FSH levels may be related to a decrease in total inhibin in both follicular and luteal phases of the cycle. Along with the changes in the levels of FSH, inhibin and LH, a marked decrease in estrogen concentration occurs in the menopause [5]. This disrupted ovarian function leads to changes in the pattern of menstrual bleeding during the perimenopausal phase.

Estrogen is the major reproductive hormone in the female body and promotes the development of female secondary sex characteristics. In women, naturally occurring estrogen is produced from androgens via enzymatic reactions which yield three major forms: estradiol, estriol, and estrone. In the perimenopausal years, 17β-estradiol, is the most potent and predominant estrogen, whereas the weaker form, estrone, is the predominant estrogen in the postmenopausal phase. The synthesis of estrogen is stimulated by FSH and LH and takes place primarily in developing follicles in the ovaries and the corpus luteum. Estrogen is also produced in small amounts by the liver, adrenal glands, fat cells, and breasts. In postmenopausal women, estrone is formed as a result of the peripheral conversion of androstenedione in both adipose tissue and the liver. Estrogen metabolites have been proven to exert both antioxidant [6, 7] and pro-oxidant effects [7]. Methoxyestrogen is seen to have the most potent antioxidant properties of the various forms of estrogen [7]. Some believe that estrogen's antioxidant properties are derived from the phenolic ring in its structure [5]. Markides et al. [7] proposed that estrogen has antioxidant activity through the inhibition of 8-hydroxylation of guanine bases of DNA. Estrogen metabolites significantly increased the concentrations of 8-hydroxyguanine bases by 54–66 % [7]. The concentration and chemical structure of estrogen metabolites determines whether it will have an antioxidant or pro-oxidant effect. At high concentrations, estrogen metabolites tend to produce antioxidant effects—whereas at lower concentrations, estrogen metabolites are more likely to produce pro-oxidant effects. Estrogen metabolites that possess a catechol structure act in a pro-oxidant manner [7]. In one study, estrogen supplementation led to a decrease of the oxidation of LDL cholesterol in postmenopausal women [8]. According to Pansini et al. [9], supplementing postmenopausal women with estrogen can improve their lipid profile, by increasing HDL levels and decreasing LDL and lipoprotein A levels. However, further studies are needed to assess the direct implications of this finding on the cardiovascular complications often seen in postmenopausal women [10].

11.3 The Role of Oxidative Stress in the Menopause

Oxidative stress, which is defined as an imbalance between oxidants and antioxidants, plays a well-established role in normal aging and has been implicated in the pathogenesis of a number of disease processes, including age-related degenerative processes such as atherosclerotic cardiovascular disease [11], non-alcoholic liver cirrhosis, and various pathologies afflicting the female reproductive system. Various studies have shown that vasomotor disturbances [12], osteoporosis [13] and cardiovascular diseases [14] significantly correlate with the progressive loss of estrogen and its protective effects, combined with deficient antioxidant defense leading to a pronounced redox imbalance.

Vural et al. [15] compared follicular phase levels of serum TNF-α, IL-4, IL-10, and IL-12 in premenopausal women, ages 19–38, to the levels seen in postmenopausal women, ages 37–54. Higher serum concentrations of TNF-α, IL-4, IL-10, and IL-12 were seen in postmenopausal women compared to premenopausal women [15]. Levels of TNF-α and inflammatory cytokines have been established to be elevated in the presence of oxidative stress. Therefore, it can be speculated that oxidative stress is present in increased amounts in postmenopausal women. This study also demonstrated a compensatory relationship between TNF-α and IL-4. Elevated levels of IL-4, with its anti-inflammatory effects, may act to counter the pro-inflammatory state induced by increased TNF-α levels [15].

Signorelli et al. [16] also reported findings that show a high degree of oxidative stress is experienced by postmenopausal women. Blood serum levels assessing for malonaldehyde (MDA), 4-hydroxynenal (4-HNE), oxidized LDL, and glutathione peroxidase (GSH-Px) were compared in two groups of women: fertile women, between the ages of 30–35 and postmenopausal women, between the ages of 45–55. The postmenopausal group demonstrated significantly higher levels of the pro-oxidant biomarkers MDA, 4-HNE, and oxidized LDL, whereas levels of the antioxidant GSH-Px were significantly decreased when compared to premenopausal control subjects.

Estrogen is involved in a number of physiological processes in the tissues of the cardiovascular system. It is known to be protective against cardiovascular disease by way of endothelial and non-endothelial mediated effects, favorable effects on lipoprotein, glucose, and insulin homeostasis, changes in extracellular matrix composition, atherosclerotic plaque destabilization and the facilitation of collateral vessel formation [9]. Postmenopausal estrogen deficiency is associated with higher blood levels of free fatty acids, which contribute to the pathogenesis of the metabolic syndrome and insulin resistance. Menopause complicated by poorly controlled diabetes is linked to an elevated risk of atherosclerosis and cardiovascular disease. The risk of cardiovascular disease is present even in non-diabetic postmenopausal women in the presence of recognized risk factors such as elevated lipid and glucose concentrations in plasma [17]. Atherogenesis is considered to be an inflammatory, fibroproliferative process [18]. The incidence of atherosclerosis is increased in menopause, as the antioxidant influence of estrogen is lost, leading to increased oxidation of LDL cholesterol. Moreau et al. [19, 20] demonstrated elevated levels of plasma oxidized LDL in postmenopausal women compared to premenopausal women. The administration of antioxidant vitamin C was shown to reverse this effect, with the decrease in oxidized LDL concentrations leading to an improvement in parameters of vascular health such as blood flow and vascular conductance [20].

Elevated cholesterol coupled with vascular endothelial injury contributes to the development of atherosclerotic plaques. Angiotensin type I (AT-1) receptor activation is thought to be a predominant source of free radical production in vasculature. In a study conducted by Wassmann et al. [21], treatment of spontaneously hypertensive rats with the AT-I receptor antagonist irbesartan normalized the vascular production of free radicals and reverse endothelial dysfunction. These findings suggest that menopause-induced oxidative stress may be mediated by overexpressed AT-I receptor, resulting in an enhanced vasoconstriction and endothelial dysfunction. Increased breakdown of nitric oxide (NO) may be another mechanism by which oxidative stress contributes to the pathogenesis of cardiovascular disease in postmenopausal women [22]. NO, which is derived from the endothelium, is an important physiological regulator of blood flow and regulates blood pressure by inducing vascular relaxation [23–25]. It also demonstrates anti-aggregative, anti-inflammatory, fibrinolytic, thrombolytic, cardio-protective, and cyto-protective properties [23, 25, 26]. NO acts to suppress smooth muscle proliferation, and exerts an anti-atherogenic influence on the vasculature. NO levels in men and postmenopausal women are found to exist at lower levels than those measured in premenopausal women [27, 28].

Leal et al. [29] implicated oxidative stress in the pathogenesis of menopausal symptoms including hot flashes. Hot flashes are characterized by a generalized, transient increase in metabolic rate which may manifest clinically as sweating, irritability, and panic, as well as cardiovascular alterations which cause an increase in blood flow and heart rate. Repetitive increases in metabolic activity are thought to contribute to the development of oxidative stress, possibly by exhausting the antioxidant capacity to regulate reactive oxygen species production. Postmenopausal women experiencing vasomotor symptoms were shown to have lower plasma antioxidant activity than postmenopausal women of the same age without hot flashes [29].

Postmenopausal osteoporosis is a progressive loss of bone density which results in pathological fracture within 10–20 years of the onset of menopause [13]. However, the reason why the incidence of osteoporosis is higher in postmenopausal women and the mechanism by which osteoporosis occurs is not yet completely understood. Iqbal et al. [30] analyzed various markers and cells present in bone marrow samples from mice to characterize the mechanism of osteoporosis development in postmenopausal women. Results demonstrated that mice deficient in the β subunit of FSH are protected from excessive bone turnover despite experiencing a state of severe estrogen deficiency. Furthermore, these FSH-β deficient mice were found to have significantly lower levels of TNF-α. Thus, TNF-α production may be regarded as being dependent on FSH. Decreased TNF-α appears to render mice resistant to hypogonadal bone loss, suggesting TNF-α may be critical to the action of FSH on bone. Estrogen normally prevents bone loss by way of multiple effects on bone marrow and bone cells which cause decreased osteoclast formation, increased osteoclast apoptosis, and decreased capacity of mature osteoclasts to resorb bone [13]. In estrogen deficiency, TNF-α is most likely produced from macrophages and granulocytes, and induces osteoclast and osteoblast formation leading to increased bone turnover [30].

A study conducted by Vural et al. [13] demonstrated that the plasma cytokines—TNF-α, IL-4, IL-10, and IL-12, and markers of bone turnover-urinary hydroxyproline and calcium were elevated in postmenopausal women compared to premenopausal controls. A weak but significant correlation was found between IL-4 and TNF-α, suggesting that anti-inflammatory cytokines such as IL-4, IL-10, and IL-12 serve to counteract pro-inflammatory TNF-α, helping to balance oxidative stress and osteoclast activity. TNF-α contributes to increased osteoclast formation by direct stimulation of osteoclast precursor proliferation and enhancement of pro-osteoclastogenic activity of stromal cells [13]. The role of pro-inflammatory cytokine TNF-α in bone resorption implicates oxidative stress as a key factor in the age-related decline of bone mass density.

The high FSH level in menopause stimulates osteoclast differentiation and TNF-α production from bone marrow macrophages and granulocytes. This leads to the activation of three mechanistic pathways: an increase in oxidative stress, increased M-CSF levels, and M-CSF receptor expression which increase osteoclast precursors and macrophages inducing the proliferation of activated T lymphocytes, leading to an increase in receptor activator of nuclear factor kappa B ligand (RANK-L) expression, resulting in a further increase in TNF-α production. This cycle of increased TNF-α production results in a greater number of osteoclast precursors, giving rise to the bone resorption characteristic of osteoporosis. This process may be inhibited by various substrates. Selective estrogen receptor modulators (SERMs) can prevent an increase in FSH and interact selectively with either α or β estrogenic receptors to activate protective estrogen-signaling pathways in skeletal tissue. The antioxidant vitamin C can block TNF-α production from macrophages and granulocytes, while suppressing high levels of FSH to halt and reverse increased bone turnover. A recombinant RANK-L antagonist or osteoprotegerin can block the RANK-L expression [30] and bisphosphonates, such as alendronate and risedronate, inhibit resorption and are mainstays in the treatment of osteoporosis [9]. Another prophylactic measure or treatment is the synthetic steroid tibilone, which has been reported to decrease urinary markers of bone resorption [15].

Based on the evidence which shows a strong relationship between oxidative stress and estrogen deficiency, hormone replacement and antioxidant supplementation have been investigated for their efficacy and safety in the treatment and prevention of menopause-related symptoms and complications.

11.4 Medical Management of Menopause

The medical treatment of menopause has been extensively studied. It is difficult to clearly distinguish which compounds may be superior in alleviating OS and menopause-associated symptoms and diseases. Several pharmacotherapeutic agents and compounds have been evaluated for their efficacy in alleviating oxidative stress and menopause-related symptoms and associated disease, with the aim to provide clinicians with evidence-based treatment options.

11.4.1 Hormone Replacement Therapy

Estrogen supplementation has been thoroughly investigated as a treatment for the myriad of symptoms and long-term degenerative effects of menopause. The use of Hormone Replacement Therapy (HRT) to improve the redox status in postmenopausal women has been debated by the clinical and research community, as estrogen can exhibit both antioxidant and pro-oxidant properties. Many studies have attempted to determine the relationship between HRT and oxidative stress, and although the studies' conclusions may differ, more studies favor the use of HRT.

Unfer et al. [5] compared the serum levels of superoxide dismutase (SOD), catalase (CAT), GPx, and thiobarbituric acid reactive substances (TBARS) in premenopausal women with levels in postmenopausal women, both with and without HRT. HRT consisted of differing regimens containing conjugated estrogens, estradiol or estrogen plus progestin. Postmenopausal women without HRT demonstrated significantly lower SOD activity, not related to aging, and similar levels of CAT, GPx, and TBARS activity compared with premenopausal women and postmenopausal women on HRT. Therefore, HRT estrogen supplementation may boost SOD activity, thereby antagonizing oxidative stress. Leal et al. [29] compared 6 postmenopausal women without hot flashes to 12 menopausal women with hot flashes. All subjects were administered transdermal estradiol (17-β E2; 50 µg per day, twice a week) and medroxyprogesterone acetate (MPA) (5 mg per day for the first 12 days of each month). Postmenopausal women with hot flashes has lower baseline total antioxidant status (TAS) and higher baseline levels of lipoperoxides compared with women without hot flashes. After 4 months treatment with HRT, postmenopausal women with and without hot flashes experienced a significant increase in TAS and decrease in lipoperoxides. However, the correlation of vasomotor symptoms with increased oxidative stress was seen to persist, as the subjects with hot flashes continued to display lower TAS and higher lipoperoxide levels even after HRT administration. Therefore, in addition to decreasing oxidative stress in postmenopausal women, HRT is effective in reducing the frequency and severity of hot flushes [29].

Estrogen is hypothesized to increase NO levels, by stimulating NO synthase [31, 32] or through other indirect mechanisms. The antioxidant properties of estrogen are also thought to modulate the levels of NO [27, 33–35]. However, the precise mechanism by which estrogen affects NO levels remains unclear. A study by Cincinelli et al. [36] provided evidence that estrogen modulates NO concentration as higher NO levels were demonstrated during the follicular phase compared to the secretory phase of the menstrual cycle [36].

The effect of estrogen/estrogen-progestin therapy (ET/EPT) on plasma NO was studied in 80 postmenopausal women, including 26 with surgically induced menopause and 54 with physiological menopause were compared with 40 healthy premenopausal women [37]. The group with surgically induced menopause was treated with 4 months of ET and those with physiological menopause were given 4 months of EPT. Transdermal E2 (50 μgm twice weekly) and oral MPA (5 mg daily for 12 days) were used in the treatment. The pre- and post-treatment levels of serum E2, NO, lipid peroxide, and FSH were measured and compared to the controls. The pretreatment NO levels were lower in the postmenopausal women compared with controls, with these levels increasing significantly after hormonal therapy. As a result of treatment, the levels of total cholesterol, LDL cholesterol, triglycerides, and apolipoprotein B levels decreased to the levels seen in the control group. Interestingly, there was no correlation between increased levels of NO and the improvement in lipid profile (especially LDL) in postmenopausal women taking ET/EPT. This finding is in disagreement with the hypothesis that an improved lipid profile may promote the generation of NO. No significant difference in NO levels was observed between the ET and EPT treated groups, suggesting that progesterone does not have a significant action in the regulation of NO levels [37]. Furthermore, some studies have suggested that the addition of progesterone may actually antagonize the beneficial NO-mediated effects of estrogen on blood flow [38, 39].

Kurtay et al. [40] studied the effects of transdermal infusion of estradiol hemi-hydrate (2 mg) and norethisterone acetate (NETA) (0.25 mg) in 80 postmenopausal women. Plasma NO levels were monitored at 1, 3, 6, and finally at 12 months. A significant increase in NO levels was observed in postmenopausal women receiving HRT transdermally over a 12 month period. However, no significant change in serum NO was seen in postmenopausal women that were given oral HRT. Therefore, the route of administration of HRT may have a direct bearing on the mechanism by which supplemental hormones are metabolized by the body and influence the degree to which oxidative stress is counteracted [40].

Many researchers have also assessed the specific effects of progestin as part of HRT. In a study by Rosselli et al. [39], 26 postmenopausal women were randomized into a group that received HRT in the form of a transdermal patch of 17β-estradiol and an oral progestin supplement of 1 mg of NETA and another group which served as a control. The levels of NO were not significantly altered from baseline levels when measured at 6, 12, and 24 month intervals. Therefore, progestin supplementation did not appear to have a favorable effect on the NO levels and the redox status of postmenopausal women [39].

The use of progestin supplementation in HRT to prevent and improve cardiovascular disease in postmenopausal women was further investigated by Imthurn et al. [41]. A subject group of 26 postmenopausal women received orally administered estradiol valerate tablets for 21 continuous days. On days 12 through 21 of the treatment cycle, this treatment was supplemented with one of two chemically distinct progestins: cyproterone acetate (CPA) or MPA. Following day 21, treatment was followed by a 7-day treatment-free interval. Blood samples of the postmenopausal women receiving HRT were collected while the subjects were being treated with estradiol valerate alone and estradiol valerate plus CPA or MPA. After 12 months of treatment with estradiol valerate alone, NO levels were significantly increased. However, when estradiol valerate was supplemented with CPA or MPA, no significant difference in NO levels was seen. Therefore, progestin supplementation may have reversed the cardioprotective effects provided by estrogen in postmenopausal women [41]. The conflicting results of the above studies regarding the interaction of progesterone with the beneficial effects of estrogen on the NO-mediated blood may be attributed to the fact that various studies tested different types of progestin.

Vasodilation is also mediated by the effect of estrogen on the synthesis of prostacyclin and endothelin, blocking calcium channels and interfering with the potassium conductance [18]. Estrogen may oppose atherosclerosis by downregulating inflammatory markers, such as cell adhesion molecules and chemokines. In addition, estrogen inhibits smooth muscle cell proliferation and downregulates angiotensin receptor gene expression. Estrogens may also stabilize atherosclerotic plaques, by reducing the expression of matrix metalloproteinases, and may decrease the thrombogenic potential of ruptured plaques by downregulating the synthesis of plasminogen activator inhibitor-1 [18].

A study by Archer et al. [42] randomized 1,147 postmenopausal subjects to groups who received either 1 mg of estradiol alone or in combination with 0.5, 1, 2, or 3 mg of drospirenone. Drospirenone is a progestin derived from spironolactone with anti-minerocorticoid and anti-androgen actions. The combination treatment group had decreased incidence of endometrial hyperplasia as compared to the group treated with estradiol alone. Furthermore, endometrial thickness remained stable over time in the combination treatment group. The combination regimen had a favorable effect on lipid profile as it reduced the total cholesterol, triglyceride and LDL levels. Due to the anti-aldosterone action of drospirenone, these patients were able to maintain or even lose weight. Urogenital and vasomotor symptoms improved in all treatment groups. Combination treatment was able to achieve an increase in the bone mineral density, thus lowering the risk of osteoporosis. Interestingly, a post-hoc analysis of a subgroup of hypertensive women in this study demonstrated a significant reduction in blood pressure in women receiving drospirenone and estradiol, in combination [43]. This finding may be attributable to the anti-mineralocorticoid action of drospirenone. Drospirenone and estradiol combination treatment was reported to improve the quality of life in postmenopausal women. Overall, combination therapy was considered more effective in treating menopause-related symptoms and complications compared to estrogen monotherapy [43]. Since the use of progestin provides varied results, further studies are required to arrive at a general consensus.

There have been several studies which have failed to find a relationship between HRT and oxidative stress. Maffei et al. [44] randomized 15 postmenopausal women to receive either 2 mg oral micronized 17β-estradiol daily or transdermal estradiol therapy (1.5 mg 17β estradiol gel) The oxidative stress biomarker 8-epi $PGF_{2\alpha}$ were evaluated over 12 months and was not found to be significantly altered in response to treatment. However, the sample size in this study was considerably small and the reliability and significance of 8-epi $PGF_{2\alpha}$, as a biomarker of oxidative stress, is not confirmed. Another form of HRT, tibilone, is a synthetic steroid with combined progesterogenic, weak estrogenic and androgenic properties. In a study by Vural et al., postmenopausal women were treated with oral tibilone daily for 6 months. Treatment failed to demonstrate any modifying effect on the levels of cytokines TNF-α, IL-4, IL-10, and IL-12 in postmenopausal women [15]. Vassalle et al. [45] confirmed the idea that tibolone has no effect on the biochemical parameters of oxidative stress, as 2.5 mg per day for 3 months did not significantly alter the levels of IL-6, C-reactive protein or antioxidant status in both pre- and postmenopausal women. However, treatment was reported to significantly lower diastolic and systolic BP, TNF-α and glucose, and HDL. Despite the fact that HDL was reduced, tibolone may lower the overall cardiovascular risk in postmenopausal women because of a beneficial effect on blood pressure, inflammation, and glycemic control [45].

There are a considerable number of risks and side-effects associated with HRT use, including higher incidence of estrogen-dependent breast, ovarian, and endometrial cancers, and increased risk of thromboembolism, cardiovascular, and cerebrovascular events [22]. There is thought to be a certain time frame which is a window of opportunity in postmenopausal life, during which HRT is beneficial, and outside of which harm may be caused. The timing of HRT is relevant, as longer periods of estrogen deficiency lead to reduced number and activity of estrogen receptors which contributes to more extensive atherosclerotic damage or endothelial dysfunction, resulting in decreased vascular responsiveness and lowered efficacy of HRT. If HRT is given early enough, it may protect postmenopausal women by maintaining their vascular health, improving vascular reactivity to estrogen's effects and delaying the clinical manifestations of artherosclerosis [44].

According to the International Menopause Society (IMS) [46], women who start late HRT may have a transient, slightly increased risk of cardiovascular events [46]. Thus, age after menopause may be considered an important factor in determining the individualized risk–benefit ratio of HRT use. Mares et al. [47] studied the relationship between the risk of heart disease and HRT. This prospective cohort study compared 2,693 women currently taking HRT or stopped HRT 5 years or less with an unexposed group of 2,256 women who had never taken HRT or stopped taking HRT for more than 5 years. After 2 years, no significant increased risk of heart disease was observed in the exposed group as compared with the unexposed group. The authors concluded that the time to menopause is a crucial factor [47].

The current recommendation is that postmenopausal women should use the lowest possible dose of HRT, with treatment being based on clear indications. The long-term data regarding fracture risk and cardiovascular implications is considered insufficient. However, HRT may prevent cardiovascular disease if started in young women at the onset of menopause, with long-term administration. The benefits of HRT are thought to generally outweigh the risks for women under the age of 60 years [46].

11.4.2 Selective Estrogen Receptor Modulators

SERMs are a class of compounds that act on the estrogen receptor, with the possibility to selectively stimulate or inhibit the effects of estrogen in various tissues. Raloxifene was the first SERM to be used to prevent and treat osteoporosis [9]. The compound functions in the breast and uterus as an estrogen antagonist [9, 48]. Raloxifene shares properties similar to those of estrogen, particularly in its capacity to reduce oxidative stress. The antioxidant activity of raloxifene is attributed to the presence of phenolic rings in its structure [9, 49]. The mechanism of action targets NADPH oxidase, an enzyme responsible for generating free radicals [9]. In normal physiologic conditions, NADPH oxidase requires activation of a particular subunit by GTPase rac 1. Raloxifene was shown to downregulate rac1 protein expression in the aortic membrane, further reducing the activity of GTPase rac1. The effects of raloxifene to ultimately decrease NADPH oxidase activity result in a less oxidative stress due to hindered ROS production [50].

Raloxifene was shown to reduce blood pressure and improve endothelial dysfunction in male spontaneously hypertensive rats. Furthermore, treatment was seen to cause a significant increase in SOD levels and the release of NO and upregulation of endothelial NOS in spontaneously hypertensive rats [48]. Raloxifene has also been shown to prevent the accumulation of cholesterol in ovariectomized cholesterol-fed rabbits and inhibit macrophage lipid oxidation [51]. A recent study by Ozbasar et al. [52] studied the effect of daily raloxifene administration in a group of 24 postmenopausal women who were undergoing long-term hemodialysis for the treatment of chronic renal failure. A regimen of 60 mg per day for 3 months lead to significantly lower levels of serum MDA and NO levels, with favorable effects on the lipid profile. The results of these studies illustrate the protective effect of raloxifene on the vascular endothelium.

Oviedo et al. [49] reported that levels of myeloperoxidase and $F_{2\alpha}$-isoprostane, markers of oxidative stress, did not change in a cohort of 30 postmenopausal women treated with raloxifene, at a dose of 60 mg per day for a 6 month period. However, the results of this study should be taken with caution as myeloperoxidase and $F_{2\alpha}$-isoprostane have not yet been proven to be reliable indicators of oxidative stress [49].

Based on the evidence, raloxifene is thought to serve a vasoprotective role by decreasing blood pressure levels and improving endothelial function as well as providing preventing hypogonadal bone loss. These effects are mediated via estrogen-receptor pathways and may result in protection against oxidative stress.

11.5 Exercise

Exercise training is thought to modulate oxidative stress by suppressing the production of free radicals and upregulating antioxidant production, resulting in an augmented antioxidant capacity [53]. Campbell et al. [54] assessed the impact of regular aerobic exercise on the levels of F2-isoprostane, a specific marker of lipid peroxidation and general oxidative stress. After 12 months of intervention, previously sedentary postmenopausal women who exercised exhibited marked gains in aerobic fitness and decreased oxidative stress compared with non-exercising control subjects. Menopause is generally accompanied by an increase in body weight, particularly in the upper body [55]. In an observational clinical study of 90 women, total body fat mass of postmenopausal women was significantly increased by 22 %, compared to premenopausal control subjects. Furthermore, both antioxidant status and hydroperoxide levels were significantly correlated with trunk fat mass [55]. In a study by Mittal et al. [56], postmenopausal women were found to have greater body weight and a higher degree of oxidative stress compared with menstruating and perimenopausal control subjects. There was a highly significant association between weight greater than 60 kg and increased levels of SOD and MDA and decreased CAT. Karolkiewicz et al. [57] reported that an 8-week intervention of moderate intensity physical workout enhanced insulin sensitivity and improved the redox balance in healthy, postmenopausal women.

Exercise training is conservative, cost-effective strategy that may have a beneficial role in the treatment of menopausal symptoms such as hot flashes, sweating, anxiety, and depression [58, 59]. Exercise training may be useful in alleviating the symptoms of menopause, without the potential risks associated with long-term HRT use. Attipoe et al. [60] evaluated the combined effect of HRT and exercise training on oxidative stress. The study included 48 previously sedentary postmenopausal women placed into two groups: 21 women using HRT and 27 women not using HRT. Pre-exercise training and post-exercise training levels of plasma TBARS, a sensitive biomarker of lipid peroxidation and oxidative stress, were measured to assess exercise intensity. The results demonstrated a significant decrease in the plasma TBARS levels in both groups; however, no significant difference existed between the two groups. The authors concluded that a 24-week aerobic exercise training regimen significantly decreased oxidative stress in postmenopausal women regardless of HRT use [60]. However, in this study the HRT administration was not standardized, dietary intake of antioxidants was not strictly assessed in the present study, and the independent effects of HRT and exercise on oxidative stress were not assessed.

11.6 Dietary Factors and Antioxidant Supplementation

As oxidative stress has been implicated in the pathophysiology of various menopause-associated disorders, supplementing postmenopausal women with substances with antioxidant properties may serve as a useful adjunct to enhance the beneficial effect of pharmacological treatments often prescribed to postmenopausal patients. Furthermore, postmenopausal women predisposed to developing estrogen-dependent cancers based on either personal or family history, and women who suffer harsh side effects of HRT may instead benefit from dietary changes. Supplementing the diet of postmenopausal women might serve to prevent antioxidant deficiency, preserving the health of women who are exposed to high levels of oxidative stress due to either genetic factors, lifestyle elements such as poor diet, smoking, excessive alcohol intake, and psychological stress.

11.6.1 Vitamin C and Vitamin E

Vitamins C (ascorbic acid) and E (α-tocopherol) are well-known antioxidants that can be obtained through one's diet. They are thought to counteract oxidative stress through their ability to scavenge free radicals, and this effect can be harnessed to prevent and reverse the symptoms and disorders associated with age-related estrogen decline. Vitamins C and E are thought to protect against and alleviate the damaging effects of oxidative stress on the cardiovascular system of postmenopausal women. In a study conducted by Naziroglu et al. [17], 40 postmenopausal women were studied in comparison to 20 postmenopausal women with type 2 diabetes. Diabetic postmenopausal women had increased plasma and RBC lipid peroxide levels and decreased activity of key antioxidants, such a GSH-Px. Six weeks supplementation of vitamins C and E plus HRT resulted in significant decreases in levels of MDA, LDL-cholesterol, total cholesterol, and triglyceride levels in both diabetic and non-diabetic postmenopausal women. Furthermore, treatment improved fasting glucose levels. Therefore, vitamin C and E might help in lowering the risk of cardiovascular disease (with or without diabetes) in postmenopausal women by inhibiting the biosynthesis of cholesterol and oxidation of LDL-cholesterol as well as by improving glycemic balance and lipid profiles [17].

Kushi et al. [61] studied 34,486 postmenopausal women to assess the effect of vitamin E on the risk of acquiring cardiovascular diseases. After 7 years, 242 of these women died of coronary heart disease. Kushi et al. [61] reported an inverse relationship between vitamin E consumption and cardiovascular mortality and morbidity. Therefore, vitamin E obtained through dietary intake may have a significant antioxidant effect which may be helpful in decreasing cardiovascular risk [61].

Moreau et al. [62] assessed the effect of ascorbic acid on large elastic arteries in postmenopausal women. The compliance of large arteries in the cardiothoracic region decreases with age and has an important role in the increased prevalence of cardiovascular disease in postmenopausal women. The study demonstrated the ability of ascorbic acid to selectively improve large elastic artery compliance, increasing vascular conductance and blood flow in postmenopausal women, suggesting that oxidative stress might contribute to the reduced large elastic artery compliance in sedentary, estrogen-deficient postmenopausal women [62].

Furthermore, Moreau et al. [20] analyzed the relationship of oxidative stress with lower limb vasoconstriction in estrogen-deficient postmenopausal women. It should be noted that this study was limited by a small sample size as it compared a group of only 20 postmenopausal women with 9 premenopausal women. The subjects were administered an oral pharmacological dose of ascorbic acid followed by a drip infusion of ascorbic acid and saline. Lower limb vascular conductance increased by 15 % in postmenopausal women while no effect was seen on the lower limb vascular conductance of premenopausal women [20]. Vitamin C is thought to improve vascular function through its activation of the endothelial L-arginine-NO pathway. In a study by McSorley et al. [63], a 1.5 g dose of vitamin C was sufficient to induce relaxation of vascular smooth muscle via release of NO, resulting in improved vascular function [63].

Conversely, some studies failed to yield results which validate the adjuvant use of antioxidants with HRT to prevent postmenopausal women from acquiring coronary atherosclerosis. A large randomized, controlled, double-blind clinical trial evaluating the effects of HRT and antioxidant vitamin supplementation on coronary atherosclerosis in 423 postmenopausal women having baseline coronary stenosis at angiogram, reported both fatal and non-fatal myocardial infarctions during the first 2 years of treatment in patients with cardiovascular disease [64].

Low intake of ascorbic acid has been linked to increased rates of bone loss via enhanced osteoblast and osteoclast function, which result in accelerated bone turnover. This property of vitamin C has prompted investigation of its potential role in the prevention and treatment of osteoporosis in postmenopausal women. According to Iqbal et al. [30], ascorbic acid may prevent FSH-induced hypogonadal bone loss by modulating the destructive action of TNF-α, limiting its stimulatory effects on osteoclast formation. The efficacy and safety of vitamin C and E in the prevention and treatment of postmenopausal cardiovascular disease and osteoporosis should be further investigated in large-scale, double-blinded randomized, controlled trials.

11.6.2 Phytoestrogens

Phytoestrogens are weakly estrogenic compounds contained in soybeans. They are derived from the diet in the form of soymilk, soy protein, and beverages. Dietary phytoestrogen are also known as isoflavones, a broad group of polyphenolic compounds that are distributed widely among foods of plant origin. Isoflavones may be considered as natural SERMs [60] due to their structural similarity with 17β-estradiol which allows binding to both types of estrogen receptors: Erα and Erβ [65].

Isoflavones have been thought to act protectively against cardiovascular disease, osteoporosis, and cancers of the breast and prostate through their prevention of LDL oxidation and inhibition of DNA damage. Phytoestrogens may decrease the risk of cardiovascular disease by lowering the levels of oxidized LDL and decreasing the frequency of hot flushes in postmenopausal women [12]. Furthermore, phytoestrogens have been shown to exhibit defensive immunoprotective properties, such as their role in B cell stimulation and in the inhibition of oxidative damage of DNA in postmenopausal women [66].

Engelman et al. [67] evaluated the effect of isoflavone treatment in a 55 postmenopausal women. The subjects were administered varying proportions of soy proteins and isoflavones. After 6 weeks of supplementation, neither phytate nor isoflavone demonstrated any effect on redox status. Hence, additional studies employing higher doses of soyflavones in a greater sample size should be conducted to arrive at a conclusion [67].

Another study reported that the consumption of soy milk and supplemental isoflavones in 52 postmenopausal women led to decreased plasma levels of 8-hydroxydeoxyguanosine (8-OHdg) and 8-isoprostane [66]. Hallund et al. [65] verified the benefits of phytoestrogens in postmenopausal women by examining the effects of soy cereal bar consumption for an 8 week period. Specific markers of cardiovascular health, including plasma nitrate concentrations, the nitrate: endothelin-1 ratio, and the amount of nitroglycerine-mediated endothelium-independent vasodilatation, were found to be significantly increased in postmenopausal women who consumed soy cereal bars in comparison to the control group that received a placebo. Flow-mediated endothelium-dependent vasodilation was not affected [65]. Isoflavone supplementation was reported to be beneficial, in conjunction with regular exercise, in regulating weight gain, lipid profiles, and oxidative stress in the ovariectomized rat model [68]. After 12 weeks of intervention, isoflavone treatment, both alone and with exercise, led to a significant decrease in total cholesterol, triglycerides and LDL-cholesterol compared to ovariectomized control subjects [68].

A recent study by Beavers et al. [69] conducted a single-blind, randomized, controlled trial that found no significant alteration in markers of inflammation or oxidative stress in 16 postmenopausal women who consumed soymilk 3 times a day for 4 weeks, compared with 15 postmenopausal control subjects that consumed reduced fat dairy. The duration and dosage of isoflavone treatment in this study was comparable to that studied in the literature. However, the indices used to measure oxidative stress were not, which may explain the contradictory findings. The results may have been confounded by lifestyle factors that influence the expression of plasma markers of oxidative stress. It is possible that soy supplementation is efficacious only in those having significantly elevated biomarkers of oxidative stress. However, the patients in this study were not selected according to baseline oxidative stress status [69].

The role of isoflavones in reducing the risk of cardiovascular disease through oxidative stress-induced pathways must be further assessed. Studies have suggested that phytoestrogens may have a protective effect against osteoporosis through their intrinsic growth-promoting activity which stimulates osteoblasts. This action of phytoestrogens could be a new therapeutic approach toward prevention and treatment of osteoporosis. More research is required to arrive at a consensus on the use of isoflavones in the therapeutics of menopause.

11.6.3 Curcuma longa

C. longa is an herbal extract with phenolic antioxidant properties. The compound has powerful free radical-neutralizing properties and was shown to decrease the levels of oxidized HDL and LDL in women (40–90 years) without inducing hepatic or renal problems [14].

Apolipoprotein A (Apo A) is involved with the metabolism of HDL-cholesterol and is a component of the body's anti-atherogenic defense. Conversely, Apo B has pro-atherogenic effects as it induces the formation of LDL cholesterol. In a study analyzing apolipoproteins in relation to postmenopausal subjects, the ratio of apo A and B was significantly altered after treatment with curcuma longa and it was concluded that *C. longa* may normalize the apo B/apo A ratio [70]. *C. longa* extract has also been reported to decrease abnormally high levels of plasma fibrinogen to normal values [71].

11.6.4 Lycopene

LycoRed, a form of lycopene, is thought to decrease the risk of cardiovascular diseases in postmenopausal women. In healthy women ranging from 31 to 75 years, circulating lycopene levels were seen to exhibit an inverse relationship with arterial stiffness, as measured by brachial-ankle pulse wave velocity [72]. This effect may be mediated by lycopene's capacity to reduce the oxidative modification of LDL. Misra et al. [73] reported that supplementation led to a decrease in serum HDL, LDL, MDA and an increase in GSH compared to the pretreatment serum levels. The decrease in levels of MDA and LDL (a risk factor for atherosclerosis) and the increase in protective antioxidant glutathione suggest an overall decline in oxidative stress as a result of LycoRed administration [73].

11.6.5 Grape Polyphenols

Grape polyphenols have also been considered to be used as an alternative treatment to reduce oxidative stress. In both premenopausal and postmenopausal women, grape polyphenols was reported to reduce indices of oxidative stress such as plasma F_2-isoprostane and plasma TNF-α, as well as resulting in reduced triglycerides, LDL and apo-B levels [74].

11.6.6 Acanthopanax senticosus

A. *senticosus* is a common Asian herb also referred to as "Siberian Ginseng" or "Eleutherococcus senticosus". It has been shown to have antioxidant effects in rats [75]. Lee et al. [76] studied the effects of A. *senticosus* supplementation on serum lipid profiles, biomarkers of oxidative stress, and lymphocyte DNA damage in postmenopausal women. A significant decrease in the concentration of LDL, LDL/HDL ratio, serum MDA concentration, serum protein carbonyl levels, and lymphocyte DNA damage was observed. Additionally, no side effects were reported [76].

11.6.7 Vitamin A

Behr et al. [77] conducted a recent, inaugural study of low-dose retinol palmitate, a vitamin A supplement, in the treatment of menopause symptoms and associated oxidative stress. The subjects of this study, Wistar rats, were bilaterally ovariectomized and subsequently exhibited characteristics of menopause, including increases in body weight, uterine atrophy, altered lipid profile, increased blood peroxidase activity and decreased plasma antioxidant status. Low-dose supplementation with vitamin A was shown to reverse some of these effects, by restoring the levels of enzymatic and non-enzymatic antioxidant defense and decreasing the degree of oxidative damage incurred by proteins [77]. The results of this study are compelling and should promote further research to elucidate whether vitamin A is safe and effective in the treatment of menopause and associated oxidative stress. Safety is a concern as high doses of vitamin A may have embryotoxic and teratogenic effects [78].

11.6.8 Klamin

Klamin is an algae extract that is rich in potent algal antioxidant, Aphanizomenon flos-aquae (AFA) phycocyanin, and natural neuromodulators, such as phenylethylamine and selective monoamine oxidase inhibitors. Klamin has been proposed as an alternative treatment for psychological, somatic, and vasomotor symptoms related to menopause. Scoglio et al. [79] investigated the effect of Klamath algae on the general and psychological health of 21 postmenopausal women that did not take HRT. Treatment led to significantly reduced MDA levels, indicating decreased plasma lipid peroxidation. An increase in antioxidants such as carotenoids, tocopherols, and retinols was observed. Furthermore, treatment was reported to improve the overall and psychological well-being of subjects, as indicated objectively by a decreased average Green Scale score. A favorable side-effect profile was suggested by the fact that Klamin did not exhibit any steroid like effects on hormonal parameters. Therefore, Klamin may be having a role as a complementary treatment or as a plausible, natural alternative for patients who wish to avoid hormonal therapy [79].

11.6.9 Melatonin

Melatonin is secreted by the pineal gland and exhibits anti-oxidant properties. Melatonin is thought to arrest lipid peroxidation and protein oxidation in a dose-dependent manner. Up until now, the effect of melatonin on oxidative stress and symptoms in the postmenopausal state has only been studied in the ovariectomized rat. Baeza et al. [80] reported that, as part of a combination with growth hormone, estrogens, and phytoestrogens, melatonin supplementation led to a significant reduction in oxidative stress, represented by a decrease in MDA levels and the degree of glutathione depletion [80]. Melatonin was shown to influence oxidative stress in the blood and brain of ovariectomized rats [81]. In comparison with the non-treated, ovariectomized control group, melatonin supplementation for 30 days decreased lipoperoxide levels, while increasing erythrocyte glutathione, vitamin A, C, and E levels, and the concentration of the 2B subunit of the hippocampal N-methyl-D-aspartate receptor (NMDA) [81]. Therefore, by boosting antioxidant defense and upregulating the NMDA receptor, melatonin may prevent the excess oxidative stress seen in the postmenopausal state. The results of these preliminary animal studies warrant further investigation into the efficacy of melatonin supplementation in the treatment of postmenopausal women.

11.7 Conclusion

Estrogen is an established antioxidant; therefore, in menopause, estrogen deficiency leads to the development of oxidative stress. Various studies have demonstrated increased oxidative stress marker levels and decreased antioxidant levels in postmenopausal women. Oxidative stress has been linked to the development of osteoporosis and increased cardiovascular risk in these women. HRT decreases oxidative stress in women with menopause by increasing the TAS and preventing the breakdown of NO. HRT can be effective in reducing the frequency and severity of hot flushes and may be protective against osteoporosis and cardiovascular complications during menopause. HRT may also delay the clinical manifestations of artherosclerosis.

In addition to HRT, various dietary changes, exercise training, and SERMs are potential therapeutic alternatives which have been assessed in postmenopausal women for their potential role in alleviating the oxidative stress underlying the symptoms and complications of menopause. The use of antioxidant vitamins and herbal extracts may prove to be beneficial in postmenopausal women by normalizing the redox status of the cell. Further investigations are required to study their efficacy and safety before they can be implemented for clinical use in postmenopausal women. Wide varieties of treatment options are now available to prevent and reverse the effects of oxidative stress associated with reproductive aging in postmenopausal women, and treatment should be tailored according to personal circumstances with periodic reviews.

References

1. Arredondo FLJ (2007) Menopause. In: Falcone THW (ed) Clinical reproductive medicine and surgery. Mosby Elsevier, Philadelphia, pp 353–370
2. Lee SJ, Lenton EA, Sexton L, Cooke ID (1988) The effect of age on the cyclical patterns of plasma LH, FSH, oestradiol and progesterone in women with regular menstrual cycles. Hum Reprod 3(7):851–855
3. Lenton EA, Sexton L, Lee S, Cooke ID (1988) Progressive changes in LH and FSH and LH: FSH ratio in women throughout reproductive life. Maturitas 10(1):35–43
4. Welt CK, McNicholl DJ, Taylor AE, Hall JE (1999) Female reproductive aging is marked by decreased secretion of dimeric inhibin. J Clin Endocrinol Metab 84(1):105–111
5. Unfer TC, Conterato GM, da Silva JC, Duarte MM, Emanuelli T (2006) Influence of hormone replacement therapy on blood antioxidant enzymes in menopausal women. Clin Chim Acta 369(1):73–77
6. Ayres S, Tang M, Subbiah MT (1996) Estradiol-17beta as an antioxidant: some distinct features when compared with common fat-soluble antioxidants. J Lab Clin Med 128(4):367–375
7. Markides CS, Roy D, Liehr JG (1998) Concentration dependence of prooxidant and antioxidant properties of catecholestrogens. Arch Biochem Biophys 360(1):105–112
8. Mendelsohn ME, Karas RH (1999) The protective effects of estrogen on the cardiovascular system. N Engl J Med 340(23):1801–1811
9. Pansini F, Mollica G, Bergamini CM (2005) Management of the menopausal disturbances and oxidative stress. Curr Pharm Des 11(16):2063–2073
10. Liehr JG (1996) Antioxidant and prooxidant properties of estrogens. J Lab Clin Med 128(4):344–345
11. Becker BN, Himmelfarb J, Henrich WL, Hakim RM (1997) Reassessing the cardiac risk profile in chronic hemodialysis patients: a hypothesis on the role of oxidant stress and other non-traditional cardiac risk factors. J Am Soc Nephrol 8(3):475–486
12. Witteman JC, Grobbee DE, Kok FJ, Hofman A, Valkenburg HA (1989) Increased risk of artherosclerosis in women after the menopause. BMJ 298:642–644
13. Vural P, Akgul C, Canbaz M (2006) Effects of hormone replacement therapy on plasma pro-inflammatory and anti-inflammatory cytokines and some bone turnover markers in postmenopausal women. Pharmacol Res 54(4):298–302
14. Bittner V (2009) Menopause, age, and cardiovascular risk: a complex relationship. J Am Coll Cardiol 54(25):2374–2375
15. Vural P, Canbaz M, Akgul C (2006) Effects of menopause and postmenopausal tibolone treatment on plasma TNFalpha, IL-4, IL-10, IL-12 cytokine pattern and some bone turnover markers. Pharmacol Res 53(4):367–371
16. Signorelli SS, Neri S, Sciacchitano S, Pino LD, Costa MP, Marchese G et al (2006) Behaviour of some indicators of oxidative stress in postmenopausal and fertile women. Maturitas 53(1):77–82

17. Naziroglu M, Simsek M, Simsek H, Aydilek N, Ozcan Z, Atilgan R (2004) The effects of hormone replacement therapy combined with vitamins C and E on antioxidants levels and lipid profiles in postmenopausal women with type 2 diabetes. Clin Chim Acta 344(1–2):63–71
18. Mueck AO, Seeger H (2004) Estrogens acting as cardiovascular agents: direct vascular actions. Curr Med Chem Cardiovasc Hematol Agents 2(1):35–42
19. Moreau KL, DePaulis AR, Gavin KM, Seals DR (2007) Oxidative stress contributes to chronic leg vasoconstriction in estrogen-deficient postmenopausal women. J Appl Physiol 102(3):890–895
20. Moreau KL, Gavin KM, Plum AE, Seals DR (2005) Ascorbic acid selectively improves large elastic artery compliance in postmenopausal women. Hypertension 45(6):1107–1112
21. Wassmann S, Baumer AT, Strehlow K, van Eickels M, Grohe C, Ahlbory K et al (2001) Endothelial dysfunction and oxidative stress during estrogen deficiency in spontaneously hypertensive rats. Circulation 103(3):435–441
22. Arnal JF, Scarabin PY, Tremollieres F, Laurell H, Gourdy P (2007) Estrogens in vascular biology and disease: where do we stand today? Curr Opin Lipidol 18(5):554–560
23. Moncada S, Palmer RM, Higgs EA (1991) Nitric oxide: physiology, pathophysiology, and pharmacology. Pharmacol Rev 43(2):109–142
24. Gryglewski RJ (1994) Prostacyclin and nitric oxide. Acta Haematol Pol 25(2 Suppl 2):75–81
25. Cohen RA (1995) The role of nitric oxide and other endothelium-derived vasoactive substances in vascular disease. Prog Cardiovasc Dis 38(2):105–128
26. Tinker AC, Wallace AV (2006) Selective inhibitors of inducible nitric oxide synthase: potential agents for the treatment of inflammatory diseases? Curr Top Med Chem 6(2):77–92
27. Bednarek-Tupikowska G, Tupikowski K, Bidzinska B, Bohdanowicz-Pawlak A, Antonowicz-Juchniewicz J, Kosowska B et al (2004) Serum lipid peroxides and total antioxidant status in postmenopausal women on hormone replacement therapy. Gynecol Endocrinol 19(2):57–63
28. Bednarek-Tupikowska G, Tworowska U, Jedrychowska I, Radomska B, Tupikowski K, Bidzinska-Speichert B et al (2006) Effects of oestradiol and oestroprogestin on erythrocyte antioxidative enzyme system activity in postmenopausal women. Clin Endocrinol (Oxf) 64(4):463–468
29. Leal M, Diaz J, Serrano E, Abellan J, Carbonell LF (2000) Hormone replacement therapy for oxidative stress in postmenopausal women with hot flushes. Obstet Gynecol 95(6 Pt 1): 804–809
30. Iqbal J, Sun L, Kumar TR, Blair HC, Zaidi M (2006) Follicle-stimulating hormone stimulates TNF production from immune cells to enhance osteoblast and osteoclast formation. Proc Nat Acad Sci USA 103(40):14925–14930
31. Weiner CP, Lizasoain I, Baylis SA, Knowles RG, Charles IG, Moncada S (1994) Induction of calcium-dependent nitric oxide synthases by sex hormones. Proc Nat Acad Sci USA 91(11):5212–5216
32. Hishikawa K, Nakaki T, Marumo T, Suzuki H, Kato R, Saruta T (1995) Up-regulation of nitric oxide synthase by estradiol in human aortic endothelial cells. FEBS Lett 360(3):291–293
33. Sugioka K, Shimosegawa Y, Nakano M (1987) Estrogens as natural antioxidants of membrane phospholipid peroxidation. FEBS Lett 210(1):37–39
34. Liao JK, Shin WS, Lee WY, Clark SL (1995) Oxidized low-density lipoprotein decreases the expression of endothelial nitric oxide synthase. J Biol Chem 270(1):319–324
35. Wang MY, Liehr JG (1995) Induction by estrogens of lipid peroxidation and lipid peroxide-derived malonaldehyde-DNA adducts in male Syrian hamsters: role of lipid peroxidation in estrogen-induced kidney carcinogenesis. Carcinogenesis 16(8):1941–1945
36. Cicinelli E, Ignarro LJ, Lograno M, Galantino P, Balzano G, Schonauer LM (1996) Circulating levels of nitric oxide in fertile women in relation to the menstrual cycle. Fertil Steril 66(6):1036–1038
37. Bednarek-Tupikowska G, Tworowska-Bardzinska U, Tupikowski K (2008) Effects of estrogen and estrogen-progesteron on serum nitric oxide metabolite concentrations in post-menopausal women. J Endocrinol Invest 31(10):877–881

38. Jokela H, Dastidar P, Rontu R, Salomaki A, Teisala K, Lehtimaki T et al (2003) Effects of long-term estrogen replacement therapy versus combined hormone replacement therapy on nitric oxide-dependent vasomotor function. J Clin Endocrinol Metab 88(9):4348–4354
39. Rosselli M, Imthurn B, Keller PJ, Jackson EK, Dubey RK (1995) Circulating nitric oxide (nitrite/nitrate) levels in postmenopausal women substituted with 17 beta-estradiol and norethisterone acetate. A two-year follow-up study. Hypertension 25(4 Pt 2):848–853
40. Kurtay G, Ozmen B, Erguder I (2006) A comparison of effects of sequential transdermal administration versus oral administration of estradiol plus norethisterone acetate on serum NO levels in postmenopausal women. Maturitas 53(1):32–38
41. Imthurn B, Rosselli M, Jaeger AW, Keller PJ, Dubey RK (1997) Differential effects of hormone-replacement therapy on endogenous nitric oxide (nitrite/nitrate) levels in postmenopausal women substituted with 17 beta-estradiol valerate and cyproterone acetate or medroxyprogesterone acetate. J Clin Endocrinol Metab 82(2):388–394
42. Archer DF (2007) Drospirenone and estradiol: a new option for the postmenopausal woman. Climacteric 10(Suppl 1):3–10
43. Archer DF, Thorneycroft IH, Foegh M, Hanes V, Glant MD, Bitterman P et al (2005) Long-term safety of drospirenone-estradiol for hormone therapy: a randomized, double-blind, multicenter trial. Menopause 12(6):716–727
44. Maffei S, Mercuri A, Prontera C, Zucchelli GC, Vassalle C (2006) Vasoactive biomarkers and oxidative stress in healthy recently postmenopausal women treated with hormone replacement therapy. Climacteric 9(6):452–458
45. Vassalle C, Cicinelli E, Lello S, Mercuri A, Battaglia D, Maffei S (2011) Effects of menopause and tibolone on different cardiovascular biomarkers in healthy women. Gynecol Endocrinol 27(3):163–169
46. Pines A, Sturdee DW, Birkhauser MH, Schneider HP, Gambacciani M, Panay N (2007) IMS updated recommendations on postmenopausal hormone therapy. Climacteric 10(3):181–194
47. Mares P, Chevallier T, Micheletti MC, Daures JP, Postruznik D, De Reilhac P (2008) Coronary heart disease and HRT in France: MISSION study prospective phase results. Gynecol Endocrinol 24(12):696–700
48. Wassmann S, Laufs U, Stamenkovic D, Linz W, Stasch JP, Ahlbory K et al (2002) Raloxifene improves endothelial dysfunction in hypertension by reduced oxidative stress and enhanced nitric oxide production. Circulation 105(17):2083–2091
49. Oviedo PJ, Hermenegildo C, Tarin JJ, Cano A (2005) Therapeutic dosages of raloxifene do not modify myeloperoxidase and F2alpha-isoprostane levels in postmenopausal women. Fertil Steril 84(6):1789–1792
50. Miyazaki H, Oh-ishi S, Ookawara T, Kizaki T, Toshinai K, Ha S et al (2001) Strenuous endurance training in humans reduces oxidative stress following exhausting exercise. Eur J Appl Physiol 84(1–2):1–6
51. Bjarnason NH, Haarbo J, Byrjalsen I, Kauffman RF, Christiansen C (1997) Raloxifene inhibits aortic accumulation of cholesterol in ovariectomized, cholesterol-fed rabbits. Circulation 96:1964–1969
52. Ozbasar D, Toros U, Ozkaya O, Sezik M, Uzun H, Genc H, Kaya H (2010) Raloxifene decreases serum malondialdehyde and nitric oxide levels in postmenopausal women with end-stage renal disease under chronic hemodiálisis therapy. J Obstet Gynaecol Res 36(1):133–137
53. McArdle A, Jackson MJ (2000) Exercise, oxidative stress and ageing. J Anat 197(Pt 4):539–541
54. Campbell PT, Gross MD, Potter JD, Schmitz KH, Duggan C, McTiernan A, Ulrich CM (2010) Effect of exercise on oxidative stress: a 12-month randomized, controlled trial. Med Sci Sports Exerc 42(8):1448–1453
55. Pansini F, Cervellati C, Guariento A, Stacchini MA, Castaldini C, Bernardi A, Pascale G, Bonaccorsi G, Patella A, Bagni B (2008) Oxidative stress, body fat composition, and endocrine status in pre- and postmenopausal women. Menopause 15(1):112–118

56. Mittal PC, Kant R (2009) Correlation of increased oxidative stress to body weight in disease-free postmenopausal women. Clin Biochem 42(10–11):1007–1011
57. Karolkiewicz J, Michalak E, Pospieszna B, Deskur-Smielecka E, Nowak A, Pilacyznska-Szczesniak L (2009) Response of oxidative stress markers and antioxidant parameters to an 8-week aerobic physical activity program in healthy, postmenopausal women. Arch Gerontol Geriatr 49(1):e67–e71
58. Mirzaiinjmabadi K, Anderson D, Barnes M (2006) The relationship between exercise, body mass index and menopausal symptoms in midlife Australian women. Int J Nurs Pract 12(1):28–34
59. Villaverde-Gutierrez C, Araujo E, Cruz F, Roa JM, Barbosa W, Ruiz-Villaverde G (2006) Quality of life of rural menopausal women in response to a customized exercise programme. J Adv Nurs 54(1):11–19
60. Attipoe S, Park JY, Fenty N, Phares D, Brown M (2008) Oxidative stress levels are reduced in postmenopausal women with exercise training regardless of hormone replacement therapy status. J Women Aging 20(1–2):31–45
61. Kushi LH, Folsom AR, Prineas RJ, Mink PJ, Wu Y, Bostick RM (1996) Dietary antioxidant vitamins and death from coronary heart disease in postmenopausal women. N Engl J Med 334(18):1156–1162
62. Subbiah MT (2002) Estrogen replacement therapy and cardioprotection: mechanisms and controversies. Braz J Med Biol Res 35(3):271–276
63. McSorley PT, Young IS, Bell PM, Fee JP, McCance DR (2003) Vitamin C improves endothelial function in healthy estrogen-deficient postmenopausal women. Climacteric 6(3):238–247
64. Waters DD, Alderman EL, Hsia J, Howard BV, Cobb FR, Rogers WJ et al (2002) Effects of hormone replacement therapy and antioxidant vitamin supplements on coronary atherosclerosis in postmenopausal women: a randomized controlled trial. JAMA 288(19):2432–2440
65. Hallund J, Bugel S, Tholstrup T, Ferrari M, Talbot D, Hall WL et al (2006) Soya isoflavone-enriched cereal bars affect markers of endothelial function in postmenopausal women. Br J Nutr 95(6):1120–1126
66. Ryan-Borchers TA, Park JS, Chew BP, McGuire MK, Fournier LR, Beerman KA (2006) Soy isoflavones modulate immune function in healthy postmenopausal women. Am J Clin Nutr 83(5):1118–1125
67. Engelman HM, Alekel DL, Hanson LN, Kanthasamy AG, Reddy MB (2005) Blood lipid and oxidative stress responses to soy protein with isoflavones and phytic acid in postmenopausal women. Am J Clin Nutr 81(3):590–596
68. Beavers KM, Serra MC, Beavers DP, Cooke MB, Willoughby DS (2009) Soymilk supplementation does not alter plasma markers of inflammation and oxidative stress in postmenopausal women. Nutr Res 29(9):616–622
69. Oh HY, Lim S, Lee JM, Kim DY, Ann ES, Yoon S (2007) A combination of soy isoflavone supplementation and exercise improves lipid profiles and protects antioxidant defense-systems against exercise-induced oxidative stress in ovariectomized rats. BioFactors 29(4):175–185
70. Ramirez-Bosca A, Soler A, Carrion MA et al (2000) An hydroalcoholic extract of *Curcuma longa* lowers the apo B/apo A ratio. Implications for artherogenesis prevention. Mech Ageing Dev 119(1):41–47
71. Ramirez Bosca A, Soler A, Carrion-Gutierrez MA, Pamies Mira D, Pardo Zapata J, Diaz-Alperi J et al (2000) An hydroalcoholic extract of *Curcuma longa* lowers the abnormally high values of human-plasma fibrinogen. Mech Ageing Dev 114(3):207–210
72. Kim OY, Yoe HY, Kim HJ, Park JY, Kim JY, Lee SH, Lee JH, Lee KP, Jang Y, Lee JH (2010) Independent inverse relationship between serum lycopene concentration and arterial stiffness. Artherosclerosis 208(2):581–586

73. Misra R, Mangi S, Joshi S, Mittal S, Gupta SK, Pandey RM (2006) LycoRed as an alternative to hormone replacement therapy in lowering serum lipids and oxidative stress markers: a randomized controlled clinical trial. J Obstet Gynaecol Res 32(3):299–304
74. Zern TL, Wood RJ, Greene C, West KL, Liu Y, Aggarwal D et al (2005) Grape polyphenols exert a cardioprotective effect in pre- and postmenopausal women by lowering plasma lipids and reducing oxidative stress. J Nutr 135(8):1911–1917
75. Lee S, Son D, Ryu J, Lee YS, Jung SH, Kang J et al (2004) Anti-oxidant activities of Acanthopanax senticosus stems and their lignan components. Arch Pharm Res. 27(1): 106–110
76. Lee YJ, Chung HY, Kwak HK, Yoon S (2008) The effects of *A. senticosus* supplementation on serum lipid profiles, biomarkers of oxidative stress, and lymphocyte DNA damage in postmenopausal women. Biochem Biophys Res Commun 375(1):44–48
77. Behr GA, Schnorr CE, Moreira JC (2011) Increased blood oxidative stress in experimental menopause rat model: the effects of vitamin A low-dose supplementation upon antioxidant status in bilateral ovariectomized rats. Fundam Clin Pharmacol [Epub ahead of print]
78. Meyers DG, Maloney PA, Weeks D (1996) Safety of antioxidant vitamins. Arch Intern Med 156:925–935
79. Scoglio S, Benedetti S, Canino C, Santagni S, Rattighieri E, Chierchia E, Canestrari F, Genazzani AD (2009) Effect of a 2-month treatment with klamin, a klamath algae extract, on the general well-being, antioxidant profile and oxidative status of postmenopausal women. Gynecol Endocrinol 25(4):235–240
80. Baeza I, Fdez-Tresguerres J, Ariznavarreta C, De la Fuente M (2010) Effects of growth hormone, melatonin, oestrogens and phytoestrogens on the oxidized glutathione (GSSG)/reduced glutathione (GSH) ratio and lípid peroxidation in aged ovariectomized rats. Biogertontology 11(6):687–701
81. Dilek M, Naziroglu M, Baha Oral H, Suat Ovey I, Kucukayaz M, Mungan MT, Kara HY, Sutcu R (2010) Melatonin modulates hippocampus NMDA receptors, blood and brain oxidative stress levels in ovariectomized rats. J Membr Biol 233(1–3):135–142

Chapter 12
Oxidative Stress in Assisted Reproductive Technologies

Catherine M. H. Combelles and Margo L. Hennet

Abstract Assisted reproductive technologies (ART) represent a new frontier of medical knowledge regarding women's reproductive health. The success rates of procedures such as in vitro fertilization (IVF), in vitro maturation (IVM), intracytoplasmic sperm injection (ICSI), in vitro embryo culture, preimplantation genetic diagnosis (PGD), and cryopreservation simultaneously testify to significant progress in cellular and molecular research, as well as to the myriad cellular processes and phenomena that are not yet fully understood. One such phenomenon of particular interest to ART and female reproduction is oxidative stress. Oxidative stress occurs when reactive oxygen species (ROS) overwhelm cellular defenses and cause structural and functional damage. It is suspected to influence negatively the development of oocytes and embryos in ART environments, since several factors found therein are foreign to in vivo conditions. In this chapter, the major contributions of the ART environment to oxidative stress in oocytes and embryos are discussed, with a focus on the mechanisms by which oxidative stress is induced and the questions facing future ART research.

Keywords Oxidative stress · Oocyte cryopreservation · Oocyte · Embryo · Assisted reproductive technologies (ART) · In vitro maturation (IVM)

C. M. H. Combelles (✉)
Department of Biology, Middlebury College, McCardell Bicentennial Hall 346, Middlebury, VT 05753, USA
e-mail: ccombell@middlebury.edu

M. L. Hennet
Department of Biology, Middlebury College, McCardell Bicentennial Hall 348, Middlebury, VT 05753, USA
e-mail: margohennet@gmail.com

12.1 Introduction

During assisted reproductive technologies (ART), oocytes and embryos encounter a host of environmental stresses, most of which are not naturally present in vivo (Fig. 12.1). Many of these new in vitro challenges can contribute to a state of oxidative stress, in which intracellular concentrations of reactive oxygen species (ROS) increase to dangerous levels. ROS contain unpaired electrons that react readily with lipids, proteins, and nucleic acids. Although their presence at normal, physiological levels is required for some essential cell processes including blastocyst hatching [1–3], excessive ROS can lead to severe damage—including membrane malfunction, protein damage, and altered gene expression—or apoptosis [4]. When oocytes and embryos sustain oxidative damage by ROS, their development and viability are often impaired [5–7].

ART procedures include in vitro fertilization (IVF), in vitro maturation (IVM), intracytoplasmic sperm injection (ICSI), in vitro embryo culture, preimplantation genetic diagnosis (PGD), and cryopreservation (Fig. 12.2). Oxidative stress during any of these procedures can impact the quality and viability of the oocyte or embryo. In IVM, immature oocytes are retrieved from the ovary and resume meiosis in vitro. This maturation—normally achieved upon ovulation in vivo—must be completed to allow oocytes to be fertilized by sperm. IVM, but most often the retrieval of mature eggs, are followed by IVF or ICSI. IVF entails the placement of oocytes with sperm in culture medium to achieve fertilization, while ICSI is a procedure where the male gamete is delivered directly into the oocyte cytoplasm by microinjection. In vitro embryo culture maintains embryos in culture until they reach either the cleavage or blastocyst stages (by day 3 or 5 of development), at which point they are transferred to the uterus. PGD is a method to screen for genetic conditions by removing a cell(s) from cleavage stage embryos, and cryopreservation freezes oocytes or embryos to allow long-term storage. The goal of all ART procedures is always to produce viable, high quality embryos that culminate in successful pregnancies. Therefore, how oxidative stress is induced, and to what extent, during each procedure is important to understand.

In both humans and cows, fewer oocytes reach the blastocyst stage in vitro as compared to in vivo [8, 9]. Oxidative stress is suspected to play a role in promoting this discrepancy [10]. There is a direct relationship between hydrogen peroxide concentration and apoptosis or DNA fragmentation in mouse embryos in vitro [7], and it has been shown that in vivo-derived mouse embryos produce less hydrogen peroxide than those cultured in vitro do [11, 12]. Extracellular and intracellular production of ROS such as superoxide anion, hydrogen peroxide, and hydroxyl radicals during ART can be exacerbated by features of the in vitro environment, including atmospheric oxygen concentration, light, metal ions, and culture temperature (Fig. 12.3). In attempts to minimize ROS generation during ART procedures, antioxidants and chelators are commonly added to culture media. Antioxidants are enzymatic or non-enzymatic molecules that can prevent the oxidation of other molecules and include superoxide dismutase (SOD), catalase,

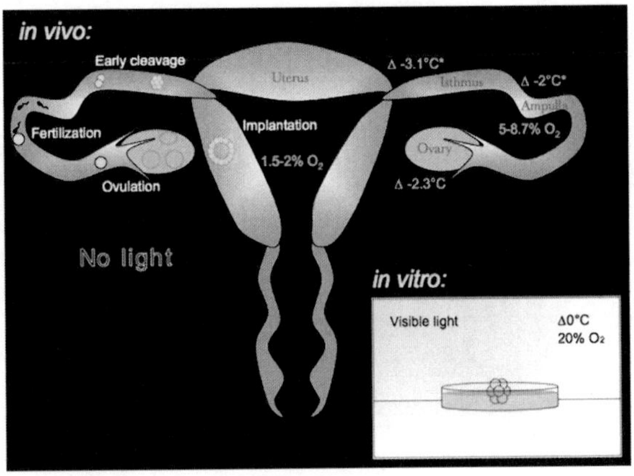

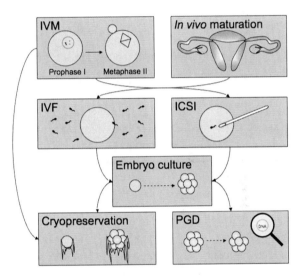

Fig. 12.1 The ART and in vivo environments of developing oocytes and embryos differ in many respects. Here, several of the differences are emphasized. The differences between core body temperature and the temperature of the ovary (human) and isthmus and ampulla (rabbit) are shown, indicated by a *delta* symbol. Oxygen tension in the oviduct and uterus are values reported in rhesus monkeys

Fig. 12.2 A schematic depicting the variety and sequence of ART procedures available for human fertility treatments

glutathione peroxidases (GPx), vitamins, glutathione, cysteamine, and pyruvate. Chelators are molecules that bind to metals and sequester them, inhibiting their ability to participate in reactions. A common chelator in embryo culture media is EDTA. Amino acids are also capable of chelation to some degree [13, 14]. The use of antioxidants to overcome oxidative stress during ART is discussed in a subsequent chapter.

In this chapter we will address the major contributions of the in vitro ART environment to promoting oxidative stress in oocytes and embryos, with special attention to the mechanisms by which these effects are exerted. Furthermore, to conclude this chapter we will briefly describe some cases of in vivo exposure to ROS before or after ovulation that may affect oocyte and embryo susceptibility to OS during ART.

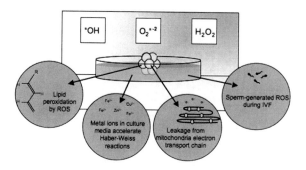

Fig. 12.3 ROS are produced by various components of the in vitro environment. Lipid peroxidation by ROS gives rise to lipid radicals, which self-propagate and generate further ROS; metal ions in culture media accelerate ROS-generating reactions; leakage from the mitochondria electron transport chain produces new intracellular ROS; sperm are known to generate ROS during IVF coincubations

12.2 Sperm Exposure During IVF

The coincubation of gametes during IVF procedures is a scenario that merits attention when discussing oxidative stress in ART. Spermatozoa are known to generate ROS and other potentially dangerous by-products—especially when dead or damaged, as in many cases of male factor infertility [15–20]. Although a physiological level of ROS is necessary for hyperactivation and capacitation in spermatozoa [21], excessive generation of ROS may be detrimental to both female and male gametes, particularly in the in vitro environment where they are exposed to higher oxygen levels than in vivo. Therefore, concentration of spermatozoa and the duration of coincubation are important stress factors for oocytes and zygotes during IVF.

Conventional IVF utilizes a coincubation period of 16–18 h to ensure that gametes have ample time to achieve fertilization. This time period was originally established because it corresponds with the observation of pronuclei in zygotes [22]. Interestingly, however, human spermatozoa enter the cumulus complex within 15 min of in vitro insemination [22], and the number of spermatozoa positioned within the cumulus mass reaches a plateau of 15 within that time [22, 23]. In most mammals, fertilization is normally completed within 2–4 h of insemination [22, 24]. Several studies in the human have confirmed that fertilization rates are similar between oocytes coincubated with normal sperm for 1–2 h and those coincubated with normal sperm for the standard 16–18 h in IVF [23, 25, 26], suggesting that 2 h is sufficient time for fertilization to occur in vitro.

Despite similar fertilization rates, human oocytes coincubated with sperm for short time periods (1–4 h) are reported to give rise to higher quality embryos and significantly improved embryo development, implantation, and pregnancy rates compared to those coincubated with sperm for 16–18 h [22, 23, 25–27].

Interestingly, in vivo-derived mouse zygotes coincubated with sperm for 12–18 h following retrieval exhibit higher levels of degradation and arrest than those coincubated with sperm for 2–4 h, suggesting that shortened exposure of embryos to sperm produces higher blastocyst rates [24]. The involvement of sperm-generated ROS in promoting these differences is supported by observations that fertilization media containing sperm—with or without oocytes—contains significantly higher levels of ROS after 18 h than after 4 h [24], and that the success of 16–18 h coincubations of human gametes is improved when the media contains EDTA and glutamine [23].

Furthermore, the number of spermatozoa encountered by oocytes during IVF (1×10^6/ml) is many times higher than what would be found in vivo [28]. This higher number helps to ensure fertilization in IVF by increasing the probability of sperm–egg collision, especially in cases of male factor infertility where a majority of spermatozoa may be immotile or defective. However, high concentrations of spermatozoa produce high levels of ROS relative to any antioxidants or protective components in the medium. This is perhaps one reason that reduced coincubation times have been shown to be beneficial by the studies described above. Indeed, a study comparing different sperm concentrations found that coincubation of zygotes with sperm negatively affects embryo development in a dose-dependent manner [29]. The question of spermatozoa concentration may also help to explain why some studies have reported no effect of reduced coincubation times on embryo development [30, 31]. In these studies, a fixed number of spermatozoa (20,000–50,000) were placed with oocytes in 0.75–1 ml drops, whereas the other studies reporting benefits of shorter incubation times used the same number of spermatozoa in microdrops. A higher ratio of spermatozoa—and therefore ROS—to volume of medium may necessitate shorter incubation times.

The ROS generated by spermatozoa during IVF may affect preimplantation embryo development in two ways. First, ROS are known to cause zona pellucida hardening [23, 32]. This may interfere with embryo hatching and consequently reduce their viability. Next, ROS can oxidize polyunsaturated fatty acids in the oocyte membrane, leading to a decrease in membrane fluidity and flexibility. Fluidity of oocyte membranes is necessary for polar body extrusion during maturation (which occurs during IVM), proper sperm–oocyte fusion (during IVF), and subsequent embryo cleavage (during embryo culture). As no negative effects of shortened coincubation time on embryo development have been reported, it may be the safer choice for IVF. In particular, dead sperm are known to generate high levels of ROS, and the probability of their presence during IVF increases with longer coincubation times. From this perspective, ICSI provides the advantage of avoiding coincubation of oocytes with sperm, potentially avoiding excess exposure to ROS during and after fertilization. Overall, it is important to be aware that any damaging effects of sperm-derived ROS during IVF will depend heavily on the duration of sperm exposure, sperm quality, environmental features such as temperature, media composition, and oxygen tension, and the type of ROS generated.

12.3 Oxygen Tension

The percent of oxygen that oocytes and embryos are exposed to during ART (be it during gamete manipulation, IVF, or embryo culture) must be considered, the reasons being twofold. First, oxygen concentrations differ significantly between the in vivo and in vitro environments, with mean values ranging between 5–9 % oxygen in the fallopian tubes and 1.5–8 % oxygen in the uteri of rhesus monkeys, hamsters, and rabbits [33]. This is in contrast to the ambient air in the laboratory and the traditional 5 % CO_2 cell culture incubators both containing about 20 % oxygen. Oxygen concentration in the oviduct is thus about 40 % lesser than atmospheric, and even less in the uterus. Second, 20 % oxygen tension is associated with increased ROS production when compared to culture in 5 % oxygen; for example, this was demonstrated in mouse, pig, and cow embryos [12, 34–36].

Mechanistically, high oxygen concentrations could activate the reactions of xanthine oxidase with hypoxanthine, in turn resulting in the increased production of free radicals. There is supporting evidence for such effects of high oxygen during bovine embryo culture [37]. Beyond the risks of increased ROS with non-physiological oxygen concentrations are changes in the expression of oxygen-sensitive genes. For instance, the hypoxia-inducible factor (HIF) transcription factors respond to external oxygen, in turn modulating the expression of a battery of genes [38–40]. Altered metabolism and energy substrate utilization of an embryo cultured under varying oxygen tension merits attention, and overall future studies remain to establish with certainty the mechanisms by which oxygen influences early embryonic development.

To prevent in the first place the generation of excessive free radicals and changes in oxygen-sensitive genes, one approach has been to perform cultures under 5 % oxygen. A myriad of animal studies have compared embryos cultured under 5 % versus 20 % oxygen, with only a growing number of them now focused on human embryos. Taken together, animal studies do not all support significant benefits of lowering oxygen during embryo culture (reviewed by [41]). Of note is the fact that it is now common practice to culture bovine embryos in reduced (5 %) oxygen from zygote through blastocyst development; yet this is not the case for mouse embryos that are routinely cultured in 20 % oxygen. It is not implied that reduced oxygen should not be used in the mouse, as some studies do report on beneficial effects, but a consensus is yet to be reached [42–44]. The following discussion focuses on human studies unless noted otherwise; a few domestic species and mouse reports are used. In this respect, it is important to note that mouse embryos may not be the best study system for human (when compared to the bovine) since the volume of a mouse embryo is four times less than the volume of a human embryo [45]. As a result, oxygen diffusion and availability would differ between embryos of the two species. The existence of potential oxygen gradients across the developing embryo also merits consideration.

Human studies testing the influence of low oxygen all focused initially on comparisons with atmospheric oxygen during pre-compaction development, namely from the time of insemination to cleavage-stage embryos. An early report examining reduced oxygen tension in human embryos dates to 1994 and while there was no comparison with ambient oxygen, it first showed encouraging blastocyst formation, pregnancy, and live birth rates [46]. A first comparison was conducted with oxygen levels manipulated from the time of insemination to day 2 (D2), with no effect on fertilization, embryonic development, pregnancy, and implantation rates [47]. In a prospective randomized IVF study exposing gametes and embryos from the time of fertilization under varying oxygen levels, 5 % oxygen resulted in similar fertilization, embryo development, pregnancy, and implantation rates with D2 or D3 transfers. However, when culture was continued post-compaction, blastocyst formation and cell numbers were improved under low oxygen [48]. Such study began pointing to the importance of timing when exposing pre-implantation embryos to modified oxygen levels. Further, the in vivo environment (following embryo transfer on D2 or D3) may have mitigated any detriments of ambient oxygen cultures beforehand, thereby attesting to the potential plasticity and recovery of embryos. A prospective randomized study comprising ICSI cycles also demonstrated no benefits of lowered oxygen concentration from the time of oocyte retrieval to embryo transfer at either D2 or D3 [49]. Interestingly, morphological evaluations of embryos prior to transfer on D3 showed significantly increased quality with low oxygen culture, a difference that was not apparent on D3. This also supports the time dependence of oxygen effects, with extended in vitro cultures (for even a single day) further benefiting from reduced oxygen. Kea et al. [50] showed no changes in fertilization and pregnancy rates for cultures (from oocyte retrieval to D3 embryo transfer) in either 5 or 20 % oxygen; this was in spite of improvements in morphological embryo scores on D3. Several studies thus fail to support the need to culture under reduced oxygen; however, there are conflicting results with other reports that document the overall benefits of 5 % oxygen, as summarized below.

An often cited early study by Catt and Henman [51] reported on higher pregnancy and implantation rates in the 5 % oxygen group; however, it is important to note that the study was preliminary in nature and little details were provided on the experimental design. In a randomized prospective study design, human sibling oocytes fertilized and cultured under 5 % oxygen developed favorably in comparison to the 20 % oxygen group; there was a significant increase in cleavage by D3, and blastocyst formation and expansion by the end of culture [52]. While an evaluation of ongoing pregnancy outcomes did not reveal significant differences between the two oxygen groups, an improvement in favor of culture in 5 % oxygen was seen in the cumulative pregnancy rates and in poor responders (<40 yo with <5 metaphase-II oocytes) [53].

Two recent studies also examined the influences of reduced oxygen from insemination to D5 in blastocyst transfer cycles. One randomized trial reported increased blastocyst formation, viable pregnancy, and live birth (with a 10 % increase) in the 5 % when compared to 19 % oxygen group [54]. Similarly, implantation and live birth rates were higher (by 15 and 11 %, respectively) for blastocysts cultured from the time of oocyte retrieval in 5 % than 21 % oxygen [55]. Interestingly, significance only existed when considering D5 transfers in low oxygen versus D3 transfers, also in low oxygen [55].

It is currently difficult to explain with certainty the discrepancy between randomized control trials concluding on the benefits, or lack thereof, of reduced oxygen. Discordant results may be explained by differences in culture media composition and conditions, such as the influences of the volume of microdrops and depth of oil overlays [56], or variations in percent CO_2 that would in turn impact the external pH and thus possibly confound effects directly attributable to oxygen alone [57]; nonetheless, further studies remain needed. Particular considerations should be given to culture conditions, with any subpar ones, or simply the inclusion of free radical scavengers, potentially masking a detrimental effect of oxygen. It is also relevant to note that all human studies (with the exception of one, [53]) that report on the success of low oxygen focused on D5 transfers and/or blastocyst quality. There may thus be an accrued benefit of extending the use of 5 % oxygen for the in vitro culture of human blastocysts, rather than merely for cleavage-stage embryos that become replaced back in a natural uterine low-oxygen environment.

In a small-scale study considering a potential temporal effect of oxygen levels during the pre-implantation period, there was an improvement in blastocyst yield and morphological quality with culture in reduced oxygen from D0 through D6 when compared to a D0 to D3 culture in ambient oxygen followed by reduced oxygen for D3 to D6 [58]. Lowering oxygen in vitro beyond D3 may thus not alleviate any detriments that are already caused from cultures in atmospheric oxygen until D3. In support of this is a recent retrospective non-sibling embryo study comparing D3 with D5 culture in either 5 or 20 % oxygen, following culture in 20 % oxygen for all groups; there were no differences in embryo quality, pregnancy, and implantation rates [59].

A study employing a fully crossed design tested the effects of oxygen and time of culture in the mouse; more specifically, it examined the temporal influence of 5 % versus 20 % oxygen during the first 48 h and/or another 48 h of culture [60]. In vivo fertilized mouse zygotes suffered developmental delays in 20 % oxygen during the first and/or second culture periods although the greatest effects were seen pre-compaction; further, early detriments from elevated oxygen were not found reversible when placing the embryos in reduced oxygen post-compaction [60]. The sensitivity of embryos to external oxygen varies during pre-implantation development, with several mouse reports pointing to a heightened sensitivity in the early stages of development [42–44]. Of caution are studies that may conclude on the insensitivity of embryos to oxygen during cleavage stages; unless tested further, one cannot eliminate the possibility that early damages do occur but with consequences not apparent until later stages of development. Of utmost importance is thus the need to support development adequately in vitro and prior to transfer, with the evidence to date supporting the involvement of oxygen tension. Yet future studies remain not only to elucidate the exact effects but also to take advantage of the optimal oxygen levels at the right time(s) during ART. Indeed, oviductal oxygen varies between follicular (5 %) and early post-ovulatory (8.2 %) or mid-luteal (8.7 %) phases in the rhesus monkey, with levels decreasing to 1.5–2 % in the uterus [33]. Similar low ranges of oxygen tension were measured more recently at the endometrial surface in human [61]. Future human studies should thus evaluate the impact of lowering oxygen down to 2 % post-compaction from the morula (rather than 8-cell) to the blastocyst stage. This precise timing of exposure has not yet been investigated in human embryos, while it has been considered in recent animal studies [40, 62].

In spite of the aforementioned plasticity and potential ability of embryos to recover from supraphysiological oxygen exposures (albeit perhaps only in vivo), there are more subtle effects that may persist and merit careful attention. Animal studies tested development to blastocyst and gene expression under 2 % oxygen after morula formation (with zygote to morula culture in 7 % oxygen). In both the mouse [62] and the cow [40], 2 % oxygen resulted in increased expression of several oxygen-sensitive genes while there were no effects on development to blastocysts when compared with 7 and 20 % oxygen. With respect to some recent interest in lowering the oxygen below 5 % post-compaction, a study supports a note of caution with respect to hypoxia; indeed, 2 % oxygen results in detriments in fetal development after blastocyst transfer in a mouse model [63]. Species difference should also be considered, given the beneficial effects of 2 % oxygen on blastocyst development reported in the cow [64], in contrast with the detrimental effects in the mouse [62, 63]. In this vein, differences in oxygen levels as measured in vivo are relevant to note, with variations in absolute values across species. Indeed, while all three species show lowest oxygen tension in the uterus at the time of implantation, that lowest point is 1.5, 5.3, and 3.5% in the rhesus monkey, hamster, and rabbit, respectively [33]. That said, the hamster is the only species not showing decreased oxygen in the uterus when compared to the oviduct, perhaps explaining the detriments of reducing oxygen in vitro post-compaction. Furthermore, a careful analysis of the oviductal and uterine environments of pregnant hamsters showed a consistent increase during early cleavage stages and before a subsequent decline; such dynamic would suggest a need for increased oxygen availability (albeit with a high of about 8 % in vivo) during a limited developmental period [33]. In vitro studies ought to consider the previously reported variations in the in vivo oxygen tension across organs, times of the reproductive cycle, during conception, and across species.

The expression of genes related to metabolism, oxidative stress, antioxidant protection, and stress response varied in a complex fashion depending on the developmental stage (from 2-cell through blastocyst) of bovine embryos cultured continuously under 5 or 20 % oxygen [65]. A gene microarray analysis also showed increased perturbations in the global pattern of gene expression of mouse blastocysts that were cultured from in vivo-derived zygotes in 20 % when compared with 5 % oxygen; of note were similar patterns of gene expression between embryos cultured either in vivo or under 5 % oxygen [66]. Even beyond such short-term effects (themselves not routinely evaluated in human ART) lie concerns following embryo transfers. This is not surprising in light of the known influences of pre-implantation development in the programming of later development and the health of the offspring. Also of caution with studies to date is a prevailing reliance on morphological parameters for the assessment of embryo quality (be it in animal or human studies). Human studies including pregnancy and live birth outcomes are undoubtedly telling; interestingly, fetal development (as assessed by fetal weight) decreased with cultures of in vivo-produced mouse zygotes to blastocysts in 20 % oxygen, while morphological assessment alone did not detect any differences [44]. The inner cell mass may also prove particularly susceptible to oxygen-related damage as shown in the mouse [44]. Taken together, there is experimental support for elevated oxygen influencing embryo viability but with effects not always apparent until later developmental stages.

It is relevant to note that human studies testing the influence of reduced oxygen have all varied oxygen exposures from the time of gamete retrieval through fertilization. This is in contrast to some of the current animal ART practices, notably in the bovine for which 5 % oxygen is the standard culture condition from the zygote stage only (with IVF taking place under 20 % oxygen). Such significant difference in experimental design ought to be considered when comparing findings, with additional studies focusing on the true impact of oxygen during insemination on subsequent embryo quality. With embryo culture in reduced oxygen, IVF in 5 % oxygen significantly benefited blastocyst formation (with no changes in cell numbers) when compared with 20 % oxygen in a sheep model [67]. Similarly, insemination in reduced oxygen resulted in increased blastocyst yield (with no changes in fertilization) in the bovine [68]. Yet, a bovine study demonstrated detrimental effects of low oxygen during IVF, notably when performed following IVM under 20 % and not 5 % oxygen [69]. It must also be noted that all these animal studies manipulating oxygen levels during IVF did so after IVM, leaving unanswered the question of potential oxygen effects during IVF for in vivo-derived mature oocytes. Regardless, gametes and embryos will be handled for varying amounts of time in the ambient air; these times will differ depending on the specific ART procedures and protocols, including routine IVF, ICSI, PGD, other micromanipulations, and/or embryo evaluations. An initial exposure to 5 % oxygen, for a period as short as 1 h, proved detrimental to the continued developmental progression of mouse in vivo-derived zygotes, particularly past the morula stage [42]. There is thus a need to minimize exposure of oocytes and embryos to atmospheric oxygen at every step during clinical ART.

Studies in animal models may also vary in the source of gametes or embryos, and such difference may impart varying responses to elevated oxygen. Interestingly, in vivo-derived pig cleavage-stage embryos proved particularly resilient to oxygen toxicity when compared with in vitro fertilized and parthenogenetically activated embryos following IVM [70]. Yet, this was not the case in a mouse study with detrimental effects of 20 % oxygen shown for in vivo produced embryos when culture was initiated at the 2-cell stage but not at the 8-cell stage [44]. The influence of prior developmental history on the effects of oxygen on embryo quality is also relevant in human ART. Indeed, while routine ART is currently performed using in vivo matured oocytes retrieved following ovarian stimulation, other sources of clinically useful oocytes are immature and cryopreserved ones (immature or mature). It is conceivable that in vivo-derived gametes or embryos, obtained under natural conditions, possess superior cellular defenses against the oxidative threats of ART. Alternatively, differences in sensitivity may reflect the stages of embryonic development at which cells are exposed to varying oxygen levels.

In spite of the higher cost associated with maintaining embryos in reduced oxygen conditions, the suggested beneficial outcomes on embryo development and cycle outcomes may justify the additional expense (most of which is incurred only initially with the purchase of new cell culture incubators). To date, no studies identify any detriments of culturing in low oxygen; and while additional studies await, a fail-safe approach may indeed be to reduce oxygen levels to 5 %. This is particularly true for prolonged cultures to D5. Beyond pursuing large-scale and independent confirmation of findings by multiple centers, there remain several other specific avenues of research for the future, many of which are informed by pertinent findings in animal studies (as highlighted herein).

Lastly, the influence of oxygen during IVM has received much less attention, likely due to the experimental nature of clinical IVM and its current use in only select cases. Yet oxygen merits consideration as a potential factor that may help explain the superior developmental competencies of oocytes matured in vivo over those matured in vitro. As for embryos, the oocyte normally matures in a microenvironment (i.e. the ovarian follicle) that contains oxygen levels much less than in air. The exact levels of oxygen in this environment is clearly dynamic and merits further characterization; nonetheless, all reports indicate a maximum of 8 % oxygen with some regions of the large follicles perhaps even approaching anoxia [71–73].

No human studies comparing oxygen tensions during IVM exist, and a consensus is not yet established from experiments in animal models. Some studies document benefits (albeit not always in all outcome measures) of lowering oxygen in pig, cow, and mouse [35, 74–78], while others show developmental improvements in 20 % when compared to 5 % oxygen in the same three species [79–82]. Differences are often not seen in overall developmental rates, but rather in further analyses such as blastocyst cell numbers. Interestingly, advantages of low oxygen IVM may be augmented by not only maturation but also fertilization and embryo culture performed under 5 % oxygen, as shown in the pig [74]. More recently, a bovine study (with a 2 × 2 design) aimed to distinguish the effects of oxygen during IVM and/or IVF [69]. All embryo cultures were conducted in 5 % oxygen, but interestingly embryo development (as assessed by overall cleavage rate and blastocyst yield) was improved with IVM in 5 % while IVF in 5 % proved detrimental. There was also an interaction between oxygen tension used in IVM and IVF, with the worst development obtained with IVM and IVF in 20 % and 5 % oxygen, respectively. As proxy for oocyte and blastocyst quality, the relative mRNA abundance of select genes indicated improvements in competence markers of the cumulus-oocyte-complex with IVM in 5 % oxygen [69]. Even once the optimal oxygen concentration is ascertained for IVM, it remains essential to consider the percent of oxygen used during IVF and embryo culture. It should be noted that aforementioned studies used for all of the groups either IVF and embryo culture under 5 % oxygen in the mouse [76, 83], or IVF in 20 % and embryo culture in 5 % oxygen in the cow [35, 80]. It also appears relevant to consider potential interactions between media composition and oxygen, with improved development only observed with low oxygen when culturing oocytes in elevated glucose for instance [35, 84]. In a mouse study, oocytes matured in vitro under increasing oxygen concentrations (between 2 and 20 %) showed corresponding decreases in blastocyst cell numbers; this was in spite of maturation, fertilization, and blastocyst formation outcomes remaining comparable across all concentrations. Interestingly, implantation and fetal outcomes following embryo transfer also failed to show significant effects of oxygen during IVM (although fetal and placental weight were reduced with IVM in 5 % oxygen) [83]. Taken together, these reports testify to the complexity of the response of oocytes to maturation under varying oxygen. There is thus a dire need for future studies on the effects of oxygen during IVM; animal and human oocytes are generally cultured in 20 % oxygen, but the exact benefits or detriments remain controversial.

12.4 Metals in Culture Media

Most laboratory reagents and chemicals contain trace levels of metals such as iron. The typical concentrations of these metals, though low (1.6–19.4 µM), are high enough to catalyze ROS generation in culture media [85, 86]. This de novo ROS synthesis occurs through the Haber-Weiss and Fenton reactions, in which metals such as iron or copper serve as oxidizing and reducing agents to generate the most reactive ROS, hydroxyl radicals, from hydrogen peroxide and superoxide (Fig. 12.4) [85]. Furthermore, some of these same metals—namely ferrous ions—can facilitate the oxidation of lipids by other oxidized lipids [87], thus propagating a chain of oxidative damage.

There is ample evidence from studies of embryo culture that metals in culture media can be detrimental beyond a certain threshold. Iron, when added to culture media during mouse embryo culture, causes hydrogen peroxide concentrations to increase. This increase is also accompanied by higher rates of block or early mortality [88]. In another study, the supplementation of culture medium with the chelator EDTA lowered ROS levels compared to unsupplemented medium [89], suggesting that EDTA's sequestration of metal ions inhibited ROS generation to some degree. EDTA has also been shown to protect mouse embryos from toxic impurities, believed to be zinc from the silicone oil used as overlay [90]. It should be noted, however, that there are physiological requirements for low levels of metal ions by embryos and that, therefore, the complete removal of metal ions from culture media would most likely be detrimental to development [88, 91, 92]. In fact, mammalian oocytes and embryos in vivo normally encounter metals such

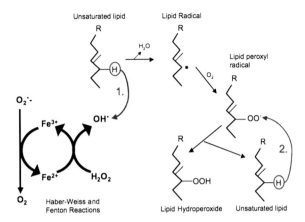

Fig. 12.4 ROS generation is a complex and cyclic process. Here, the mechanism by which the Haber-Weiss and Fenton reactions interact with lipid peroxidation is shown. Hydroxyl radicals are generated in Haber-Weiss and Fenton reactions, facilitated by metal ions such as iron. If hydroxyl radicals encounter unsaturated lipids, they will react together to form lipid radicals and eventually lipid hydroperoxides. *Red arrows* indicate radical-forming reactions. In reaction (*1*), an ROS forms a lipid radical. In reaction (*2*), an unstable lipid peroxyl radical reacts with an unsaturated lipid to form a lipid hydroperoxide and a new lipid radical

as potassium, calcium, magnesium, zinc, and other heavy metals in the form of protein nucleotides, in the reproductive tract [93, 94].

12.5 Visible Light

Yet another unnatural stress factor for gametes and embryos during ART is visible light. Light, or electromagnetic radiation, is energy that travels through space in oscillating waves. Types of electromagnetic radiation are categorized based on the frequency of their waves (their wavelengths). The wavelength of visible light ranges from approximately 380 to 780 nm, and oocytes and embryos are inevitably exposed to it during ART when they are handled and inspected. A particular concern for ART is how visible light affects oocyte and embryo viability.

Light can generate ROS, and is therefore a potential aggravator of oxidative stress. When light is absorbed by a molecule, its energy excites electrons. Excited electrons weaken chemical bonds, which can lead to the degradation of the molecule into radicals and other molecular fragments. Alternatively, molecules can absorb light energy and transfer it to other molecules that then degrade. Whether this process occurs depends on which wavelength of light a molecule can absorb and the strength of its bonds. For example, molecules such as riboflavin and tryptophan are sources of ROS both intracellularly and in cell culture media when they are irradiated by solar light or fluorescent light [95–98]. Light is either measured as units of intensity (lux) or by level of irradiation.

Negative effects of visible light on embryo development in vitro have been reported. Short exposure (increments of 0–10 min) to visible light from a microscope generates ROS in mouse and hamster embryos. The amount of ROS produced increases with the amount of time the embryos are exposed [12, 99]. Studies monitoring the developmental progress of oocytes and embryos exposed to fluorescent or incandescent light during various stages of ART, including prior to IVF, have shown that cleavage and survival rates are reduced compared to low- or no-light controls in mice, hamsters, and rabbits. These effects are observed after as little as 5 min of exposure to 2,400 lux fluorescent light or up to 30–60 min of exposure to light intensities varying from 400 to 4,000 lux [100–103]. Hamster embryos respond differently to varied light intensities (200, 500, 900 lux), with the lowest intensity, 200 lux, allowing the highest relative rates of embryo development [104]. Interestingly, other studies have reported no effect of fluorescent light on DNA ploidy abnormalities and embryo development in rabbit and mouse embryos, respectively [105, 106]. These experimental data suggest that visible light is a component of the ART environment that merits further attention and regulation with respect to oocyte and embryo viability.

In an IVF laboratory, light is generated by microscopes, fluorescent lighting, and indirect sunlight if shaded windows are present. Therefore, Ottosen et al. [107] evaluated the intensity and spectral composition of light reaching embryos during in vitro manipulations in active IVF laboratories. They found that microscopes, at settings appropriate for embryo inspection, produced light at 2,500–5,000 lux, while ambient light levels (room lighting and sunlight) were over tenfold lower, at 200–400 lux. A comparison of these intensities with the results of experimental studies (summarized above) suggests that microscope light exposure is a more dangerous source of light for oocyte and embryo viability than ambient light is. In fact, microscopic light was determined to produce 95 % of the light radiation experienced by oocytes and embryos in an IVF laboratory [107]. ICSI and PGD procedures together require ~700 more seconds of microscope time, which translates into an average of 14.2 kJ/m2 more energy exposure, than IVF procedures do (167 s, 3.2 kJ/m2). They may therefore be particularly impacted by visible light radiation.

In addition to intensity, the spectral composition of light that reaches oocytes or embryos during ART is also relevant to consider. Short-wavelength light, particularly blue light (400–500 nm) tends to generate the most ROS and to be most harmful to embryos [104], as measured by TE and ICM apoptosis and gene expression. In hamster eggs, meiosis was also negatively affected by short-wavelength visible light [108]. Conversely, green (500–575 nm), yellow (575–585 nm), and red light (620–750 nm) were shown not to affect morula compaction in hamsters, and blastulation rates were increased by red light as compared to full-spectrum visible light [104]. In the study by Ottosen et al. [107], blue light was found to contribute 5 % of the energy produced by microscope light.

The spectral composition of light that reaches oocytes and embryos can be managed by using certain fluorescent lights in laboratory equipment and filters on microscopes. A 2007 study showed that warm white fluorescent lights or incandescent lights produce significantly fewer ROS in mouse and hamster zygotes than do cool white fluorescent lights [109]. Cool white fluorescent lights are particularly rich in short-wavelength light, while warm white fluorescent lights and incandescent lights are not. Furthermore, the use of microscope filters has also been proposed to shield oocytes and embryos from the most damaging wavelengths of visible light (400–500 nm) during ART manipulations [107]. In one study, a green pass filter of 498–563 nm wavelengths was used during bovine oocyte collection and other in vitro embryo procedures. The no-filter control group exposed to microscopic light expressed the inducible stress-response form of Hsp72/73 protein, while the filter group did not. Therefore, although this study observed no differences in embryo development rates, total cell counts, or morphological quality up to day 8 between the control and experimental groups, light-induced stress was evident at the gene expression level in controls [110]. To our knowledge, this is the only study directly testing the effects of a microscope filter on oxidative stress during ART.

In considering the effect of visible light on oxidative stress during ART, there are several additional factors to consider. First, there is evidence that sensitivity of an embryo to light varies by species. No studies to date have examined the effects of light on human oocyte or embryo development during ART procedures, however, studies in the mouse, hamster, and rabbit have shown that, at least between these three species, significant differences in light tolerance exist. In the rabbit, oocytes and embryos can tolerate strong light, while in the hamster they are extremely sensitive [109]. As information regarding the sensitivity of human embryos to light is lacking, it is difficult to precisely delineate safe light exposures for human ART. Furthermore, in vitro- versus in vivo-derived embryos may vary in their tolerance of light as well. Lastly, any exposure to direct sunlight will include exposure to ultraviolet (UV) light (300–400 nm). With even shorter wavelengths than blue light, UV light has been shown to exert the most damage of any other light source available during ART [111], and even transient exposures should be avoided entirely.

12.6 Temperature

In an effort to approximate in vivo conditions, human ART procedures are conducted at core body temperature, or 37 °C. This practice was established by 1969, at a time when few studies had been done to elucidate whether temperature gradients exist in the female reproductive tract [112]. Interestingly, temperatures in the oviduct and within follicles—ovarian structures that comprise the microenvironment of preovulatory oocytes—are now known to be lower than core body temperature in humans and other mammals [113–116]. Human follicles, for example, are up to 2.3 °C cooler than the ovarian stroma. The ovarian stroma, in turn, is approximately 35–36.75 °C, slightly cooler than core body temperature [113]. Although no data exist for human oviduct temperatures, in the rabbit the isthmus and ampulla are approximately 3.1 °C and 2 °C cooler than rectal (core) temperature, respectively [116]. Similar gradients are reported in other mammals [114], and the presence of a similar gradient in human oviducts is assumed [115, 117]. The degree of difference between core body temperature and that of follicles or oviducts fluctuates slightly with respect to stages of the estrous cycle.

The temperature discrepancy between in vivo conditions for oocyte maturation, fertilization, and early embryo development and the in vitro environment are a potential concern for oxidative stress during ART. Heat shock, a phenomenon that occurs when cells are subjected to temperatures higher than those found under ideal in vivo conditions, can lead to cellular damage and apoptosis through a variety of mechanisms. Of particular relevance to this chapter, heat shock has been shown to directly increase the flux of ROS in live rat intestinal epithelial cells [118]. Therefore, it is possible that oxidative stress may play a role in the cellular consequences of heat shock.

The temperature difference between the in vivo environment and in vitro conditions during ART might be sufficient to establish a state of mild heat shock. At the very least, it is a departure from natural conditions that oocytes and embryos have evolved to develop in. As such, temperature during ART has been identified by some as a topic in need of further investigation [115]. A clinical study measuring the follicle temperatures of women found one patient (out of 13) whose follicle temperature was warmer—by 0.2 °C—than the ovarian stroma. This patient was infertile, and in vitro observations revealed that her oocytes could not divide in culture [113]. In another clinical study, human pregnancy rates increased when the incubator environment was up to 0.2 °C less than core body temperature rather than up to 0.2 °C above it [57]. Although indirect, the potential link between temperature and normal oocyte or embryo development, and the direct link between temperature and ROS generation (described earlier; [118]), merits attention. One study comparing core body temperature (38.5 °C) and 37 °C during IVM, IVF, and in vitro embryo culture in the bovine found that a decrease from 38.5 to 37 °C during the last 14 h of maturation resulted in slightly (though not significantly) higher cleavage, morula, and blastocyst development [119]. To date, however, studies of heat shock in embryos have primarily focused on comparisons between core body temperature and 2 °C above that temperature. Although these studies do not compare the in vivo and in vitro conditions, they offer insight into the effects that small temperature changes—those roughly equivalent to the in vivo–in vitro temperature difference—can have on development and ROS concentrations in oocytes or embryos.

In mouse and bovine models, temperatures approximately 2 °C higher than core body temperature increase the production of intracellular ROS in pre-implantation embryos after as little as 6 h of culture [120–123]. This rise in ROS is accompanied by increased DNA fragmentation, decreased blastocyst rates, and reduced blastocyst cell numbers [122, 124–126]. Similar results have been reported in somatic cells [127, 128] and in maternally heat-stressed mouse embryos [120]. These results suggest that small temperature increases might affect the ROS production in oocytes or embryos, which could in turn influence the course of development in vitro.

Antioxidants provide protection against heat shock during in vitro embryo development, once again supporting a role for ROS in heat-induced cellular damage. Antioxidants such as vitamin E, glutathione, and beta-mercaptoethanol minimize the negative effects of heat on embryo development [125, 126, 129, 130]. Interestingly, a 2002 study reported that hydrogen-peroxide-induced inhibition of cellular antioxidants, specifically reduced glutathione, was exacerbated by 5 °C heat shock in Chinese hamster ovary cells [131]. Therefore, it is possible that in embryos, heat shock can "inactivate" some of the antioxidant defense system, allowing ROS concentrations to increase to dangerous levels. Conflicting results have also been reported regarding the relationship between oxygen and heat shock. Although one study reports that the negative effects of heat shock are evident in bovine embryos only when cultured under high oxygen tension [130], another study reports no oxygen–temperature interaction [132].

The question of temperature during ART may be relevant to creating an in vitro environment similar to the in vivo one. It is generally thought that departures from in vivo conditions can cause stress to developing oocytes and embryos, however, further studies are needed to elucidate the precise relationship between temperature and developmental quality. Furthermore, some features of oocytes or embryos may be especially sensitive to ART temperatures. It is known that meiotic spindles, for example, are vulnerable to temperature fluctuations and heat [133]. In particular, spindle integrity is important during IVM and IVF, during which oocytes complete various stages of meiosis as they mature. Therefore, questions have also been raised regarding the temperature fluctuations induced by the opening and closing of incubator doors, as well as the consistency and reliability of stage warmers [134, 135]. As the studies in this section may suggest, the negative effects of temperature discrepancies and changes might be induced or exacerbated by changing ROS levels.

12.7 Metabolism

Animal cells require the consumption of oxygen and nutrients to produce energy. In the course of this aerobic metabolism, oxygen is used as a terminal acceptor in the mitochondrial electron transport chain and thus facilitates a highly efficient mode of ATP production. However, inevitable leakage from the electron transport chain, combined with the presence of oxygen, contributes to the formation of intracellular ROS [2, 38, 136]. For this and other reasons, cells regulate their metabolism to maintain viability and quality.

In oocytes and embryos, the regulation of metabolism is particularly important as it ultimately impacts the fate of the organism as a whole. In 2002, Leese et al. [137] presented a hypothesis of embryo metabolism based on data from rodent and domestic species. This hypothesis, termed the "quiet embryo hypothesis" posits that viable embryos will consume lower levels of oxygen and nutrients (a "quieter" metabolism) than less viable embryos will. This difference exists because less viable embryos have experienced significant damage to their genomes, transcriptomes, and proteomes, and consequently undertake repair or rescue operations that have a high energy price. The upregulation of their metabolisms, in turn, leads to increased production of ROS [115, 138]. In summary, the quiet embryo hypothesis presents another possible mechanism by which various stress factors in the in vitro environment can impact intracellular ROS production.

During ART, embryo metabolism may be up-regulated directly by environmental factors or indirectly via damage to cellular components. High oxygen tension, high concentration of energy substrates in the media, radiation, preexisting ROS, and temperature extremes have been identified by Leese et al. [115] as features of the in vitro environment that favor non-quiet metabolisms, and are addressed elsewhere in this chapter. Often, direct and indirect effects on metabolism can be induced by the same stress factor. For instance, the difference between core body temperature (ART incubation temperature) and the temperature in the human oviduct is theoretically enough to increase the metabolic rate of oocytes and early embryos in vitro by at least 15 % compared to those in vivo [115]. In fact, some unpublished results (reviewed in [115]) suggest that when bovine blastocysts are cultured at physiological oviduct temperatures rather than core body temperature, metabolic activity is reduced with no negative impact on blastocyst rates. Additionally, higher-than-physiological temperatures experienced by oocytes and embryos in vitro may generate ROS directly, which, in turn, could up-regulate embryo metabolism and lead to further ROS production (see Sect. 12.6).

Excessive or inadequate concentrations of energy substrates in culture media can alter the normal course of metabolic development as embryos grow. During early cleavage stages, embryos have no net growth and therefore their energy requirements are fairly low. Low oxygen consumption, and therefore low ROS production, might be important during cleavage stage development, as it entails the activation of the zygotic genome at the 4–8 cell stage in humans [115]. Energy demands and metabolism increase later, during compaction and cavitation. High concentrations of energy substrates such as glucose can promote inappropriate or premature metabolic activity [51, 139]. Therefore, Leese et al. [137] advocate the use of culture media that are low in nutrients (like the female reproductive tract), to encourage early embryos to use endogenous stores of nutrients for energy production. Alternatively, endogenous compounds in culture media may autooxidize to produce ROS [140]. The autooxidation of such compounds may deplete levels of antioxidants and other molecules with antioxidant function. One such molecule is pyruvate—although it is included in culture media as an energy substrate, it also reduces hydrogen peroxide [140] and has been shown to have protective effects against these ROS intracellularly in somatic cells [131]. Excessive ROS may lead to some media deficiencies by invoking the antioxidant functions of molecules intended for other purposes. This, in turn, could adversely affect the metabolism and cell function of oocytes or embryos.

12.8 Oocyte Cryopreservation

The cryopreservation of oocytes is currently an intense area of study, due to its strong potential as a method for fertility preservation in patients at risk of losing their fertility from cancer treatment or those of advanced maternal age. The success of oocyte cryopreservation does not yet match success rates with embryo freezing; relevant culprits that may explain such differences could lie in the significantly different cell cycle state and properties of oocytes and early embryos. There are multiple damages associated with cryopreservation, and oxidative stress numbers among one of many potential influencing factors.

Evidence in other study systems supports the relevance of considering oxidative stress in cryopreservation injury, and notably when cellular materials are cooled to low temperatures. For example, liver tissues kept cool for transplantation are known to suffer from oxidative damage. When early and late products of lipid peroxidation (conjugated dienes and malondialdehyde, respectively) were measured after liver tissue storage at cold temperatures, peroxidation occurred at -20 °C while not at -196 °C. Free radical activity, as reflected by changes in ratios of reduced to oxidized glutathione, also increased with freezing at -20 °C, with a proposed stabilization once the tissue reached ultra-low temperatures [141]. Interestingly, cold storage of kidney tissue at only 4 °C augmented ROS levels concomitantly with reduced glutathione defenses, while Mn-SOD expression increased [142]. Supplementation with antioxidants and free radical scavengers also abrogated the cold-induced ROS formation [141] and cellular damages [142]. All of these effects of cold temperatures were time-dependent, thus buttressing the likely improved oxidative outcomes with an ultra-low rapid freezing protocol. Further, yeast cells that lacked some antioxidants, notably Cu, Zn-SOD, and Mn-SOD, proved the least tolerant to freeze–thaw stress; cryopreservation also resulted in an oxidative burst of superoxide radicals [143]. Together, there is thus an effect of cold temperature exposures on the pro- and anti-oxidant balances of cells, as well as an involvement of endogenous antioxidants. Conversely, the generation of free radicals is also relevant during the rewarming of frozen tissues [144] (reviewed by [145]); cellular damage may thus be further exacerbated upon warming and when compared to freezing injury alone.

To date, studies examining oxidative stress during oocyte cryopreservation are grossly lacking. This is a clear gap in the field, one that demands attention given a precedent for the involvement of oxidative stress during both sperm and embryo cryopreservation. Further, it is conceivable that the oocyte, when frozen individually without its natural microenvironment, may prove particularly sensitive to oxidative stress as induced by cryopreservation. Oocytes possess stores of antioxidants [146–148], but whether these stores equip the oocytes with sufficient defenses against exogenously induced oxidative stress is unclear. In vivo, the oocyte also develops in a microenvironment that is rich in antioxidant defenses (reviewed by [149] and [150]), and cryopreservation conditions and solutions may merit antioxidant supplementations in order to make up for any deficiencies.

Experimental studies using mammalian oocytes support the involvement of OS during oocyte cryopreservation, even if indirectly. The most direct evidence to date includes a measurement of ROS concentrations during the vitrification of porcine oocytes. When compared to never frozen, metaphase-II oocytes that underwent vitrification contained increased intracellular ROS concentration [151, 152]; there was no difference in ROS depending on the percent oxygen (5 or 20 %) these oocytes had been in vitro matured in [151]. Species differences cannot be ignored, and it is relevant to note that oocytes from pigs are particularly susceptible to cryopreservation, largely due to its very high lipid content. With very few studies focused on the oocyte, further insight may be gained from experimentation in early embryos. First demonstrated in mouse 8-cell embryos was a benefit of pre-culture prior to freezing in media containing inhibitors of lipid peroxidation [153]. Benefits were more pronounced with slow-freezing than with vitrification, perhaps indicating the particular needs for uncompromised membrane permeability for the penetration of cryoprotectants during slow-freezing. In this vein, membrane characteristics were measured in slow-frozen mouse 2-cell embryos. When compared with unfrozen controls, cryopreserved embryos had decreased membrane fluidity (presumably due to lipid peroxidation damage), showed changes in membrane depolarization (reflecting functional damage), and increased hydrogen peroxide levels. Cell death was further demonstrated in frozen embryos, with thus a proposed involvement of free radical generation and damage with an induction of apoptosis [154]. Mouse 2-cell embryos showed a 50 and 20 % increase in hydrogen peroxide after cryopreservation with slow-freezing and vitrification, respectively; vitrification thus resulted in less hydrogen peroxide formation than slow-freezing [155]. Further evidence points to the influences of specific cryopreservation protocols, with the fastest cooling method resulting in reduced ROS production during the vitrification of human ovarian tissue [156].

Also pertinent are indications that antioxidant defenses become compromised during cryopreservation. In porcine in vitro matured M-II oocytes, glutathione levels become decreased following vitrification, together with the decreased fertilization and developmental potential of cryopreserved embryos [151]. This is however, the only study to date reporting on compromised antioxidants in the frozen oocyte.

In contrast, there are several studies indirectly buttressing the involvement of oxidative stress during oocyte cryopreservation. All of these entail the addition of antioxidants during various phases of the cryopreservation protocol; an improvement in developmental outcomes is in turn interpreted as due to the mitigation of cryopreservation-induced oxidative stress by these exogenous antioxidants. The survival and fertilization rates of slow-frozen mouse metaphase-II oocytes were improved when the freezing and thawing media were supplemented with SOD, with further benefits obtained with the dual addition of SOD and Catalase [157]. Further insight may be gained from studies manipulating the pro- and/or antioxidants during the freezing of early embryos. For 2-cell mouse embryos, supplementation of the cryopreservation solutions with the antioxidant Vitamin C (ascorbate) resulted in some benefits, namely in reduced hydrogen peroxide, increased number of cells in the inner cell mass, and decreased damage to the membrane [155]. Improvements were consistently observed after slow-freezing rather than vitrification, likely reflecting an increased oxidative damage from slow-freezing. When added to the culture medium following cryopreservation, the ability of exogenous antioxidants to alleviate deficiencies in antioxidant defenses that resulted from freezing injury was considered. This was tested for after the vitrification of porcine zygotes, for which there was no significant improvement of GSH addition on in vitro developmental outcomes [158]. However, when ROS are presumably reduced experimentally (with beta-mercaptoethanol, BME) after the vitrification and warming of bovine blastocysts, survival and re-expansion were improved after a 6 h culture period; hatching and total cell numbers were also increased with post-freezing culture up to 48 h in BME [159]. A more recent study also vitrifying bovine blastocysts demonstrated the usefulness of not only supplementing the culture medium with BME from the zygote to the blastocyst stage but also during post-warming for 48 h on the resulting quality of blastocysts (based on hatching, total cell numbers, and DNA fragmentation) [160]. The timing of antioxidant supplementation thus proves influential with studies in early embryos revealing important lessons; notably, studies should also focus on the post-warming period, rather than only aim to neutralize ROS prior to freezing. Most recently, BME treatment during the vitrification and warming of metaphase-II porcine oocytes significantly diminished intracellular ROS, but to levels that remained higher than that of control oocytes. Viability and fertilization was not affected by BME exposures, but cleavage rates improved with BME although there was no increase in blastocyst formation [152]. Interestingly, the inclusion of BME during vitrification, warming, and in vitro culture of porcine oocytes that were frozen 4 h after IVF resulted in increased blastocyst formation and total cell numbers [152]. BME supplementation can thus partially neutralize ROS induced by freezing and warming. Further consideration should also be given to the most optimal type of exogenous free radical scavenger to use (e.g. GSH, BME, or others), and the developmental periods when vitrified/warmed oocytes or embryos may be most susceptible to oxidative damage.

The future thus awaits an improved understanding of the exact involvement and influences of oxidative stress during oocyte cryopreservation. Past work supports the relevance of oxidative stress, but strategies to prevent and/or remedy oxidative damage during cryopreservation are not yet available. Notably, ways to preserve and/or restore the antioxidant protection systems of the oocyte should be addressed. Future consideration should also be given to potential differences in the challenges posed by oxidative stress during the cryopreservation of either immature or mature oocytes. Of relevance is a recent study by Somfai et al. [161] demonstrating no differences in GSH content following the IVM of vitrified porcine GV immature oocytes, while in vitro matured M-IIs showed decreased GSH with vitrification [151]; it has been proposed that GSH levels may recover to normal levels during IVM, although direct supporting evidence is needed. Even among mature oocytes, the particular susceptibility of those matured in vitro merit comparison with those matured in vivo. The stage and origin of oocytes are thus variables that remain to be tested when examining the roles of oxidative stress during oocyte cryopreservation.

12.9 Conclusion

In this chapter we have addressed features of the in vitro environment that are known or suspected to contribute to oxidative stress during ART. However, it is also important to note that patient history—lifestyle factors, diet and nutrition, age, medical history—may influence the vulnerability of oocytes and embryos to oxidative threats that exist in vitro. Smoking, for instance, alters the balance of ROS and antioxidants in the in vivo developmental milieu of oocytes (i.e. follicles) to favor oxidative stress conditions [162, 163], potentially making oocytes more sensitive to the in vitro stresses they later encounter during ART [164].

Patient age is a prominent concern in ART, as advanced maternal age is strongly correlated with low oocyte and embryo quality. Oxidative stress is suspected to play a role in this aetiology. Notably, follicle cells in women and mice of advanced age exhibit low antioxidant expression [165–167]. The "free radical theory of ageing" proposes that oxidative damage gradually accumulates in oocytes that spend an extended number of years in arrest, as they are metabolically active throughout that time [168, 169]. Lowered antioxidant expression in the corresponding follicle cells may exacerbate that condition.

A number of medical conditions are associated with oxidative stress to oocytes and embryos. Polycystic ovary syndrome (PCOS), for example, is a common female endocrine disorder that causes infertility. Evidence including elevated concentrations of oxidized protein in sera [170], increased systemic oxidative stress [171, 172], and increased DNA susceptibility to oxidative damage [173] suggest that oxidative stress is intimately tied to the aetiology of this syndrome. Oocytes and embryos from PCOS patients are likely exposed to higher ROS levels in vivo than normal, thus possibly altering their tolerance of further stress. Other relevant medical conditions include obesity, diabetes, and nutrition disorders [174–177].

Each of these scenarios bring to light yet another important question for women's reproductive health and ART success—to what degree do oocytes and embryos rely on external antioxidant sources during development? The fluid inside follicles contains high levels of many different antioxidants [178], suggesting that oocyte protection against oxidative stress requires extracellular components. Similarly, the early embryo develops in the oviduct, an environment naturally rich in antioxidants [179, 180]. Attention to the in vitro environment and patient history is thus of the utmost importance in achieving successful ART outcomes. Furthermore, it is evident that ROS levels in follicle fluid vary [178, 181–183], and that both minimum and maximum ROS thresholds for oocyte quality likely exist. Identifying and replicating that balance in vitro is an important task facing ART advancement today.

12.10 Summary Statements

- During ART, oocytes and embryos encounter an artificial environment that differs significantly from in vivo. Developing under such conditions may contribute to oxidative stress.

- Oxidative stress can lead to severe cellular damage that impairs the viability and quality of oocytes and embryos. It occurs when oxygen-containing radicals, ROS overwhelm a cell's defense system.
- Coincubation with sperm during IVF exposes oocytes to ROS. The consequences of this exposure might depend on the duration of coincubation.
- Oxygen tension influences ROS formation and gene expression. Culturing embryos under low oxygen tension (5 %) rather than atmospheric oxygen tension (20 %) appears to benefit development; however, evidence in human models is still tentative. As for the optimal oxygen tension during the IVM of oocytes, even data in animal models are still preliminary and controversial.
- Most laboratory reagents and chemicals contain trace levels of metals. Metal ions can accelerate ROS generation in cell culture media through Haber-Weiss and Fenton reactions.
- Visible light can radiate molecules and generate ROS. Short wavelengths are more dangerous in this respect, and the use of microscope filters during ART is thus supported by several research groups.
- Human ART incubations occur at core body temperature, however, the temperatures found inside the oviduct and uterus are consistently cooler. Inappropriate temperatures could contribute to ROS generation and metabolism disruption.
- Oocyte and embryo metabolisms may be up-regulated during ART procedures due to environmental factors or cellular damage. Increased electron leakage from the mitochondria—a consequence of aerobic metabolism—subsequently promotes higher intracellular ROS generation.
- Cryopreservation appears to affect the balance of antioxidants, as well as concentrations of ROS, in oocytes.
- Maternal lifestyle factors can influence the vulnerability of oocytes and embryos to oxidative stress during ART.

References

1. Kamata H, Hirata H (1999) Redox regulation of cellular signalling. Cell Signal 11:15
2. Halliwell B (2006) Reactive species and antioxidants. Redox biology is a fundamental theme of aerobic life. Plant Physiol 141:312–322
3. Thomas M, Jain S, Kumar GP, Laloraya M (1997) A programmed oxyradical burst causes hatching of mouse blastocysts. J Cell Sci 110(Pt 14):1597–1602
4. Halliwell B (2007) Biochemistry of oxidative stress. Biochem Soc Trans 35:1147–1150
5. Favetta LA, Madan P, Mastromonaco GF, St John EJ, King WA, Betts DH (2007) The oxidative stress adaptor p66Shc is required for permanent embryo arrest in vitro. BMC Dev Biol 7:132
6. Jurisicova A, Varmuza S, Casper RF (1996) Programmed cell death and human embryo fragmentation. Mol Hum Reprod 2:93–98
7. Yang HW, Hwang KJ, Kwon HC, Kim HS, Choi KW, Oh KS (1998) Detection of reactive oxygen species (ROS) and apoptosis in human fragmented embryos. Hum Reprod 13:998–1002
8. Suikkar AM (2008) In vitro maturation: its role in fertility treatment. Curr Opin Obstet Gynecol 20:242–248

9. Mermillod P, Oussaid B, Cognie Y (1999) Aspects of follicular and oocyte maturation that affect the developmental potential of embryos. J Reprod Fertil Suppl 54:12
10. Bedaiwy MA, Falcone T, Mohamed MS, Aleem AA, Sharma RK, Worley SE, Thornton J, Agarwal A (2004) Differential growth of human embryos in vitro: role of reactive oxygen species. Fertil Steril 82:593–600
11. Nasr-Esfahani MH, Aitken JR, Johnson MH (1990) Hydrogen peroxide levels in mouse oocytes and early cleavage stage embryos developed in vitro or in vivo. Development 109:501–507
12. Goto Y, Noda Y, Mori T, Nakano M (1993) Increased generation of reactive oxygen species in embryos cultured in vitro. Free Radical Biol Med 15:69–75
13. Lindenbaum A (1973) A survey of naturally occurring chelating ligands. Adv Exp Med Biol 40:67–77
14. Jeppsen RB (2001) Toxicology and safety of ferrochel and other iron amino acid chelates. Arch Latinoam Nutr 51:26–34
15. Aitken RJ, Clarkson JS (1987) Cellular basis of defective sperm function and its association with the genesis of reactive oxygen species by human spermatozoa. J Reprod Fertil 81:459–469
16. Iwasaki A, Gagnon C (1992) Formation of reactive oxygen species in spermatozoa of infertile patients. Fertil Steril 57:409–416
17. Aitken RJ (1994) Pathophysiology of human spermatozoa. Curr Opin Obstet Gynecol 6:128–135
18. Mazzilli F, Rossi T, Marchesini M, Ronconi C, Dondero F (1994) Superoxide anion in human semen related to seminal parameters and clinical aspects. Fertil Steril 62:862–868
19. Plante M, de Lamirande E, Gagnon C (1994) Reactive oxygen species released by activated neutrophils, but not by deficient spermatozoa, are sufficient to affect normal sperm motility. Fertil Steril 62:387–393
20. Baker HW, Brindle J, Irvine DS, Aitken RJ (1996) Protective effect of antioxidants on the impairment of sperm motility by activated polymorphonuclear leukocytes. Fertil Steril 65:411–419
21. de Lamirande E, Gagnon C (1995) Impact of reactive oxygen species on spermatozoa: a balancing act between beneficial and detrimental effects. Hum Reprod 10(Suppl 1):15–21
22. Gianaroli L, Cristina Magli M, Ferraretti AP, Fiorentino A, Tosti E, Panzella S, Dale B (1996) Reducing the time of sperm-oocyte interaction in human in vitro fertilization improves the implantation rate. Hum Reprod 11:166–171
23. Quinn P, Lydic ML, Ho M, Bastuba M, Hendee F, Brody SA (1998) Confirmation of the beneficial effects of brief coincubation of gametes in human in vitro fertilization. Fertil Steril 69:399–402
24. Enkhmaa D, Kasai T, Hoshi K (2009) Long-time exposure of mouse embryos to the sperm produces high levels of reactive oxygen species in culture medium and relates to poor embryo development. Reprod Domest Anim 44:634–637
25. Kattera S, Chen C (2003) Short coincubation of gametes in in vitro fertilization improves implantation and pregnancy rates: a prospective, randomized, controlled study. Fertil Steril 80:1017–1021
26. Gianaroli L, Fiorentino A, Magli MC, Ferraretti AP, Montanaro N (1996) Prolonged sperm-oocyte exposure and high sperm concentration affect human embryo viability and pregnancy rate. Hum Reprod 11:2507–2511
27. Dirnfeld M, Shiloh H, Bider D, Harari E, Koifman M, Lahav-Baratz S, Abramovici H (2003) A prospective randomized controlled study of the effect of short coincubation of gametes during insemination on zona pellucida thickness. Gynecol Endocrinol 17:397–403
28. Trounson AO (1994) The choice of the most appropriate microfertilization technique for human male factor infertility. Reprod Fertil Dev 6:37–43
29. Dumoulin JC, Bras M, Land JA, Pieters MH, Enginsu ME, Geraedts JP, Evers JL (1992) Effect of the number of inseminated spermatozoa on subsequent human and mouse embryonic development in vitro. Hum Reprod 7:1010–1013
30. Swenson K, Check JH, Summers-Chase D, Choe JK, Check ML (2000) A randomized study comparing the effect of standard versus short incubation of sperm and oocyte on subsequent

pregnancy and implantation rates following in vitro fertilization embryo transfer. Arch Androl 45:73–76
31. Lundqvist M, Johansson U, Lundkvist O, Milton K, Westin C, Simberg N (2001) Reducing the time of co-incubation of gametes in human in vitro fertilization has no beneficial effects. Reprod Biomed Online 3:21–24
32. Dirnfeld M, Bider D, Koifman M, Calderon I, Abramovici H (1999) Shortened exposure of oocytes to spermatozoa improves in vitro fertilization outcome: a prospective, randomized, controlled study. Hum Reprod 14:2562–2564
33. Fischer B, Bavister BD (1993) Oxygen tension in the oviduct and uterus of rhesus monkeys, hamsters and rabbits. J Reprod Fertil 99:673–679
34. Kwon HC, Yang HW, Hwang KJ, Yoo JH, Kim MS, Lee CH, Ryu HS, Oh KS (1999) Effects of low oxygen condition on the generation of reactive oxygen species and the development in mouse embryos cultured in vitro. J Obstet Gynaecol Res 25:359–366
35. Hashimoto S, Minami N, Takakura R, Yamada M, Imai H, Kashima N (2000) Low oxygen tension during in vitro maturation is beneficial for supporting the subsequent development of bovine cumulus-oocyte complexes. Mol Reprod Dev 57:353–360
36. Kitagawa Y, Suzuki K, Yoneda A, Watanabe T (2004) Effects of oxygen concentration and antioxidants on the in vitro developmental ability, production of reactive oxygen species (ROS), and DNA fragmentation in porcine embryos. Theriogenology 62:1186–1197
37. Iwata H, Akamatsu S, Minami N, Yamada M (1999) Allopurinol, an inhibitor of xanthine oxidase, improves the development of IVM/IVF bovine embryos (>4 cell) in vitro under certain culture conditions. Theriogenology 51:613–622
38. Harvey AJ, Kind KL, Thompson JG (2002) REDOX regulation of early embryo development. Reproduction 123:479–486
39. Harvey AJ, Kind KL, Pantaleon M, Armstrong DT, Thompson JG (2004) Oxygen-regulated gene expression in bovine blastocysts. Biol Reprod 71:1108–1119
40. Harvey AJ, Navarrete Santos A, Kirstein M, Kind KL, Fischer B, Thompson JG (2007) Differential expression of oxygen-regulated genes in bovine blastocysts. Mol Reprod Dev 74:290–299
41. Harvey AJ (2007) The role of oxygen in ruminant preimplantation embryo development and metabolism. Anim Reprod Sci 98:113–128
42. Pabon JE Jr, Findley WE, Gibbons WE (1989) The toxic effect of short exposures to the atmospheric oxygen concentration on early mouse embryonic development. Fertil Steril 51:896–900
43. Umaoka Y, Noda Y, Narimoto K, Mori T (1992) Effects of oxygen toxicity on early development of mouse embryos. Mol Reprod Dev 31:28–33
44. Karagenc L, Sertkaya Z, Ciray N, Ulug U, Bahceci M (2004) Impact of oxygen concentration on embryonic development of mouse zygotes. Reprod Biomed Online 9:409–417
45. Byatt-Smith JG, Leese HJ, Gosden RG (1991) An investigation by mathematical modelling of whether mouse and human preimplantation embryos in static culture can satisfy their demands for oxygen by diffusion. Hum Reprod 6:52–57
46. Noda Y, Goto Y, Umaoka Y, Shiotani M, Nakayama T, Mori T (1994) Culture of human embryos in alpha modification of eagle's medium under low oxygen tension and low illumination. Fertil Steril 62:1022–1027
47. Dumoulin JC, Vanvuchelen RC, Land JA, Pieters MH, Geraedts JP, Evers JL (1995) Effect of oxygen concentration on in vitro fertilization and embryo culture in the human and the mouse. Fertil Steril 63:115–119
48. Dumoulin JC, Meijers CJ, Bras M, Coonen E, Geraedts JP, Evers JL (1999) Effect of oxygen concentration on human in vitro fertilization and embryo culture. Hum Reprod 14:465–469
49. Bahceci M, Ciray HN, Karagenc L, Ulug U, Bener F (2005) Effect of oxygen concentration during the incubation of embryos of women undergoing ICSI and embryo transfer: a prospective randomized study. Reprod Biomed Online 11:438–443
50. Kea B, Gebhardt J, Watt J, Westphal LM, Lathi RB, Milki AA, Behr B (2007) Effect of reduced oxygen concentrations on the outcome of in vitro fertilization. Fertil Steril 87:213–216

51. Catt JW, Henman M (2000) Toxic effects of oxygen on human embryo development. Hum Reprod 15(Suppl 2):199–206
52. Kovacic B, Vlaisavljevic V (2008) Influence of atmospheric versus reduced oxygen concentration on development of human blastocysts in vitro: a prospective study on sibling oocytes. Reprod Biomed Online 17:229–236
53. Kovacic B, Sajko MC, Vlaisavljevic V (2010) A prospective, randomized trial on the effect of atmospheric versus reduced oxygen concentration on the outcome of intracytoplasmic sperm injection cycles. Fertil Steril 94:511–519
54. Waldenstrom U, Engstrom AB, Hellberg D, Nilsson S (2009) Low-oxygen compared with high-oxygen atmosphere in blastocyst culture, a prospective randomized study. Fertil Steril 91:2461–2465
55. Meintjes M, Chantilis SJ, Ward DC, Douglas JD, Rodriguez AJ, Guerami AR, Bookout DM, Barnett BD, Madden JD (2009) A randomized controlled study of human serum albumin and serum substitute supplement as protein supplements for IVF culture and the effect on live birth rates. Hum Reprod 24:782–789
56. McKiernan SH, Bavister BD (1990) Environmental variables influencing in vitro development of hamster 2-cell embryos to the blastocyst stage. Biol Reprod 43:404–413
57. Higdon HL 3rd, Blackhurst DW, Boone WR (2008) Incubator management in an assisted reproductive technology laboratory. Fertil Steril 89:703–710
58. Ciray HN, Aksoy T, Yaramanci K, Karayaka I, Bahceci M (2009) In vitro culture under physiologic oxygen concentration improves blastocyst yield and quality: a prospective randomized survey on sibling oocytes. Fertil Steril 91:1459–1461
59. Nanassy L, Peterson CA, Wilcox AL, Peterson CM, Hammoud A, Carrell DT (2010) Comparison of 5 % and ambient oxygen during days 3–5 of in vitro culture of human embryos. Fertil Steril 93:579–585
60. Wale PL, Gardner DK (2010) Time-lapse analysis of mouse embryo development in oxygen gradients. Reprod Biomed Online 21:402–410
61. Ottosen LD, Hindkaer J, Husth M, Petersen DE, Kirk J, Ingerslev HJ (2006) Observations on intrauterine oxygen tension measured by fibre-optic microsensors. Reprod Biomed Online 13:380–385
62. Kind KL, Collett RA, Harvey AJ, Thompson JG (2004) Oxygen-regulated expression of GLUT-1, GLUT-3, and VEGF in the mouse blastocyst. Mol Reprod Dev 70:37–44
63. Feil D, Lane M, Roberts CT, Kelley RL, Edwards LJ, Thompson JG, Kind KL (2006) Effect of culturing mouse embryos under different oxygen concentrations on subsequent fetal and placental development. J Physiol 572:87–96
64. Thompson JG, McNaughton C, Gasparrini B, McGowan LT, Tervit HR (2000) Effect of inhibitors and uncouplers of oxidative phosphorylation during compaction and blastulation of bovine embryos cultured in vitro. J Reprod Fertil 118:47–55
65. Balasubramanian S, Son WJ, Kumar BM, Ock SA, Yoo JG, Im GS, Choe SY, Rho GJ (2007) Expression pattern of oxygen and stress-responsive gene transcripts at various developmental stages of in vitro and in vivo preimplantation bovine embryos. Theriogenology 68:265–275
66. Rinaudo PF, Giritharan G, Talbi S, Dobson AT, Schultz RM (2006) Effects of oxygen tension on gene expression in preimplantation mouse embryos. Fertil Steril 86: 1252–1265, 1265 e1251-1236
67. Leoni GG, Rosati I, Succu S, Bogliolo L, Bebbere D, Berlinguer F, Ledda S, Naitana S (2007) A low oxygen atmosphere during IVF accelerates the kinetic of formation of in vitro produced ovine blastocysts. Reprod Domest Anim 42:299–304
68. Takahashi Y, Kanagawa H (1998) Effect of oxygen concentration in the gas atmosphere during in vitro insemination of bovine oocytes on the subsequent embryonic development in vitro. J Vet Med Sci 60:365–367
69. Bermejo-Alvarez P, Lonergan P, Rizos D, Gutierrez-Adan A (2010) Low oxygen tension during IVM improves bovine oocyte competence and enhances anaerobic glycolysis. Reprod Biomed Online 20:341–349

70. Booth PJ, Holm P, Callesen H (2005) The effect of oxygen tension on porcine embryonic development is dependent on embryo type. Theriogenology 63:2040–2052
71. Shalgi R, Kraicer PF, Soferman N (1972) Gases and electrolytes of human follicular fluid. J Reprod Fertil 28:335–340
72. Fischer B, Kunzel W, Kleinstein J, Gips H (1992) Oxygen tension in follicular fluid falls with follicle maturation. Eur J Obstet Gynecol Reprod Biol 43:39–43
73. Clark AR, Stokes YM, Lane M, Thompson JG (2006) Mathematical modelling of oxygen concentration in bovine and murine cumulus-oocyte complexes. Reproduction 131: 999–1006
74. Karja NW, Wongsrikeao P, Murakami M, Agung B, Fahrudin M, Nagai T, Otoi T (2004) Effects of oxygen tension on the development and quality of porcine in vitro fertilized embryos. Theriogenology 62:1585–1595
75. Iwamoto M, Onishi A, Fuchimoto D, Somfai T, Takeda K, Tagami T, Hanada H, Noguchi J, Kaneko H, Nagai T, Kikuchi K (2005) Low oxygen tension during in vitro maturation of porcine follicular oocytes improves parthenogenetic activation and subsequent development to the blastocyst stage. Theriogenology 63:1277–1289
76. Preis KA, Seidel GE Jr, Gardner DK (2007) Reduced oxygen concentration improves the developmental competence of mouse oocytes following in vitro maturation. Mol Reprod Dev 74:893–903
77. Hashimoto S (2009) Application of in vitro maturation to assisted reproductive technology. J Reprod Dev 55:1–10
78. Pereira MM, Machado MA, Costa FQ (2010) Effect of oxygen tension and serum during IVM on developmental competence of bovine oocytes. Reprod Fertil Dev 22:9
79. Pinyopummintr T, Bavister BD (1995) Optimum gas atmosphere for in vitro maturation and in vitro fertilization of bovine oocytes. Theriogenology 44:471–477
80. Watson AJ, De Sousa P, Caveney A, Barcroft LC, Natale D, Urquhart J, Westhusin ME (2000) Impact of bovine oocyte maturation media on oocyte transcript levels, blastocyst development, cell number, and apoptosis. Biol Reprod 62:355–364
81. Hu Y, Betzendahl I, Cortvrindt R, Smitz J, Eichenlaub-Ritter U (2001) Effects of low O2 and ageing on spindles and chromosomes in mouse oocytes from pre-antral follicle culture. Hum Reprod 16:737–748
82. Park JI, Hong JY, Yong HY, Hwang WS, Lim JM, Lee ES (2005) High oxygen tension during in vitro oocyte maturation improves in vitro development of porcine oocytes after fertilization. Anim Reprod Sci 87:133–141
83. Banwell KM, Lane M, Russell DL, Kind KL, Thompson JG (2007) Oxygen concentration during mouse oocyte in vitro maturation affects embryo and fetal development. Hum Reprod 22:2768–2775
84. Oyamada T, Fukui Y (2004) Oxygen tension and medium supplements for in vitro maturation of bovine oocytes cultured individually in a chemically defined medium. J Reprod Dev 50:107–117
85. Halliwell B, Gutteridge JM (1989) Free radicals in biology and medicine, 2nd edn. University Press, Oxford
86. Gutteridge JM (1987) A method for removal of trace iron contamination from biological buffers. FEBS Lett 214:362–364
87. Minotti G (1993) Sources and role of iron in lipid peroxidation. Chem Res Toxicol 6:134–146
88. Nasr-Esfahani M, Johnson MH, Aitken RJ (1990) The effect of iron and iron chelators on the in vitro block to development of the mouse preimplantation embryo: BAT6 a new medium for improved culture of mouse embryos in vitro. Hum Reprod 5:997–1003
89. Martin-Romero FJ, Miguel-Lasobras EM, Dominguez-Arroyo JA, Gonzalez-Carrera E, Alvarez IS (2008) Contribution of culture media to oxidative stress and its effect on human oocytes. Reprod Biomed Online 17:652–661
90. Erbach GT, Bhatnagar P, Baltz JM, Biggers JD (1995) Zinc is a possible toxic contaminant of silicone oil in microdrop cultures of preimplantation mouse embryos. Hum Reprod 10:3248–3254

91. Nasr-Esfahani MH, Johnson MH (1992) How does transferrin overcome the in vitro block to development of the mouse preimplantation embryo? J Reprod Fertil 96:41–48
92. Matsukawa T, Ikeda S, Imai H, Yamada M (2002) Alleviation of the two-cell block of ICR mouse embryos by polyaminocarboxylate metal chelators. Reproduction 124:65–71
93. Aguilar J, Reyley M (2005) The uterine tubal fluid: secretion, composition and biological effects. Anim Reprod 2:15
94. Orsi NM, Leese HJ (2001) Protection against reactive oxygen species during mouse preimplantation embryo development: role of EDTA, oxygen tension, catalase, superoxide dismutase and pyruvate. Mol Reprod Dev 59:44–53
95. Grzelak A, Rychlik B, Bartosz G (2001) Light-dependent generation of reactive oxygen species in cell culture media. Free Radical Biol Med 30:8
96. Stoien JD, Wang RJ (1974) Effect of near-ultraviolet and visible light on mammalian cells in culture II. Formation of toxic photoproducts in tissue culture medium by blacklight. Proc Nat Acad Sci USA 71: 3961–3965
97. Cunningham ML, Krinsky NI, Giovanazzi SM, Peak MJ (1985) Superoxide anion is generated from cellular metabolites by solar radiation and its components. J Free Radical Biol Med 1:381–385
98. Wang RJ, Nixon BR (1978) Identification of hydrogen peroxide as a photoproduct toxic to human cells in tissue-culture medium irradiated with "daylight" fluorescent light. In vitro. 14:715–722
99. Nakayama T, Noda Y, Goto Y, Mori T (1994) Effects of visible light and other environmental factors on the production of oxygen radicals by hamster embryos. Theriogenology 41:499–510
100. Umaoka Y, Noda Y, Nakayama T, Narimoto K, Mori T, Iritani A (1992) Effect of visual light on in vitro embryonic development in the hamster. Theriogenology 38:1043–1054
101. Barlow P, Puissant F, Van der Zwalmen P, Vandromme J, Trigaux P, Leroy F (1992) In vitro fertilization, development, and implantation after exposure of mature mouse oocytes to visible light. Mol Reprod Dev 33:297–302
102. Daniel JC (1964) Cleavage of mammalian ova inhibited by visible light. Nature 201:2
103. Takahashi M, Saka N, Takahashi H, Kanai Y, Schultz RM, Okano A (1999) Assessment of DNA damage in individual hamster embryos by comet assay. Mol Reprod Dev 54:1–7
104. Oh SJ, Gong SP, Lee ST, Lee EJ, Lim JM (2007) Light intensity and wavelength during embryo manipulation are important factors for maintaining viability of preimplantation embryos in vitro. Fertil Steril 88:1150–1157
105. Schumacher A, Fischer B (1988) Influence of visible light and room temperature on cell proliferation in preimplantation rabbit embryos. J Reprod Fertil 84:197–204
106. Kruger TF, Stander FS (1985) The effect of fluorescent light on the cleavage of two-cell mouse embryos. S Afr Med J 68:744–745
107. Ottosen LD, Hindkjaer J, Ingerslev J (2007) Light exposure of the ovum and preimplantation embryo during ART procedures. J Assist Reprod Genet 24:99–103
108. Hirao Y, Yanagimachi R (1978) Detrimental effect of visible light on meiosis of mammalian eggs in vitro. J Exp Zool 206:5
109. Takenaka M, Horiuchi T, Yanagimachi R (2007) Effects of light on development of mammalian zygotes. Proc Nat Acad Sci USA 104:14289–14293
110. Korhonen K, Sjovall S, Viitanen J, Ketoja E, Makarevich A, Peippo J (2009) Viability of bovine embryos following exposure to the green filtered or wider bandwidth light during in vitro embryo production. Hum Reprod 24:308–314
111. Kielbassa C, Roza L, Epe B (1997) Wavelength dependence of oxidative DNA damage induced by UV and visible light. Carcinogenesis 18:811–816
112. Brinster RL (1969) In vitro cultivation of mammalian ova. Adv Biosci 4:35
113. Grinsted J, Kjer JJ, Blendstrup K, Pedersen JF (1985) Is low temperature of the follicular fluid prior to ovulation necessary for normal oocyte development? Fertil Steril 43:34–39
114. Hunter RH, Nichol R (1986) A preovulatory temperature gradient between the isthmus and ampulla of pig oviducts during the phase of sperm storage. J Reprod Fertil 77:599–606

115. Leese HJ, Baumann CG, Brison DR, McEvoy TG, Sturmey RG (2008) Metabolism of the viable mammalian embryo: quietness revisited. Mol Hum Reprod 14:667–672
116. Bahat A, Eisenbach M, Tur-Kaspa I (2005) Periovulatory increase in temperature difference within the rabbit oviduct. Hum Reprod 20:2118–2121
117. Eisenbach M, Giojalas LC (2006) Sperm guidance in mammals—an unpaved road to the egg. Nat Rev Mol Cell Biol 7:276–285
118. Flanagan SW, Moseley PL, Buettner GR (1998) Increased flux of free radicals in cells subjected to hyperthermia: detection by electron paramagnetic resonance spin trapping. FEBS Lett 431:285–286
119. Shi DS, Avery B, Greve T (1998) Effects of temperature gradients on in vitro maturation of bovine oocytes. Theriogenology 50:667–674
120. Ozawa M, Hirabayashi M, Kanai Y (2002) Developmental competence and oxidative state of mouse zygotes heat-stressed maternally or in vitro. Reproduction 124:683–689
121. Matsuzuka T, Ozawa M, Nakamura A, Ushitani A, Hirabayashi M, Kanai Y (2005) Effects of heat stress on the redox status in the oviduct and early embryonic development in mice. J Reprod Dev 51:281–287
122. Sakatani M, Kobayashi S, Takahashi M (2004) Effects of heat shock on in vitro development and intracellular oxidative state of bovine preimplantation embryos. Mol Reprod Dev 67:77–82
123. Sakatani M, Yamanaka K, Kobayashi S, Takahashi M (2008) Heat shock-derived reactive oxygen species induce embryonic mortality in in vitro early stage bovine embryos. J Reprod Dev 54:496–501
124. Ealy AD, Howell JL, Monterroso VH, Arechiga CF, Hansen PJ (1995) Developmental changes in sensitivity of bovine embryos to heat shock and use of antioxidants as thermoprotectants. J Anim Sci 73:1401–1407
125. Malayer JR, Pollard JW, Hansen PJ (1992) Modulation of thermal killing of bovine lymphocytes and preimplantation mouse embryos by alanine and taurine. Am J Vet Res 53:689–694
126. Arechiga CF, Ealy AD, Hansen PJ (1995) Evidence that glutathione is involved in thermotolerance of preimplantation murine embryos. Biol Reprod 52:1296–1301
127. Loven DP (1988) A role for reduced oxygen species in heat induced cell killing and the induction of thermotolerance. Med Hypotheses 26:39–50
128. Lord-Fontaine S, Averill DA (1999) Enhancement of cytotoxicity of hydrogen peroxide by hyperthermia in chinese hamster ovary cells: role of antioxidant defenses. Arch Biochem Biophys 363:283–295
129. Ealy AD, Drost M, Barros CM, Hansen PJ (1992) Thermoprotection of preimplantation bovine embryos from heat shock by glutathione and taurine. Cell Biol Int Rep 16:125–131
130. de Castro e Paula LA, Hansen PJ (2008) Modification of actions of heat shock on development and apoptosis of cultured preimplantation bovine embryos by oxygen concentration and dithiothreitol. Mol Reprod Dev 75: 1338–1350
131. Lord-Fontaine S, Averill-Bates DA (2002) Heat shock inactivates cellular antioxidant defenses against hydrogen peroxide: protection by glucose. Free Radic Biol Med 32: 752–765
132. Rivera RM, Hansen PJ (2001) Development of cultured bovine embryos after exposure to high temperatures in the physiological range. Reproduction 121:107–115
133. Wang WH, Meng L, Hackett RJ, Odenbourg R, Keefe DL (2001) Limited recovery of meiotic spindles in living human oocytes after cooling-rewarming observed using polarized light microscopy. Hum Reprod 16:2374–2378
134. Fujiwara M, Takahashi K, Izuno M, Duan YR, Kazono M, Kimura F, Noda Y (2007) Effect of micro-environment maintenance on embryo culture after in vitro fertilization: comparison of top-load mini incubator and conventional front-load incubator. J Assist Reprod Genet 24:5–9
135. Cooke S, Tyler JP, Driscoll G (2002) Objective assessments of temperature maintenance using in vitro culture techniques. J Assist Reprod Genet 19:368–375

136. Burton GW (2002) Oxygen, early embryonic metabolism and free radical-mediated embryopathies. BioMedicine Online 6:84–96
137. Leese HJ (2002) Quiet please, do not disturb: a hypothesis of embryo metabolism and viability. BioEssays 24:845–849
138. Leese HJ, Sturmey RG, Baumann CG, McEvoy TG (2007) Embryo viability and metabolism: obeying the quiet rules. Hum Reprod 22:3047–3050
139. Crabtree HG (1929) Observations on the carbohydrate metabolism of tumours. Biochem J 23:536–545
140. Long LH, Halliwell B (2009) Artefacts in cell culture: pyruvate as a scavenger of hydrogen peroxide generated by ascorbate or epigallocatechin gallate in cell culture media. Biochem Biophys Res Commun 388:700–704
141. Whiteley GS, Fuller BJ, Hobbs KE (1992) Deterioration of cold-stored tissue specimens due to lipid peroxidation: modulation by antioxidants at high subzero temperatures. Cryobiology 29:668–673
142. Salahudeen AK, Huang H, Patel P, Jenkins JK (2000) Mechanism and prevention of cold storage-induced human renal tubular cell injury. Transplantation 70:1424–1431
143. Park J, Grant CM, Davies MJ, Dawes IW (1998) The cytoplasmic Cu, Zn superoxide dismutase of saccharomyces cerevisiae is required for resistance to freeze-thaw stress. J Biol Chem 273:22921–22928
144. Iyengar J, George A, Russell JC, Das DK (1990) The effects of an iron chelator on cellular injury induced by vascular stasis caused by hypothermia. J Vasc Surg 12:545–551
145. Bhaumik G, Srivastava KK, Selvamurthy W, Purkayastha SS (1995) The role of free radicals in cold injuries. Int J Biometeorol 38:171–175
146. Combelles CM, Holick EA, Paolella LJ, Walker DC, Wu Q (2010) Profiling of superoxide dismutase isoenzymes in compartments of the developing bovine antral follicles. Reproduction 139:871–881
147. El Mouatassim S, Guerin P, Menezo Y (1999) Expression of genes encoding antioxidant enzymes in human and mouse oocytes during the final stages of maturation. Mol Hum Reprod 5:720–725
148. Leyens G, Knoops B, Donnay I (2004) Expression of peroxiredoxins in bovine oocytes and embryos produced in vitro. Mol Reprod Dev 69:243–251
149. Guerin P, El Mouatassim S, Menezo Y (2001) Oxidative stress and protection against reactive oxygen species in the pre-implantation embryo and its surroundings. Hum Reprod Update 7:175–189
150. Combelles CM, Gupta S, Agarwal A (2009) Could oxidative stress influence the in vitro maturation of oocytes? Reprod Biomed Online 18:864–880
151. Somfai T, Ozawa M, Noguchi J, Kaneko H, Kuriani Karja NW, Farhudin M, Dinnyes A, Nagai T, Kikuchi K (2007) Developmental competence of in vitro-fertilized porcine oocytes after in vitro maturation and solid surface vitrification: effect of cryopreservation on oocyte antioxidative system and cell cycle stage. Cryobiology 55:115–126
152. Gupta MK, Uhm SJ, Lee HT (2010) Effect of vitrification and beta-mercaptoethanol on reactive oxygen species activity and in vitro development of oocytes vitrified before or after in vitro fertilization. Fertil Steril 93:2602–2607
153. Tarin JJ, Trounson AO (1993) Effects of stimulation or inhibition of lipid peroxidation on freezing-thawing of mouse embryos. Biol Reprod 49:1362–1368
154. Ahn HJ, Sohn IP, Kwon HC, Jo DH, Park YD, Min CK (2002) Characteristics of the cell membrane fluidity, actin fibers, and mitochondrial dysfunctions of frozen-thawed two-cell mouse embryos. Mol Reprod Dev 61:466–476
155. Lane M, Maybach JM, Gardner DK (2002) Addition of ascorbate during cryopreservation stimulates subsequent embryo development. Hum Reprod 17:2686–2693
156. Rahimi G, Isachenko E, Sauer H, Isachenko V, Wartenberg M, Hescheler J, Mallmann P, Nawroth F (2003) Effect of different vitrification protocols for human ovarian tissue on reactive oxygen species and apoptosis. Reprod Fertil Dev 15:343–349

157. Dinara S, Sengoku K, Tamate K, Horikawa M, Ishikawa M (2001) Effects of supplementation with free radical scavengers on the survival and fertilization rates of mouse cryopreserved oocytes. Hum Reprod 16:1976–1981
158. Somfai T, Ozawa M, Noguchi J, Kaneko H, Nakai M, Maedomari N, Ito J, Kashiwazaki N, Nagai T, Kikuchi K (2009) Live piglets derived from in vitro-produced zygotes vitrified at the pronuclear stage. Biol Reprod 80:42–49
159. Nedambale TL, Du F, Yang X, Tian XC (2006) Higher survival rate of vitrified and thawed in vitro produced bovine blastocysts following culture in defined medium supplemented with beta-mercaptoethanol. Anim Reprod Sci 93:61–75
160. Hosseini SM, Forouzanfar M, Hajian M, Asgari V, Abedi P, Hosseini L, Ostadhosseini S, Moulavi F, Safahani Langrroodi M, Sadeghi H, Bahramian H, Eghbalsaied S, Nasr-Esfahani MH (2009) Antioxidant supplementation of culture medium during embryo development and/or after vitrification-warming; which is the most important? J Assist Reprod Genet 26:355–364
161. Somfai T, Noguchi J, Kaneko H, Nakai M, Ozawa M, Kashiwazaki N, Egerszegi I, Ratky J, Nagai T, Kikuchi K (2010) Production of good-quality porcine blastocysts by in vitro fertilization of follicular oocytes vitrified at the germinal vesicle stage. Theriogenology 73:147–156
162. Paszkowski T, Clarke RN, Hornstein MD (2002) Smoking induces oxidative stress inside the Graafian follicle. Hum Reprod 17:921–925
163. Tiboni GM, Bucciarelli T, Giampietro F, Sulpizio M, Di Ilio C (2004) Influence of cigarette smoking on vitamin E, vitamin A, beta-carotene and lycopene concentrations in human preovulatory follicular fluid. Int J Immunopathol Pharmacol 17:389–393
164. Younglai EV, Holloway AC, Foster WG (2005) Environmental and occupational factors affecting fertility and IVF success. Hum Reprod Update 11:43–57
165. Tatone C, Carbone MC, Falone S, Aimola P, Giardinelli A, Caserta D, Marci R, Pandolfi A, Ragnelli AM, Amicarelli F (2006) Age-dependent changes in the expression of superoxide dismutases and catalase are associated with ultrastructural modifications in human granulosa cells. Mol Hum Reprod 12:655–660
166. Tarin JJ, Gomez-Piquer V, Pertusa JF, Hermenegildo C, Cano A (2004) Association of female aging with decreased parthenogenetic activation, raised MPF, and MAPKs activities and reduced levels of glutathione S-transferases activity and thiols in mouse oocytes. Mol Reprod Dev 69:402–410
167. Hamatani T, Falco G, Carter MG, Akutsu H, Stagg CA, Sharov AA, Dudekula DB, VanBuren V, Ko MS (2004) Age-associated alteration of gene expression patterns in mouse oocytes. Hum Mol Genet 13:2263–2278
168. Tarin JJ (1995) Aetiology of age-associated aneuploidy: a mechanism based on the 'free radical theory of ageing'. Hum Reprod 10:1563–1565
169. Tarin JJ (1996) Potential effects of age-associated oxidative stress on mammalian oocytes/embryos. Mol Hum Reprod 2:717–724
170. Palacio JR, Iborra A, Ulcova-Gallova Z, Badia R, Martinez P (2006) The presence of antibodies to oxidative modified proteins in serum from polycystic ovary syndrome patients. Clin Exp Immunol 144:217–222
171. Sabuncu T, Vural H, Harma M (2001) Oxidative stress in polycystic ovary syndrome and its contribution to the risk of cardiovascular disease. Clin Biochem 34:407–413
172. Fenkci V, Fenkci S, Yilmazer M, Serteser M (2003) Decreased total antioxidant status and increased oxidative stress in women with polycystic ovary syndrome may contribute to the risk of cardiovascular disease. Fertil Steril 80:123–127
173. Dinger Y, Akcay T, Erdem T, Ilker Saygili E, Gundogdu S (2005) DNA damage, DNA susceptibility to oxidation and glutathione level in women with polycystic ovary syndrome. Scand J Clin Lab Invest 65:721–728
174. Forsberg H, Borg LA, Cagliero E, Eriksson UJ (1996) Altered levels of scavenging enzymes in embryos subjected to a diabetic environment. Free Radical Res 24:451–459

175. Ornoy A (2007) Embryonic oxidative stress as a mechanism of teratogenesis with special emphasis on diabetic embryopathy. Reprod Toxicol 24:31–41
176. Yilmaz M, Biri A, Karakoc A, Toruner F, Bingol B, Cakir N, Tiras B, Ayvaz G, Arslan M (2005) The effects of rosiglitazone and metformin on insulin resistance and serum androgen levels in obese and lean patients with polycystic ovary syndrome. J Endocrinol Invest 28:1003–1008
177. Wu LL, Dunning KR, Yang X, Russell DL, Lane M, Norman RJ, Robker RL (2010) High-fat diet causes lipotoxicity responses in cumulus-oocyte complexes and decreased fertilization rates. Endocrinology 151:5438–5445
178. Pasqualotto EB, Agarwal A, Sharma RK, Izzo VM, Pinotti JA, Joshi NJ, Rose BI (2004) Effect of oxidative stress in follicular fluid on the outcome of assisted reproductive procedures. Fertil Steril 81:973–976
179. Lapointe S, Sullivan R, Sirard MA (1998) Binding of a bovine oviductal fluid catalase to mammalian spermatozoa. Biol Reprod 58:747–753
180. Lapointe J, Bilodeau JF (2003) Antioxidant defenses are modulated in the cow oviduct during the estrous cycle. Biol Reprod 68:1157–1164
181. Jana SK, K NB, Chattopadhyay R, Chakravarty B, Chaudhury K (2010) Upper control limit of reactive oxygen species in follicular fluid beyond which viable embryo formation is not favorable. Reprod Toxicol 29:447–451
182. Das S, Chattopadhyay R, Ghosh S, Goswami SK, Chakravarty BN, Chaudhury K (2006) Reactive oxygen species level in follicular fluid–embryo quality marker in IVF? Hum Reprod 21:2403–2407
183. Jozwik M, Wolczynski S, Szamatowicz M (1999) Oxidative stress markers in preovulatory follicular fluid in humans. Mol Hum Reprod 5:409–413

Chapter 13
Antioxidant Strategies to Overcome OS in IVF-Embryo Transfer

Mitali Rakhit, Sheila R. Gokul, Ashok Agarwal and Stefan S. du Plessis

Abstract During in vitro fertilization (IVF) procedures, the transferred embryo is exposed to many sources of reactive oxygen species (ROS) originating both internally (in vivo from the male or female reproductive tracts and secretes) and externally (in vitro during the IVF procedures). In order to prevent oxidative damage various antioxidant strategies may be employed. Several antioxidants may be used in the prevention of oxidative stress (OS) in an assisted reproductive technology setting. Most studies involve enzymatic or non-enzymatic antioxidants, and these same antioxidants may be effective in therapy individually or in combination with another treatment. Throughout this chapter, the most studied antioxidants are discussed. The experiments that provide evidence for or against the effectiveness of these compounds to decrease ROS and protect the embryo from OS in an IVF-embryo transfer setting are evaluated based on the strength of evidence in order to give a recommendation for or against the use of an antioxidant or combination of antioxidants in OS-preventing therapies.

Keywords: Antioxidant strategies · Oxidative stress · IVF-embryo transfer · Reactive oxygen species · In vitro fertilization procedures

---

M. Rakhit · S. R. Gokul · A. Agarwal
Cleveland Clinic Center for Reproductive Medicine, 9500 Euclid Avenue, Cleveland, OH 44195, USA
e-mail: mitali.rakhit@gmail.com

S. R. Gokul
e-mail: Sheila.gokul@gmail.com

A. Agarwal
e-mail: agarwaa@ccf.org

S. S. du Plessis (✉)
Medical Physiology, Faculty of Health Sciences, University of Stellenbosch, Francie van Zijl Drive, Tygerberg, Western Cape 7505, South Africa
e-mail: ssdp@sun.ac.za

13.1 Introduction

Up to 15 % of all reproductive age couples are unable to achieve a natural pregnancy after one year of regular unprotected intercourse and are thus classified as infertile [1]. As both sexes must undergo a series of complex physiological processes in order to achieve pregnancy male and female factors (endogenous or exogenous origin) can contribute equally toward couple infertility. With the advent of assisted reproductive technologies (ART) many of these couples seek clinical intervention in order to achieve pregnancy. Various ART procedures such as in vitro fertilization (IVF) and intracytoplasmic sperm injection (ICSI) are available; however these interventions are both emotionally and physically demanding. As these are highly specialized techniques, they are extremely costly and therefore place a further financial burden and stress on the couple [2]. Due to the significant amount of energy and resources invested in such ART treatment, it is paramount to reduce any factors that may potentially prevent a successful outcome.

Various endogenous and exogenous factors can impede on ART outcome [3]. First, the quality of the gametes plays a vital role, but due to the nature of the individual male or female factors, any of these endogenous and or exogenous factors will have an even more pronounced effect on the outcome. Factors such as oxidative stress (OS) that result from an imbalance between reactive oxygen species (ROS) and antioxidant capacity is just one such contributing factor. Both in vivo and in vitro (during ART procedures) OS have been implicated to be a major role player in pregnancy failure, especially when it develops in the environment surrounding the gametes [4].

In the recent past various research studies have focused on studying antioxidant strategies to prevent OS damage to gametes and developing embryos, in order to optimize ART outcome. Therefore, the aim of this chapter is to review the current understanding of antioxidant treatments and strategies employed in the clinical setting with the intention of preventing OS during IVF and embryo transfer.

13.2 ROS, Antioxidants, and Oxidative Stress

Most ROS are free radicals involved in various signaling processes throughout the body. While necessary for normal body function, they are harmful in larger amounts and are involved in the pathophysiology of various diseases. Three of the most common ROS members include hydrogen peroxide (H_2O_2), the superoxide anion (O_2^-), and the hydroxyl radical (OH^-) [5]. Free radicals, by definition, are molecules, which contain one or more unpaired electrons in their valence shell. Therefore, they are inclined to remove an electron from a surrounding molecule in order to complete their octet. However, in the process of stabilizing itself, a former free radical ends up also transforming the other molecule into a free radical, thereby causing a chain reaction. ROS result primarily as a by-product of cellular

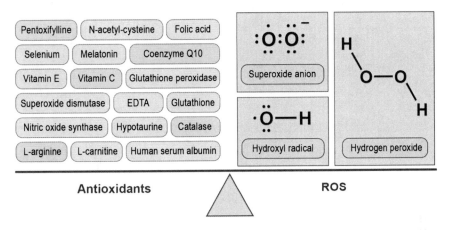

Fig. 13.1 *ROS* (reactive oxygen species) and antioxidants are contained in balanced levels throughout the body in order to maintain homeostasis and prevent oxidative stress. Reprinted with permission, Cleveland Clinic Center for Medical Art & Photography © 2010. All Rights Reserved

respiration. As oxygen is the final electron acceptor in the electron transport chain, failure of these ions to bind to hydrogen ions in order to form water molecules leads them to be released as free radicals from the mitochondria. Although this process accounts for the greatest percentage of ROS, there are many other mechanisms of ROS production in the body, such as through immature spermatozoa and leukocytes in the seminal fluid [6].

The characteristic chain reaction of ROS is usually disrupted by the presence of an antioxidant. Antioxidants are molecules that are able to donate an electron to a free radical in order to stabilize it. Antioxidants are not affected by this mechanism because they are stable with or without the electron that they have donated. There are two recognized groups of antioxidants: enzymatic and non-enzymatic. Enzymatic antioxidants such as superoxide dismutase (SOD) and its variants (Cu-SOD, Zn-SOD, Mn-SOD) can be found in cytoplasm, mitochondria, and endometrial glandular cells [7]. These enzymes scavenge O_2^- radicals and protect tissues from damage. Similarly, catalase and glutathione peroxidase both scavenge H_2O_2. Some examples of non-enzymatic antioxidants include vitamins A, C, and E as well as carotene, ascorbate, and hypotaurine. These antioxidants found in the seminal plasma, tubal fluid, epithelium of the endometrium, follicular fluid, and ovary exhibit the typical chain breaking property. Antioxidants are endogenous and acquired exogenously via dietary intake in both males and females [7]. ROS and antioxidants are contained in balanced levels throughout the body in order to maintain homeostasis (Fig. 13.1). However, if this delicate harmony is disrupted and comes into discord due to excess buildup of ROS generation or inadequate antioxidant levels, OS develops. OS is potentially harmful because elevated

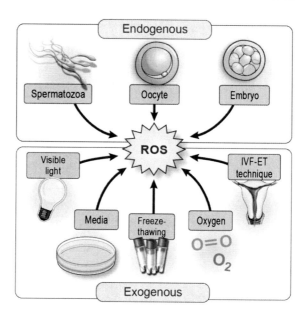

Fig. 13.2 In an IVF setting ROS can originate from both endogenous as well as exogenous sources. Reprinted with permission, Cleveland Clinic Center for Medical Art & Photography © 2010. All Rights Reserved

amounts of ROS can cause changes in basic regulatory molecules such as proteins, DNA, lipids, in addition to causing ATP depletion. The prevention of OS is key to maintaining normal reproductive function [7].

13.3 Sources of ROS in an IVF Setting

Due to the nature of an IVF setting, risk of exposure to ROS and OS is much higher than in vivo [5]. It is also important to note that the natural antioxidants found in vivo during fertilization are not present in a laboratory setting, thereby further exacerbating the situation [7]. Endogenous as well as exogenous sources of ROS exist in an IVF setting (Fig. 13.2). Both of these causes must be considered in ART procedures in order to ensure successful fertilization and pregnancy.

13.3.1 Internal

OS that develops in an in vitro setting can originate from ROS that was generated in vivo. In case of male factor, it is known that human spermatozoa requires low levels of ROS in order to function normally as ROS are involved in the processes of capacitation, and the successful completion of the acrosome reaction. Seminal ROS can originate from both spermatozoa and leukocytes. ROS can be produced in spermatozoa through the cytoplasmic cytochrome b5 reductase system in the mitochondria, the NADH-dependent oxidoreductase system in the mitochondria, and the NADPH oxidase system in the sperm plasma membrane [8]. Men with confirmed teratozoospermia and/or leukocytospermia are likely to have higher levels of ROS than normal men. This is due to the fact that increased levels of ROS correlate with higher incidences or morphologically abnormal spermatozoa and an increased white blood cell count in the semen [8]. It is estimated by Saleh et al. [9] that spermatozoa used in at least 50 % of all IVF cases are obtained from a setting that has experienced OS. Due to the magnitude of this finding it is therefore vital to address this issue and explore alternative means to minimize gamete exposure to OS prior to undergoing IVF.

Some consequences of OS, such as lipid peroxidation, can decrease the fluidity of the plasma membrane, thereby leading to reduced or possibly even complete loss of motility. As high levels of ROS can lead to a decrease in mitochondrial membrane potential a drop in membrane potential could be harmful because it is likely to activate apoptosis signaling pathways in the spermatozoa [10]. Furthermore, ROS may also cause DNA damage in spermatozoa which manifest as shifts, breaks within the nitrogenous bases, and changes in chromosome arrangements [11]. These alterations can ultimately result in increased abortion rates and the augmented incidence of various fetal abnormalities. During the use of the ICSI technique, the risk of negative outcomes is of greater concern because only one spermatozoon is selected and placed directly inside the oocyte. Therefore, since there is no barrier by natural selection, this could result in detrimental consequences for the developing embryo if the selected spermatozoon was exposed to OS-related DNA damage.

Gametes provide a source of ROS in both in vivo and in vitro settings. OS contributed by male gametes has great significance in procedures involving ART, as shown in a study by Baker et al. [12]. From this study it was concluded that there was a negative correlation between elevated levels of sperm generated ROS and fertilization rates, subsequent embryo development, and clinical pregnancies. Female factors such as follicular fluid may also contribute to increased ROS levels during ART, especially in the IVF setting. Low levels of ROS are required to ensure and maintain proper folliculogenesis, ovarian steroidogenesis, oocyte maturation, and luteolysis in the female reproductive system [13]. A study by Pasqualotto et al. [14] determined that ROS at low concentrations may actually be used as a predictive indicator of future success in terms of IVF outcomes. Certain amounts of ROS-mediated lipid peroxidation, a bioindicator of metabolic activity, are necessary in order to establish pregnancy. In addition, Bedaiwy et al. [15] was able to show that an increase in total antioxidant capacity in follicular fluid correlated with higher rates of pregnancy. Alternatively, exceedingly high levels of ROS can have negative effects on the oocyte. Free radicals are able to impair microtubule function and the cytoskeleton while also causing aneuploidy and chromosomal scattering. A study by Sugino et al. [16] also found that measuring 8-hydroxy-2-deoxyguanosine levels, a biomarker for DNA damage caused by OS, in granulosa cells can predict oocyte quality and subsequent embryo development during IVF treatment cycles.

ROS can also be produced by the embryo post fertilization. Some of the pathways that contribute to the levels of ROS include oxidative phosphorylation, NADPH, and xanthine oxidase [17]. A study conducted by Goto et al. [18] showed that ROS production was higher in embryos cultured in vitro compared to in vivo. The evidence from this study was not conclusive about whether or not the results were affected by the procedural constraints of IVF. However, the negative effects of ROS on a developing embryo have been observed in other studies. Decreased development and growth of normal blastocysts as well as high rates of fragmentation have been observed and associated with increased levels of ROS in day 1 embryo culture. These results eventually correlated with lower rates of pregnancy in clinical settings [6].

13.3.2 External Sources of ROS

In addition to the internal sources there are also many external sources and factors that contribute to the production of ROS in the clinical ART setting that should be taken into account. When considering the IVF procedure it is evident that there are a multitude of modes during IVF procedures that can lead to increased ROS production. These include everything from the specific techniques used to complete the process to freeze–thawing, visible light, oxygen concentration, and culture medium. The contribution of ROS generation varies between IVF and ICSI due to the difference in techniques employed. The main discrepancy between the two techniques is that multiple sperm are conventionally used during IVF as compared to a single selected sperm in ICSI. The greatest danger of increasing the levels of ROS in ICSI is due to the risk of transferring some of the ROS containing culture media into the oocyte along with the single spermatozoon. This procedure may put the DNA material located within the oocyte in danger of being damaged [19]. However, there is a significant decrease in the amount of incubation time required between IVF and ICSI. The incubation period is shorter in ICSI compared to IVF and this leads to a decrease in the risk of ROS generation from exposure to abnormal spermatozoa. During IVF procedures, both the spermatozoa and oocyte, along with its cumulus cells, are able to generate ROS. However, during ICSI procedures, the cumulus cells are removed from the oocyte and therefore there is an immediate reduction in the rate of ROS production [6]. Additionally, spermatozoa being prepared for either IVF or ICSI must undergo centrifugation. A study by Lampiao et al. [20] has shown that there is an increase in levels of ROS in spermatozoa that have been centrifuged, which can lead to OS.

Cryopreservation is another technique commonly used during IVF treatment cycles. Cryopreserved spermatozoa may sometimes suffer membrane damage as a result of lipid peroxidation as well as a decrease in SOD levels, which are bioindicators of OS [21]. Although oocyte cryopreservation methods such as vitrification are still novel and actively researched, a study by Rahimi et al. [22] has shown that techniques which use a slower method of thawing increase levels of apoptosis and production of ROS. When considering this observation, it is likely that more gradual freeze–thaw techniques make the oocytes more susceptible to damage from subsequent OS compared to more rapid alternatives. Therefore, it can be concluded that there are various benefits and costs of conducting IVF as well as ICSI techniques in terms of risk of exposure to ROS.

It is important to recognize that it is very difficult to simulate in vivo reproductive conditions in an in vitro setting. This is especially true of oxygen concentrations. Oxygen concentration in the culture medium is significantly greater than the concentration within the female reproductive tract [6]. This is dangerous because elevated concentrations of oxygen can lead to increased amounts of H_2O_2 production, an agent of DNA fragmentation [7]. A study by Jones et al. [23] was able to quantify the values and found that the atmospheric oxygen concentration is approximately 20 times greater than the concentration within the body. Placing gametes in an environment that is so highly concentrated with oxygen greatly increases the chances for generation of ROS as well as OS. Therefore, exposure to high atmospheric concentrations of oxygen is often presumed to be the main cause of a rise in ROS levels and the impairment of subsequent embryo development under in vitro conditions [6]. The presence of visible light during the execution of IVF procedures can also have potentially harmful consequences in terms of ROS production. Girotti et al. [24] conducted a study about the effects of visible light and found that it could cause damage to cholesterol and unsaturated lipids contained in cell membranes through OS. This phenomenon would increase overall generation of ROS and simultaneously contribute to the amount of DNA damage. Nakayama et al. [25] reported that due to the negative effects of visible light exposure during IVF, many laboratories have installed fluorescent light filters in order to reduce additional damage from this source.

Another important source of OS can result from overproduction of ROS in the culture medium used during IVF procedures. The presence of ROS within the culture medium itself can cause direct damage to the oocyte and impair embryo development [3]. This is especially likely to occur if the media contains certain metallic ions such as iron or copper, which may unintentionally be transferred to the oocyte or embryo during standard IVF techniques and initiate the generation of ROS. Bedaiwy et al. [15] reported that increased levels of ROS in the culture medium correlated with impaired blastocyst development, cleavage, and fertilization, as well as increased fragmentation rates. In a review by Du Plessis et al. [6] it was advocated that some supplements, added to enhance the culture medium, may actually lead to increasing ROS levels. For example the addition of a serum containing the amine oxidase enzyme actually contributed to an increase in the amount of H_2O_2 molecules present in the media, thereby inducing a greater likelihood of OS-related damage to the embryo.

13.4 Antioxidant Treatments in an IVF Setting

Due to the presence of so many internal and external sources of ROS that can contribute to the development of OS during IVF, it is essential to explore ways to curb the potentially negative effects this may have on successful outcomes.

In vivo, the body contains many natural antioxidant defense systems to protect a developing embryo from ROS. These mechanisms are eliminated during the in vitro state. Li et al. [26] reported that the environment surrounding an oocyte or embryo contains antioxidants from both enzymatic (such as SOD, glutathione peroxidase, and catalase), as well as non-enzymatic (including taurine, hypotaurine, vitamin C, and glutathione) origin. These antioxidants provide protection against both internal and external sources of ROS production. One possible way to facilitate the management of OS during IVF procedures includes providing the potential parents with oral antioxidant supplements to improve the quality of the gametes prior to collection for the IVF treatment cycle [27]. Another conceivable method might be to add various dosages of antioxidants directly into the culture medium during IVF to reduce the effects of OS from endogenous and exogenous factors.

13.5 Antioxidant Therapies and Studies

Antioxidant therapies have been studied extensively in the ART setting. Oral antioxidants and those used to supplement culture media are already employed in improving fertilization and pregnancy rates. Both enzymatic and non-enzymatic antioxidants have been and are being studied in an IVF setting. While non-enzymatic antioxidant supplementation is more common than enzymatic, both have their value in improving fertility in men and women, as well as minimizing oxidative damage during embryo transfer.

In the following section, we will discuss multiple studies that support or rebuff the use of specific antioxidants, along with the dosage information and methods of antioxidant treatment. We will provide a recommendation for each antioxidant as a treatment for use in IVF based on the results of the studies. The assigned grades and classification of evidence for each study can be seen in Table 13.1, organized by antioxidant. The grading system, including the methods of classifying evidence from each study and the recommended grades are presented and explained in Table 13.2. The system used in this review was employed in order to be as objective as possible in determining the effectiveness of each study. While the system assigns grades based on how well controlled and randomized a study is, it does not take into account factors such as number of participants, length of study, and type of subject. The system also does not distinguish between the supplementation type, and whether supplementation success was based on gametes or embryos. It is acknowledged that these parameters are important in reviewing the studies; however, it is difficult to assign a grade based on all these complex parameters while comparing across different types of subjects and methods of supplementation.

Table 13.1 Grading of evidence and recommendations of antioxidant studies. Under type of effect, (+) = positive effect of antioxidant, (−) = negative or no significant effect. EDTA = ethylenediaminetetraacetic acid; SOD = superoxide dismutase; GSH = glutathione

Antioxidant	Study	Classification of evidence	Grading of recommendation	Type of effect
Superoxide dismutase	Kobayashi et al. [30]	IIa, III	B	+
	Hammadeh et al. [31]	III	C	+
	Li et al. [82]	III	C	±
Catalase and EDTA	Chi et al. [34]	IIa	B	+
Glutathione peroxidase	Alvarez et al. [83]	IIb	B	+
Vitamin C (ascorbic acid)	Verma et al. [79]	IIa	B	±
	Griesinger et al. [36]	Ib	A	−
	Mostafa et al. [37]	III	C	+
Vitamin E	Kessopolou et al. [41]	Ib	A	+
	Geva et al. [39]	IIb	B	+
	Suleiman et al. [38]	Ib	A	+
Vitamins C and E	Kodama et al. [43]	IIa	B	+
	Hughes et al. [44]	IIa	B	±
	Chappell et al. [48]	Ib	A	+
	Rolf et al. [27]	Ib	A	−
	Wang et al. [10]	IIa	B	±
	Greco et al. [45]	IIb	B	±
	Greco et al. [84]	Ib	A	±
	Mier-Cabrera et al. [46]	III, Ib	A	+
Vitamin E and selenium	Keskes-Ammar et al. [47]	Ib	A	+
Selenium and N-acetyl-cysteine	Safarinejad et al. [49]	Ib	A	+
N-acetyl-cysteine	Elgindy et al. [53]	Ib	A	±
Selenium	Iwanier et al. [52]	IIb	B	−
Folic Acid (Vitamin B9)	Symanski et al. [54]	Ib	A	+

(continued)

Table 13.1 (continued)

Antioxidant	Study	Classification of evidence	Grading of recommendation	Type of effect
Glutathione	Berker et al. [55]	III	C	+
	Lenzi et al. [58]	Ib	A	+
	Griveau et al. [29]	IIa	B	+
Hypotaurine	Guerin et al. [59]	III	C	+
	Guyader-Joly et al. [56]	IIa	B	+
	Donnelly et al. [57]	IIa	B	±
L-carnitine	Costa et al. [61]	IIb	B	+
	Lenzi et al. [62]	Ib	A	+
	Abdelrazik et al. [64]	Ib	A	+
	Mansour et al. [63]	Ib	A	+
Human serum albumin	Armstrong et al. [67]	III	C	+
Pentoxifylline	Okada et al. [80]	IIa	B	±
	Cook et al. [66]	IIa	B	+
	Ohn et al. [65]	IIa	B	+
	Zhang et al. [60]	Ib	A	+
Pentoxifylline and tocopherol	Ledee-Bataille et al. [69]	IIb	B	+
	Letur-Konirsch et al. [68]	III	C	+
Melatonin	Tamura et al. [4]	III, Ib, IIb	A	+
	Farahavar et al. [81]	IIa	B	+
	Du Plessis et al. [70]	IIa	B	–
Coenzyme Q10	Lewin et al. [72]	IIa	B	+
Combination antioxidants	Tremellen et al. [71]	Ib	A	±
	Orsi et al. [73]	Ib	A	±
L-arginine and nitric oxide synthase	Battaglia et al. [74]	Ib	A	+
	Battaglia et al. [76]	Ib	A	–

Table 13.2 Key for Table 13.1. Classification of evidence and grading recommendations based on objective analysis of each article [76]

	Evidence classification
Ia	Evidence from meta-analysis of randomized controlled trials
Ib	Evidence from at least one randomized controlled trial
IIa	Evidence from at least one controlled study without randomization
IIb	Evidence from at least one other type of quasi-experimental design
III	Evidence from non-experimental descriptive studies, such as comparative studies, correlation studies, and case–control studies
IV	Evidence from expert committee reports or opinions and/or clinical experience of respected authorities
	Recommendation grading
A	Directly based on category I evidence
B	Directly based on category II evidence, or extrapolated recommendation from category I evidence
C	Directly based on category III evidence, or extrapolated recommendation from category I or II evidence
D	Directly based on category IV evidence, or extrapolated recommendation from category I, II, or III evidence

13.5.1 Enzymatic Antioxidants

13.5.1.1 Superoxide Dismutase

SOD catalyzes the breakdown of the O_2^- anion into oxygen and H_2O_2. It is part of the group of enzymatic antioxidants naturally found within the seminal plasma [28]. Griveau et al. [29] observed that SOD helps to increase hyperactivation and acrosome reaction rates in spermatozoa when added in vitro. A study by Kobayashi et al. [30] found that SOD also has an effect on sperm motility. A sperm sample incubated for 2 h in culture medium without seminal plasma showed an increase in malondialdehyde (MDA) levels and a decrease in sperm motility. After the addition of exogenous SOD (400 U/ml) to the sperm suspension, a decrease in MDA and increase in motility levels was recorded. SOD is found inside the theca interna cells of the antral follicles and therefore protects the oocyte from ROS-mediated damage during the maturation process. Hammadeh et al. [31] conducted a study that observed the effects of antioxidants in the seminal plasma of males undergoing IVF and ICSI treatments. The group reported a positive correlation between SOD levels and normal sperm morphology and a negative correlation between ROS levels and morphology as well as motility. These findings indicate that enzymatic antioxidants play a positive role in protecting spermatozoa from damage related to ROS. In addition to the positive correlation between SOD levels and normal morphology in spermatozoa used during IVF and ICSI, a negative correlation between SOD levels and pregnancy rates was also observed in patients who underwent IVF treatments. Addition of exogenous SOD to sperm undergoing preparation for IVF resulted in a decrease of motility loss and MDA levels, indicating that supplementation of SOD may help to increase the successful IVF outcomes [32]. Liu et al. [32] also noted that patients who had undergone a failed cycle of IVF-embryo transfer (IVF-ET) tended to have lower levels of SOD present in their granulose cells. These patients also had embryos of poorer quality compared to the group of patients who were able to achieve pregnancy. SOD has been reported to increase the proportions of zygotes that undergo at least one round of cleavage while improving cleavage past the two-cell stage and overall blastocyst development. Elevated levels of SOD may help provide protection from various gynecologic diseases such as pre-eclampsia or diabetes-induced embryopathy [33]. According to Table 13.1, the three studies examined in this section were given grades of B or C. These classifications indicate that the evidence provided from these studies is not of the highest quality due to the lack of randomized, controlled trials. However, the results were predominantly positive and provide a good base for the execution of further trials investigating the role of SOD during IVF treatments.

13.5.1.2 Catalase and EDTA

Catalase is another enzymatic antioxidant present in seminal plasma [30]. It is also present within the corpus luteum and tubal fluid of females and can neutralize H_2O_2 molecules by converting it into oxygen and water. A study by Chi et al. [34] investigated the effects of various concentrations of EDTA, a metal chelating agent, and catalase on human spermatozoa. Both EDTA and catalase were added in three different concentrations to the Ham's F-10 medium. The results showed that addition of EDTA increased sperm motility while combinations of EDTA and catalase increased acrosome reaction rates and decreased levels of DNA fragmentation. Due to improved functional parameters of the spermatozoa, the study may be alluding to possible increases in successful IVF outcomes. However, since only one study by Chi et al. was found on these antioxidants and the evidence provided was from controlled trials without randomization, it was therefore only assigned a grade of "B" in Table 13.1. Further studies must be executed in patients undergoing IVF treatments before a direct recommendation can be made.

13.5.1.3 Glutathione Peroxidase

Glutathione peroxidase is an enzyme that acts in conjunction with glutathione, a tripeptide cofactor. It is able to reduce peroxides such as H_2O_2 and lipid peroxides into water and alcohol while oxidizing glutathione [34]. It is located within the glandular epithelium of the endometrium in females as well as in the seminal plasma of males [34, 35]. When examining the effects of glutathione peroxidase addition on the production of ROS in the culture media used for sperm preparation, the study reviewed showed a negative correlation between GPX levels and MDA levels, a marker of lipid peroxidation [1]. Although only one study on this antioxidant was graded, the results were positive. However, the evidence was not of the highest experimental quality as the study was not controlled or randomized and did not receive the highest grade. While it is an essential enzymatic antioxidant in both males and females, no studies have examined the effects of the addition of GPX to embryo culture media.

13.5.2 Non-Enzymatic Antioxidants

13.5.2.1 Vitamin C

Griesinger et al. [36] conducted a study to test the effects of oral vitamin C supplementation during the luteal phase in women undergoing IVF treatment cycles. The 620 women participating in the study were all under 40 years of age and were given either a placebo or three different dosages of ascorbic acid in increments of 1, 5, or 10 g/day. The study showed no significant difference in either pregnancy or implantation rates between the experimental or placebo groups. The results of this trial may indicate that vitamin C supplementation alone during the luteal phase may not be sufficient in terms of increasing the likelihood of successful IVF outcomes. From the experimental studies that were testing the effect of vitamin C, the studies that were given the most positive ratings in Table 13.1 indicated that oral supplementation increases sperm quality in smokers but does not result in any change in pregnancy or implantation rates after follicle aspiration in women undergoing IVF-ET [36]. A study with category II evidence, as classified in Table 13.2, indicated increases in sperm motility and viability along with decreases in MDA levels when vitamin C was added to the culture media [37]. However, when extremely high levels of vitamin C were tested, the motility and viability of the sperm actually decreased compared to pretreatment values and actually resulted in an increase in lipid peroxidation activity. When vitamin C is used in moderate amounts, it appears to have a positive effect on semen parameters and seems to be a viable option for infertile men undergoing IVF.

13.5.2.2 Vitamin E

A study by Suleiman et al. [38] focused on the effects of oral antioxidant supplementation on sperm motility and pregnancy rates. The group also measured MDA levels to correlate the effects of lipid peroxidation on these parameters. A group of 52 males received supplementation of 100 mg vitamin E three times daily for 6 months, while the control group received a placebo. The results showed a significant decrease in MDA levels accompanied by an increase in sperm motility. Experiments investigating the effects of vitamin E supplementation recorded increased rates of sperm motility, in vitro sperm function, and fertilization rates in IVF, with decreased levels of MDA compared to placebo groups. All studies reviewed had positive results and two of them provided category I evidence, indicating a strong recommendation for the use of a vitamin E supplement in couples undergoing IVF-ET [38–40].

13.5.2.3 Vitamin C and E

Both vitamin C and vitamin E are chain breaking antioxidants. In the male, vitamin C is highly concentrated in the seminal plasma and can preserve motility levels in sperm [41]. Vitamin E is a lipophilic antioxidant, meaning that it is fats and lipid soluble. This property allows it to protect polyunsaturated fatty acids (PUFAs) from lipid peroxidation. This is important for male infertility because germ cell and sperm membranes contain large amounts of PUFAs [42].

There have been many studies conducted on the effects of vitamins C and E on male infertility. For example, in a study by Kodama et al. [43] 19 infertile men received an oral vitamin E supplement (200 mg/day) combined with vitamin C (200 mg/day) and glutathione (400 mg/day) for a 2-month period. The results of the trial showed increased sperm concentrations and a significant reduction in oxidative DNA damage as measured by 8-hydroxy-2'-deoxyguanosine levels for patients in the experimental group [42]. Hughes et al. [44] reported that using a combination of vitamin C (600 μmol/L) and vitamin E (30 and 60 μmol/L) with urate (400 μmol/L) provided significant protection against damage to sperm DNA when added under in vitro conditions to sperm wash media or during IVF procedures. In a study by Greco et al. [45] the effects of oral antioxidant supplementation to males was evaluated with regard to ICSI outcomes. The trial consisted of 38 men who presented with more than 15 % of sperm DNA damage and had gone through at least one unsuccessful ICSI treatment cycle. The men were given daily doses of vitamin C (1 g) and vitamin E (1 g) for 2 months. A 76 % decrease in sperm DNA damage was reported in the treatment group, while pregnancy rates increased from 7 to 48 %, and implantation rates from 2 to 19 %. These studies have shown that both oral antioxidant therapy as well as antioxidant supplementation during IVF preparation methods can prevent damage caused by OS to male gametes, thereby increasing the chances of favorable IVF outcomes. In females, vitamin C is present mainly in the ovaries. It recycles and enhances the activity of vitamin E and protects lipoproteins from damage by peroxyl radicals. Supplementation of vitamin C in the female prevents follicular apoptosis [34]. Conversely, a deficiency in the levels of vitamin C in females can result in premature resumption of meiosis, follicular atresia, as well as atrophy of the ovaries [45]. Vitamin E increases the number of embryos developing into blastocysts and increases their resistance to heat shock [34]. Several studies have been conducted in order to observe the effects of vitamin C and E supplementation on both in vivo and in vitro infertility. The use of antioxidants has not been as thoroughly studied in oocytes and embryos as it has in sperm preparation. The few trials that have been conducted have mainly been in women inflicted with various gynecologic diseases in order to see if it could improve pregnancy rates. A randomized control trial by Mier-Cabrera et al. [46] revealed that the addition of vitamin C and E oral antioxidant supplementation in women with endometriosis increased concentrations of both enzymatic antioxidants, SOD and glutathione peroxidase, and non-enzymatic antioxidants, vitamins C and E, in the plasma while decreasing levels of OS markers such as lipid peroxides and MDA.

Effects of antioxidant supplementation on oocytes and embryos have mainly been studied in animal models due to the physical constraints of oocyte retrieval in humans. Wang et al.[10] studied the effects of vitamin C and vitamin E supplementation to the culture media on mouse blastocyst development. After incubation for 6 h with 50 μM vitamin C, the blastocyst development rate increased significantly from 30 to 90%. The blastocyst development rate also increased during a 6-h incubation period with 400 μM vitamin E, but the results were not significant and did not surpass those achieved with vitamin C.

Several studies were reviewed which tested the effects of vitamins C and E, when used separately or in conjunction. Both oral supplementation and in vitro supplementation of these antioxidants showed a decrease in DNA fragmentation and damage in human spermatozoa compared to pretreatment values. A combined oral supplement of these vitamins showed an increase in sperm concentration in men with oligozoospermia and/or asthenozoospermia, but these results conflicted with another study that was assigned a higher grade due to experimental parameters that showed no change in semen parameters and no change in pregnancy rates compared to pretreatment values in a similar group of men [28, 47]. Although the second study was assigned an "A" in Table 13.1 in terms of overall design, a discrepancy in results may come from the fact that a significantly smaller sample size (31 in the prior compared to 14 in the latter) was used. Many of these studies only failed to receive a category I evidence rating, classified per Table 13.2, because they were not randomized [10, 46, 48–50]. In women with gynecologic diseases such as endometriosis or pre-eclampsia, results were derived from studies containing category I evidence, and are therefore highly recommended for women inflicted with these diseases and are choosing to undergo IVF-ET in order to provide a greater likelihood of successful outcomes [50, 51].

13.5.2.4 Vitamin E and Selenium

Selenium is an important element required for normal sperm function. It is involved in testicular development, spermatogenesis, motility, and maintenance of structural integrity [46]. Keskes-Ammar et al. [47] conducted an in vivo study investigating the effect of a combination of vitamin E and selenium supplements on sperm motility. The experimental group, comprising 28 men, received vitamin E (400 mg) and selenium (225 μg) daily for 3 months. Another group of 26 men were given a 4–5 g vitamin B supplement daily for the same period of time. At the end of the experimental period the group receiving vitamin E and selenium showed a significantly greater increase in sperm motility and viability as well as a decrease in MDA levels compared to the vitamin B supplement group.

13.5.2.5 Selenium and N-Acetyl-Cysteine

N-acetyl-cysteine is a derivative of l-cysteine, a naturally occurring amino acid. It can act directly as a free radical scavenger or increase existing levels of glutathione. N-acetyl-cysteine is also involved with the survival of male germ cells in the seminiferous tubules in vivo [47, 49]. In a study by Safarinejad et al. [49] a group of infertile men ($n = 468$) received daily oral supplements of either selenium (200 μg), N-acetyl-cysteine (600 μg), or a combination of both for a 26-week period. A positive correlation was established between the combination of both antioxidants and an increase in sperm concentration, motility, and morphology. The authors subsequently recommend the use of these antioxidants for infertility treatments. A study with category II evidence testing the effects of oral selenium supplementation to infertile men showed no improvement in semen parameters [50]. However, when selenium was used in conjunction with other antioxidants such as vitamin E or N-acetyl-cysteine in category I studies, as per Table 13.1, improvements in sperm motility, morphology, and concentration as well as decreases in MDA levels compared to pretreatment were observed [52, 53]. Thus, it can be recommended to infertile male partners undergoing IVF-ET to consider the use of an antioxidant supplement containing selenium as one of the components to a successful outcome. In addition, a study by Elgindy et al. [53] found that the use of N-acetyl-cysteine in female patients undergoing ICSI treatment cycles resulted in lower early and late apoptosis rates in granulosa cells compared to the control group. The study also noted a significant negative correlation between early and late apoptosis rates and fertilization rates as well as in the number of good quality embryos. Therefore, the use of N-acetyl-cysteine may be advantageous for women planning to undergo various ART procedures.

13.5.2.6 Folic Acid

Folic acid is also known as vitamin B9. Symanski et al. [54] conducted a study that evaluated 40 female patients who qualified for IVF-ET. Half of these women were given a daily folic acid supplement. After treatment, levels of homocysteine decreased significantly in the group of women who had taken the folic acid. Additionally, this group exhibited a larger percentage of oocytes in first and second degrees of maturity. Therefore, it can be concluded that folic acid supplementation can help to increase oocyte quality prior to IVF treatment. In another study, Berker et al. [55] observed the levels of homocysteine and folic acid in oocytes retrieved from women with polycystic ovarian syndrome (PCOS) prior to undergoing ART. The results of the study showed that levels of homocysteine correlated negatively with folic acid levels and fertilization rates post IVF treatment. It was further noted that homocysteine levels were lower in embryos that were given a grade of 1–2 than in embryos with a grade of 3. In women undergoing IVF-ET, the effects of oral folic acid supplementation were classified according to the previous two experiments; one with category I evidence and the other with category III, as per Table 13.2. However, both studies noted oocytes of better quality and higher grades of maturity compared to control groups, allowing for a positive recommendation with reservation for folic acid supplementation in women undergoing IVF-ET [56, 57]

13.5.2.7 Glutathione

Glutathione is a cofactor for glutathione peroxidase, and is an antioxidant that scavenges O_2^-, H_2O_2, and OH^- [34]. Lenzi et al. [58] studied the effects of glutathione in 20 male patients suffering from unilateral varicocele or genital tract inflammation. The test subjects either received a placebo or glutathione (600 mg) injection every other day for a 2-month period after which they were switched over to receive the alternative treatment for an additional 2 months. The results showed a significant increase in sperm motility and morphology in the groups treated with the glutathione supplement compared to the placebo group. Griveau et al. [29] furthermore reported that glutathione had beneficial effects on acrosome reaction rates. Although both experiments reported positive results, only one of the trials presented category I evidence and was given a grade of A [59]. Therefore, recommendations are given with reservation since only one study of sufficient quality was evaluated.

13.5.2.8 Hypotaurine

Hypotaurine is one of the antioxidant precursors of taurine, and is involved in sperm capacitation, fertilization, and embryo development. It also provides protection from ROS-induced lipid peroxidation [59]. Guerin et al. [17] was able to show that both taurine and hypotaurine were produced by epithelial cells from the oviduct in vivo. This study confirmed that hypotaurine remains stable when it is added as an antioxidant supplement to culture media used for IVF. A study by Guyader-Joly et al. [56] on the effects of hypotaurine on bovine embryos in vitro found that the supplementation of 0.5–1 mM hypotaurine to the embryo culture medium resulted in a significant improvement in blastocyst quality and development. However, adding hypotaurine to culture media had no effect on sperm motility or DNA integrity [60]. Since the evidence from all studies reviewed on hypotaurine were only assigned grades of "B" or "C" according to Table 13.2, a recommendation is only given with significant reservation.

13.5.2.9 Glutathione and Hypotaurine

Donelly et al. [57] observed the effects of adding hypotaurine and glutathione, both separately and in combination to sperm preparation media. No significant improvement was noted in total motility, DNA integrity, or ROS production.

However, the study concluded that the addition of these antioxidant supplements may be beneficial because they can potentially protect the spermatozoa from DNA damage and ROS production that was specifically induced by H_2O_2.

13.5.2.10 L-Carnitine

L-carnitine performs anti-cytokine and anti-apoptotic actions in gametes and the developing embryo. It has a potential effect on the outcome of IVF in patients with endometriosis and leukocytospermia. Costa et al. [61] observed the effects of L-carnitine in patients with asthenozoospermia. One hundred patients were given an oral supplement containing 1 g of l-carnitine for 4 months three times daily after meals. A significant increase in their total sperm count and motility was observed. A similar study by Lenzi et al. [62] using 2 g of l-carnitine as a daily oral supplement confirmed the previous results in terms of greater sperm motility and quality compared to the placebo group. Mansour et al. [63] studied the effect of L-carnitine on embryos in patients with endometriosis. Mouse embryos were used in the experiment, but were cultured in peritoneal fluid (PF) from human female patients with endometriosis. Oocytes and embryos incubated only with the PF from endometriosis patients exhibited microtubal and chromosomal damage as well as increased levels of embryo apoptosis. However, the oocytes and embryos incubated with 0.6 mg/ml l-carnitine had a significantly improved microtubule and chromosome structure with a simultaneous decrease in embryo apoptosis. Three out of the four studies reviewed for the effects of l-carnitine supplementation were classified as category I and received grades of "A." [61, 64, 65]. In terms of male factor infertility, results from a category II study showed increased motility and sperm concentration in patients with asthenozoospermia after the use of an oral L-carnitine supplement [66]. Therefore, it seems reasonable to make a fairly strong recommendation for the use of L-carnitine orally or within the culture medium when conducting IVF treatment cycles.

13.5.2.11 Human Serum Albumin

Human serum albumin (HSA) is a protein in the body that has antioxidant potential in sperm preparation techniques. Armstrong et al. [67] found that the use of HSA during sperm preparation led to the recovery of spermatozoa with better quality compared to the Percoll method. The authors conclude that the antioxidant properties of HSA would be helpful when preparing semen samples from men with increased ROS levels in the seminal fluid, such as those with leukocytospermia. Only one study was reviewed for the effects of human serum albumin. The experiment recorded that the use of HSA during sperm preparation resulted in lower rates of decrease in motility and viability in the spermatozoa compared to the group treated with Percoll. However, since the study is classified under category III in Table 13.1, the evidence is given a "C" rating and cannot be well recommended for use in IVF-ET.

13.5.2.12 Pentoxifylline

Pentoxifylline is a 3,5 nucleotide phosphodiesterase inhibitor that protects cells from lipid peroxidation induced by the presence of ROS [60]. Pentoxyfilline also possesses various immunomodulatory properties. It is a derivative of xanthine and has the potential to reduce the embryotoxic effect of H_2O_2 as was observed in a study by Zhang et al. [68] using murine embryos. In this study, approximately 500 μM of pentoxifylline was added to concentrations of H_2O_2 in solution. This resulted in a reduction in embryotoxicity compared to controls. The rate of blastocyst development increased from 67 % in a solution containing only H_2O_2, to 82 % in a solution containing both H_2O_2 and pentoxifylline, to 94 % in a solution with only pentoxifylline. In another study Cook et al. [66] examined the effects of in vitro pentoxifylline addition to human spermatozoa prior to a treatment cycle of ICSI. The authors reported a significant increase in pregnancy and implantation rates per embryo in patients over 38 years of age with oocytes injected with pentoxifylline-treated sperm. Ohn et al. [65] confirmed these results in patients suffering from severe asthenozoospermia whose partners were undergoing an ICSI cycle. Treatment of the spermatozoa from these men with pentoxifylline showed an increase in sperm motility, fertilization, and implantation rates, as well as successful embryo transfer and pregnancy rates. Therefore, it appears that the supplementation of pentoxifylline to human gametes prior to IVF or ICSI treatment cycles ensures a greater chance of positive outcomes. Pentoxifylline treatment improved murine blastocyst development and reduced H_2O_2 induced damage in a category I study, suggesting the possible advantages of using it in the culture media during IVF treatments to improve outcomes [68]. Several category II studies have also shown the effects of pentoxifylline treatment in vitro for spermatozoa from male partners undergoing ICSI treatment [68–70]. Oral pentoxifylline supplementation has not been shown to be effective in low doses, but in high doses, around 1,200 mg daily, is able to improve sperm motility. From the evidence collected, it appears that pentoxifylline treatment would be most beneficial when added to the culture medium during IVF.

13.5.2.13 Pentoxifylline and Vitamin E

Ledee-Bataille et al. [69] evaluated 18 patients who were enrolled in oocyte donation programs. These women were treated with 800 mg pentoxifylline and 1,000 IU vitamin E daily for 6 months. Seventy two percent of patients treated responded well and showed an overall increase in endometrium thickness. Three patients conceived naturally while another three became pregnant through ART treatment methods, ultimately resulting in a total pregnancy rate of 33 %. The authors concluded that a combination of pentoxifylline and vitamin E treatment helps patients with a thin endometrium as well as improving ovarian function and pregnancy rates. Another study involved daily supplementation of 800 mg pentoxifylline and 1,000 IU vitamin E for a 9-month period to patients with premature ovarian failure. This led to an increase in endometrial thickness, which may have enhanced implantation rates and resulted in two viable pregnancies through embryo transfer. Combined pentoxifylline and vitamin E treatment in a category II study and a category III study have shown to be beneficial in patients with a thin endometrium planning to undergo IVF treatments [68, 71]. However, a definite recommendation cannot be made since the evidence reviewed was given grades of "B" and "C" as shown in Table 13.1.

13.5.2.14 Melatonin

In a study by Tamura et al. [4] 18 patients undergoing IVF-ET were treated with an oral supplement of 3 mg melatonin, 600 mg vitamin E, or a combination of both antioxidants once daily. Intracellular concentrations of 8-hydroxy-2-deoxyguanosine were significantly reduced after antioxidant treatment. The second part of the experiment involved 115 patients who had previously failed to get pregnant after one IVF-ET treatment cycle. Out of these women, 56 were treated with melatonin (3 mg) daily, while the remaining 59 patients received no antioxidant supplementation. Fertilization rates increased significantly from the previous IVF-ET cycle within the melatonin treatment group, but no significance was observed when compared to the control group. The study concluded that melatonin improved oocyte and embryo quality and marginally increased fertilization rates. Du Plessis et al. [70] studied the effects of melatonin in vitro by incubating spermatozoa with melatonin (2 mM) for 2 h. Results of the experiment indicated an increase in the number of motile sperm and simultaneous reduction in nonviable sperm compared to the control. A reduction in endogenous NO levels was also reported, however, there was no significant difference in levels of ROS. The studies examining the effect of melatonin supplementation were classified as either category I or II in Table 13.1. However, one category II study that was given a "B" rating showed no improvement in terms of oocyte maturation after in vitro supplementation of melatonin [51]. Therefore, it is likely that further studies will need to be conducted in order to confirm the positive effects of melatonin in terms of IVF-ET but it is very possible that the use of this antioxidant will become definitive soon.

13.5.2.15 Coenzyme Q10

Coenzyme Q10, also known as ubiquinone, is a substance mostly produced in the mitochondria of cells. Lewin et al. [72] performed a study on 22 patients, 16 of which were diagnosed as asthenozoospermic. Supplementation with 50microM

CoQ 10 resulted in a significant increase in motile sperm from the subgroup of asthenozoospermic men. No significant increase in motility was seen in normozoospermic men. However, after daily oral treatment with 60 mg/day CoQ 10, for a mean of 103 days, a significant improvement in fertilization rate was observed. The authors conclude that ubiquinone supplementation is effective in selected patients [73]. Unfortunately, the trials were not randomized and were therefore classified as category II evidence in Table 13.1. Positive outcomes indicate a need for more studies to investigate the effects of using this antioxidant as a supplement in IVF/ICSI patients.

13.5.2.16 Combined Antioxidants

Many antioxidants, like vitamin C and vitamin E, can work in conjunction with one another to scavenge free radicals and prevent OS. Therefore, combinations of antioxidants may be more effective in vivo than single antioxidant treatments. Tremellen et al. [71] conducted a study using Menevit, an oral antioxidant supplement containing 100 mg vitamin C, 25 mg zinc, 400 IU vitamin E, 500 μg folic acid, 6 mg lycopene, 333 μg garlic oil, and 26 μg selenium per capsule, in 60 couples suffering from severe forms of male factor infertility. In the experimental group, males were instructed to take one pill daily for 3 months prior to beginning an IVF treatment cycle. After completion of the trial, the experimental group taking Menevit had a viable pregnancy rate of 38.5 % from the total number of embryos transferred compared to only 16 % in the placebo group. The two category I studies that reviewed the effects of combination antioxidant therapies were given "A" ratings in Table 13.1 [71]. Combination antioxidant supplementation seems to have overall positive effects on IVF-ET outcomes and is recommended for use orally in both male and female factors as well as in the culture medium.

13.5.3 Additional Strategies

13.5.3.1 l-Arginine and Nitric Oxide Synthase

l-Arginine is a compound with antioxidant properties that is together with O_2 converted under the influence of nitric oxide synthase (NOS) to form L-citrulline and nitric oxide (NO) [74]. Endometrial glandular cells and surface epithelia secrete NOS in females. Low levels of NO help to reduce the occurrence of apoptosis. Conversely, higher concentrations of NOS may actually promote the production of peroxynitrite and superoxide, causing cell death and embryo fragmentation [75]. While it may seem that supplementation of l-arginine could induce more negative than positive outcomes in terms of fertility, several conflicting results have been reported which indicate that this may not be true in all cases. Battaglia et al. [76] reported that l-arginine supplementation may be beneficial in poor responder patients and detrimental in normal responder patients who are undergoing controlled ovarian hyperstimulation in order to retrieve oocytes for IVF treatment cycles. The first study involved female subjects classified as poor responders. The group that received orally administered l-arginine supplements reported low cancelation rates compared to controls, high numbers of retrieved oocytes and transferred embryos, as well as three pregnancies compared to zero in the placebo group. However, the second study observed the effects of l-arginine in normal responder patients and found lower quality and quantity of oocytes retrieved in addition to decreased pregnancy rates in the experimental group versus the control group. Further studies need to be conducted in order to determine normal l-arginine levels in the body. The studies about l-arginine and NOS as listed in Table 13.1 were assigned the highest grade of evidence, an A, as the trials were controlled and randomized [76, 77]. They showed a positive outcome in poor responder patients and no improvement in normal responders who had undergone controlled ovarian hyperstimulation prior to IVF. Although the experiments were of high quality, the results remain inconclusive. Further trials will need to be conducted before a definitive recommendation can be made for these antioxidants.

13.6 Conclusion

Increased ROS levels cause oxidative damage to gametes and embryos, impairing the fertility of couples. Sources of ROS and OS can be endogenous and be prevalent in diseases such as varicocele, leukocytospermia, endometriosis, and PCOS. Between 30 and 80 % of male subfertility cases are considered to be due to the damaging effects of OS on sperm and oral supplementation with antioxidants may improve sperm quality by reducing OS. ROS are also present outside of the body, and oxidative damage can easily occur during oocyte and sperm preparation procedures, as well as in the transfer of embryos during IVF and ICSI.

Antioxidant strategies in IVF-ET include antioxidant treatments both in vivo and in vitro. While some antioxidants are still being tested in bench research and animal models, other antioxidant treatments have been shown to significantly improve the quality of human spermatozoa, oocytes, and embryos involved in assisted reproduction techniques. In a recent Cochrane review the evidence suggests that antioxidant supplementation in subfertile males may improve the outcomes of live birth and pregnancy rate for subfertile couples undergoing ART cycles [78]. Further studies are required to determine which antioxidant treatments will provide the best outcomes in ART with the least amount of negative side effects, although some current treatments have already proven to be effective in improving fertilization and pregnancy rates. Continued research on antioxidants that are not well-studied in humans or that are only studied in gametes will add to the current knowledge about the best treatment options for men and women undergoing IVF and other assisted reproduction procedures.

References

1. Sharlip ID, Jarow JP, Belker AM et al (2002) Best practice policies for male infertility. Fertil Steril 77(5):873–882
2. Boyle KE, Thomas AJ, Marmar JL et al (2002) The cost effectiveness of cryopreservation of sperm harvested intraoperativelyduring vasectomy reversal. Fertil Steril 78(Suppl 1):S260
3. Agarwal A, Said TM, Bedaiwy MA, Banerjee J, Alvarez JG (2006) Oxidative stress in an assisted reproductive techniques setting. Fertil Steril 86(3):50–512
4. Tamura H, Takasaki A, Miwa I et al (2008) Oxidative stress impairs oocyte quality and melatonin protects oocytes from free radical damage and improves fertilization rate. J Pineal Res 44(3):280–287
5. Agarwal A, Allamaneni SS (2004) Role of free radicals in female reproductive diseases and assisted reproduction. Reprod Biomed Online 9(3):338–347
6. du Plessis SS, Makker K, Desai NR, Agarwal A (2008) Impact of oxidative stress on IVF. Expert Rev Obstet Gynecol 3(4):539–554
7. Gupta S, Malhotra N, Sharma D, Chandra A, Ashok A (2009) Oxidative stress and its role in female infertility and assisted reproduction: clinical implications. Fertil Steril 2(4):147–164
8. de Lamirande E, Gagnon C (1993) A positive role for the superoxide anion in triggering hyperactivation and capacitation of human spermatozoa. Int J Androl 16(1):21–25
9. Saleh RA, Agarwal A, Nada EA et al (2003) Negative effects of increased sperm DNA damage in relation to seminal oxidative stress in men with idiopathic and male factor infertility. Fertil Steril 79(Suppl 3):1597–1605
10. Wang X, Sharma RK, Gupta A et al (2003) Alterations in mitochondria membrane potential and oxidative stress in infertile men: a prospective observational study. Fertil Steril 80(Suppl 2): 844–850
11. Kemal DN, Morshedi M, Oehninger S (2000) Effects of hydrogen peroxide on DNA and plasma membrane integrity of human spermatozoa. Fertil Steril 74(6):1200–1207
12. Baker MA, Aitken RJ (2005) Reactive oxygen species in spermatozoa: methods for monitoring and significance for the origins of genetic disease and infertility. Reprod Biol Endocrinol 3:67
13. Agarwal A, Gupta S, Abdel-Razek H, Krajcir N, Athayde KS (2006) Impact of oxidative stress on gametes and embryos in an ART laboratory. Clin Embryol 9(3):5–22
14. Pasqualotto EB, Agarwal A, Sharma RK et al (2004) Effect of oxidative stress in follicular fluid on the outcome of assisted reproductive procedures. Fertil Steril 81(4):973–976
15. Bedaiwy MA, Falcone T, Mohamed MS et al (2004) Differential growth of human embryos in vitro: role of reactive oxygen species. Fertil Steril 82(3):593–600
16. Sugino N, Takiguchi S, Kashida S, Karube A, Nakamura Y, Kato H (2000) Superoxide dismutase expression in the human corpus luteum during the menstrual cycle and in early pregnancy. Mol Hum Reprod 6(1):19–25
17. Guérin P, El Mouatassim S, Ménézo Y (2001) Oxidative stress and protection against reactive oxygen species in the pre-implantation embryo and its surroundings. Hum Reprod Update 7(2):175–189
18. Goto Y, Noda Y, Mori T, Nakano M (1993) Increased generation of reactive oxygen species in embryos cultured in vitro. Free Radic Biol Med 15:69–75
19. Agarwal A, Saleh RA, Bedaiwy MA (2003) Role of reactive oxygen species in the pathophysiology of human reproduction. Fertil Steril 79(4):829–843
20. Lampiao F, Strijdom H, Du Plessis SS (2010) Effects of sperm processing techniques involving centrifugation on nitric oxide, reactive oxygen species generation and sperm function. Open Androl J 2:1–5
21. Alvarez JG, Storey BT (1992) Evidence for increased lipid peroxidative damage and loss of superoxide dismutase activity as a mode of sublethal cryodamage to human sperm during cryopreservation. J Androl 13(3):232–241

22. Rahimi G, Isachenko E, Sauer H, Isachenko V, Wartenberg M, Hescheler J et al (2003) Effect of different vitrification protocols for human ovarian tissue on reactive oxygen species and apoptosis. Reprod Fertil Dev 15:343–349
23. Jones DP (1985) The role of oxygen concentration in oxidative stress: hypoxic and hyperoxic models. In: Sies H (ed) oxidative stress. Academic, New York, pp 152–196
24. Girotti AW (2001) Photosensitized oxidation of membrane lipids: reaction pathways, cytotoxic effects, and cytoprotective mechanisms. J Photochem Photobiol B 63:103–113
25. Nakayama T, Noda Y, Goto Y, Mori T (1994) Effects of visible light and other environmental factors on the production of oxygen radicals by hamster embryos. Theriogenology 41:499–510
26. Li J, Foote RH, Simkin M (1993) Development of rabbit zygotes cultured in protein-free medium with catalase, taurine, or superoxide dismutase. Biol Reprod 49(1):33–37
27. Rolf C, Cooper TG, Yeung CH, Nieschlag E (1999) Antioxidant treatment of patients with asthenozoospermia or moderate oligoasthenozoospermia with high-dose vitamin C and vitamin E: a randomized, placebo-controlled, double-blind study. Hum Reprod 14(4):1028–1033
28. Zini A, de Lamirande E, Gagnon C (1993) Reactive oxygen species in semen of infertile patients: levels of superoxide dismutase- and catalase-like activities in seminal plasma and spermatozoa. Int J Androl 16(3):183–188
29. Griveau JF, Le Lannou D (1994) Effects of antioxidants on human sperm preparation techniques. Int J Androl 17(5):225–231
30. Kobayashi T, Miyazaki T, Natori M, Nozawa S (1991) Protective role of superoxide dismutase in human sperm motility: superoxide dismutase activity and lipid peroxide in human seminal plasma and spermatozoa. Hum Reprod 6(7):987–991
31. Hammadeh ME, Zenner S, Ong MF, Hamad MF, Amer AS, Schmidt W (2008) Relationship between reactive oxygen species (ROS), and enzymatic antioxidants and their effect on IVF/ICSI outcome. Fertil Steril 90:S191
32. Liu J, Li Y (2010) Effect of oxidative stress and apoptosis in granulosa cells on the outcome of IVF-ET. Zhong Nan Da Xue Xue Bao Yi Xue Ban 35(9):990–994
33. Sharma RK, Agarwal A (2004) Role of reactive oxygen species in gynecologic diseases. Reprod Med Biol 3(4):177–199
34. Chi HJ, Kim JH, Ryu CS et al (2008) Protective effect of antioxidant supplementation in sperm-preparation medium against oxidative stress in human spermatozoa. Hum Reprod 23(5):1023–1028
35. Formigari A, Irato P, Santon A (2007) Zinc, antioxidant systems and metallothionein in metal mediated-apoptosis: biochemical and cytochemical aspects. Comp Biochem Physiol C: Toxicol Pharmacol 146(4):443–459
36. Griesinger G, Franke K, Kinast C et al (2002) Ascorbic acid supplement during luteal phase in IVF. J Assist Reprod Genet 19(4):164–168
37. Mostafa T, Tawadrous G, Roaia MMF, Amer MK, Kader RA, Aziz A (2006) Effect of smoking on seminal plasma ascorbic acid in infertile and fertile males. Andrologia 38:221–224
38. Suleiman SA, Ali ME, Zaki ZM, el-Malik EM, Nasr MA (1996) Lipid peroxidation and human sperm motility: protective role of vitamin E. J Androl 17(5):530–537
39. Geva BB, Zabludovsky N, Lessing JB, Lerner-Geva L, Amit A (1996) The effect of antioxidant treatment on human spermatozoa and fertilization rate in an in vitro fertilization program. Fertil Steril 66:430–434
40. Bolle P, Evandri MG, Saso L (2002) The controversial efficacy of vitamin E for human male infertility. Contraception 65(4):313–315
41. Kessopolou E, Powers HJ, Sharma KK et al (1995) double-blind randomised placebo cross-over controlled trial using the antioxidant vitamin E to treat reactive oxygen species associated male infertility. Fert Steril 64(825):31
42. Agarwal A, Nallella KP, Allamaneni SS, Said TM (2004) Role of antioxidants in treatment of male infertility: an overview of the literature. Reprod Biomed Online 8(6):616–627
43. Kodama H, Yamaguchi R, Fukuda J, Kasai H, Tanaka T (1997) Increased oxidative deoxyribonucleic acid damage in the spermatozoa of infertile male patients. Fertil Steril 68(3):519–524

44. Hughes CM, Lewis SE, McKelvey-Martin VJ, Thompson W (1998) The effects of antioxidant supplementation during Percoll preparation on human sperm DNA integrity. Hum Reprod 13(5):1240–1247
45. Greco E, Romano S, Iacobelli M et al (2005) ICSI in cases of sperm DNA damage: beneficial effect of oral antioxidant treatment. Hum Reprod 20(9):2590–2594
46. Mier-Cabrera J, Aburto-Soto T, Burrola-Mendez S, Jimenez-Zamudio L, Tolentino MC, Casanueva E et al (2009) Women with endometriosis improved their peripheral antioxidant markers after the application of a high antioxidant diet. Reprod Biol Endocrinol 7:54
47. Keskes-Ammar L, Feki-Chakroun N, Rebai T et al (2003) Sperm oxidative stress and the effect of an oral vitamin E and selenium supplement on semen quality in infertile men. Arch Androl 49(2):83–94
48. Chappell LC, Seed PT, Briley AL et al (1999) Effect of antioxidants on the occurrence of pre-eclampsia in women at increased risk: a randomised trial. Lancet 354(9181):810–816
49. Safarinejad MR, Safarinejad S (2009) Efficacy of selenium and/or N-acetyl-cysteine for improving semen parameters in infertile men: a double-blind, placebo controlled, randomized study. J Urol 181(2):741–751
50. Gressier B, Cabanis A, Lebegue S et al (1994) Decrease of hypochlorous acid and hydroxyl radical generated by stimulated human neutrophils: comparison in vitro of some thiol-containing drugs. Methods Find Exp Clin Pharmacol 16(1):9–13
51. Erkkilä K, Hirvonen V, Wuokko E, Parvinen M, Dunkel L (1998) N-acetyl-L-cysteine inhibits apoptosis in human male germ cells in vitro. J Clin Endocrinol Metab 83(7):2523–2531
52. Iwanier K, B Zachara B (1995) Selenium supplementation enhances the element concentration in blood and seminal fluid but does not change the spermatozoal quality characteristics in subfertile men. J Androl 16:441–447
53. Elgindy EA, El-Huseiny AM, Mostafa MI, Gaballah AM, Ahmed TA (2010) N-Acetyl cysteine: could it be an effective adjuvant therapy in ICSI cycles?. A preliminary study. Reprod BioMed Online 20(6):789–796
54. Szymański W, Kazdepka-Ziemińska A (2003) Effect of homocysteine concentration in follicular fluid on a degree of oocyte maturity. Ginekol Pol 74(10):1392–1396
55. Berker B, Kaya C, Aytac R, Satıroglu H (2009) Homocysteine concentrations in follicular fluid are associated with poor oocyte and embryo qualities in polycystic ovary syndrome patients undergoing assisted reproduction. Hum Reprod 24(9):2293–2302
56. Guyader-Joly C, Guérin P, Renard JP, Guillaud J, Ponchon S, Ménézo Y (1998) Precursors of taurine in female genital tract: effects on developmental capacity of bovine embryo produced in vitro. Amino Acids 15(1–2):27–42
57. Donnelly ET, McClure N, Lewis SE (1999) Antioxidant supplementation in vitro does not improve human sperm motility. Fertil Steril 72(3):484–495
58. Lenzi A, Culasso F, Gandini L, Lombardo F, Dondero F (1993) Placebo-controlled, double-blind, cross-over trial of glutathione therapy in male infertility. Hum Reprod 8(10):1657–1662
59. Guérin P, Guillaud J, Ménézo Y (1995) Hypotaurine in spermatozoa and genital secretions and its production by oviduct epithelial cells in vitro. Hum Reprod 10(4):866–872
60. Zhang X, Sharma RK, Agarwal A, Falcone T (2005) Effect of pentoxifylline in reducing oxidative stress-induced embryotoxicity. J Assist Reprod Genet 22(11–12):415–417
61. Costa M, Canale D, Filicori M, D'Iddio S, Lenzi A (1994) L-carnitine in idiopathic asthenozoospermia: a multicenter study. Italian study group on carnitine and male infertility. Andrologia 26(3):155–159
62. Lenzi A, Lombardo F, Sgrò P, Salacone P, Caponecchia L, Dondero F, et al (2003) Use of carnitine therapy in selected cases of male factor infertility: A double-blind crossover trial. Fertil Steril 79:292–300
63. Mansour G, Abdelrazik H, Sharma RK, Radwan E, Falcone T, Agarwal A (2009) L-carnitine supplementation reduces oocyte cytoskeleton damage and embryo apoptosis induced by incubation in peritoneal fluid from patients with endometriosis. Fertil Steril 91:2079–2086 5 Suppl

64. Abdelrazik H, Sharma R, Mahfouz R, Agarwal A (2009) L-carnitine decreases DNA damage and improves the in vitro blastocyst development rate in mouse embryos. Fertil Steril 91(2):589–596
65. Ohn SJ, Su SJ, Ryul LD, Thomas K, Joo KH, Yul CK (2002) The effect of pentoxifylline (PF) in severe asthenozoospermia treatment on the results of ICSI program. Fertil Steril 78(Suppl 1):S260
66. Cook CA, Liotta D, Hariprashad J, Zaninovic N, Veeck L (2002) Is pentoxifylline treatment of sperm before ICSI detrimental to subsequent preembryo development? Fertil Steril 78(Suppl 1):S190
67. Armstrong JS, Rajasekaran M, Hellstrom WJ, Sikka SC (1998) Antioxidant potential of human serum albumin: role in the recovery of high quality human spermatozoa for assisted reproductive technology. J Androl 19(4):412–419
68. Letur-Konirsch H, Delanian S (2003) Successful pregnancies after combined pentoxifylline-tocopherol treatment in women with premature ovarian failure who are resistant to hormone replacement therapy. Fertil Steril 79(2):439–441
69. Lédée-Bataille N, Olivennes F, Lefaix JL, Chaouat G, Frydman R, Delanian S (2002) Combined treatment by pentoxifylline and tocopherol for recipient women with a thin endometrium enrolled in an oocyte donation programme. Hum Reprod 17(5):1249–1253
70. du Plessis SS, Hagenaar K, Lampiao F (2010) The in vitro effects of melatonin on human sperm function and its scavenging activities on NO and ROS. Andrologia 42(2):112–116
71. Tremellen K, Miari G, Froiland D, Thompson J (2007) A randomised control trial examining the effect of an antioxidant (Menevit) on pregnancy outcome during IVF-ICSI treatment. Aust N Z J Obstet Gynaecol 47(3):216–221
72. Lewin A, Lavon H (1997) The effect of coenzyme Q10 on sperm motility and function. Mol Aspects Med 18(Suppl):S213–S219
73. Orsi NM, Leese HJ (2001) Protection against reactive oxygen species during mouse preimplantation embryo development: role of EDTA, oxygen tension, catalase, superoxide dismutase and pyruvate. Mol Reprod Dev 59(1):44–53
74. Battaglia C, Salvatori M, Maxia N, Petraglia F, Facchinetti F, Volpe A (1999) Adjuvant L-arginine treatment for in vitro fertilization in poor-responder patients. Hum Reprod 14(7):1690–1697
75. Babu BR, Frey C, Griffith OW (1999) L-arginine binding to nitric-oxide synthase. The role of H-bonds to the nonreactive guanidinium nitrogens. J Biol Chem 274(36):25218–25226
76. Battaglia C, Regnani G, Marsella T et al (2002) Adjuvant L-arginine treatment in controlled ovarian hyperstimulation: a double-blind, randomized study. Hum Reprod 17(3):659–665
77. National Clinical Guidelines for Type 2 Diabetes. (The Royal College of Medicine of General Practitioners, Effective Clinical Practice Unit, ScHAAR University of Sheffield)
78. Showell MG, Brown J, Yazdani A, Stankiewicz MT, Hart RJ (2011) Antioxidants for male subfertility. Cochrane Database of Systematic Reviews Issue 1. Art. No.: CD007411
79. Verma A, Kanwar KC (1998) Human sperm motility and lipid peroxidation in different ascorbic acid concentrations: an in vitro analysis. Andrologia 30:325–329
80. Okada H, Tatsumi N, Kanzaki M, Fujisawa M, Arakawa S, Kamidono S (1997) Formation of reactive oxygen species by spermatozoa from asthenospermic patients: response to treatment with pentoxifylline. J Urol 157(6):2140–2146
81. Farahavar A, Shahne AZ, Kohram H, Vahedi V (2010) Effect of melatonin on in vitro maturation of bovine oocytes. Afr J Biotechnol 9(17):2579–2583
82. Li YP, Liu J, et al (2008) Effect of oxidative stress and apoptosis in granulosa cells on the quality of oocytes and the outcome of in vitro fertilization and embryo transfer (IVFET). Fertil Steril 90(Suppl):S336–S337
83. Alvarez JG, Storey BT (1989) Role of glutathione per- oxidase in protecting mammalian spermatozoa from loss of motility caused by spontaneous lipid peroxidation. Gamete Res 23:77–90
84. Greco E, Iacobelli M, Rienzi L, et al (2005) Reduction of the incidence of sperm DNA fragmentation by oral antioxidant treatment. J Androl 26:349–353

Chapter 14
Oxidative Insult After Ischemia/Reperfusion in Older Adults

Tinna Traustadóttir and Sean S. Davies

Abstract Aging is associated with significant impairment in tolerance to acute stressors such as ischemia/reperfusion and this impairment appears to contribute to the increased morbidity and mortality of older adults compared to young adults in clinical conditions associated with ischemia/reperfusion. These conditions include ischemic heart disease, stroke, and a variety of surgeries. Ischemia leads to a transition from aerobic to anaerobic metabolism with subsequent depletion of ATP, accumulation of lactate, acidosis, and loss of intracellular ion homeostasis. Reperfusion results in increased oxidative stress and calcium overload which together stimulate pathways of necrosis and apoptosis. These adverse events are exaggerated in older individuals. In particular, oxidative stress plays a central role in the age-related increase in vulnerability to ischemia/reperfusion. We outline a useful forearm ischemia/reperfusion method to assess this vulnerability in older adults and have shown that several interventions can markedly improve resistance to ischemia/reperfusion.

Keywords Oxidative insult · Ischemia/reperfusion · Older adults · Ischemic preconditioning · Myocardial ischemia · Forearm ischemia/reperfusion

T. Traustadóttir
Biological Sciences, Northern Arizona University, Building 21,
Room 227, 517 S. Beaver Street, Flagstaff, AZ 86011, USA
e-mail: tinna.traustadottir@nau.edu

S. S. Davies (✉)
Department of Pharmacology, Vanderbilt University, 2222 Pierce Avenue,
Room 506A RRB, Nashville, TN 37232-6602, USA
e-mail: sean.davies@vanderbilt.edu

14.1 Introduction

Ischemia/reperfusion is an oxidative insult requiring upregulation of enzymatic antioxidants and stress proteins if tissue is to be protected against damage. Aging is associated with increased oxidative damage as a result of ischemia/reperfusion and this increased oxidative damage may be a significant contributor to the increased morbidity and mortality of older adults compared to younger adults with similar medical conditions. This chapter will outline the processes of ischemia and reperfusion and the ensuing cascade of events in the affected tissue. We will then discuss the effects of aging and gender on the ischemia/reperfusion response and how ischemic preconditioning, a mechanism known to have a protective effect on ischemia/reperfusion injury, has a reduced efficiency in older adults. Finally, this chapter will present interventions that may reduce ischemia/reperfusion injury by increasing the individual's capacity to effectively counterbalance formation of Reactive Oxygen Species (ROS).

14.2 Ischemia/Reperfusion

Ischemia results from restricted blood flow to downstream tissues causing hypoxia relative to the metabolic demand. Reperfusion occurs when blood flow is subsequently restored and oxygen levels revert to near normal levels. Ischemia/reperfusion occurs in a variety of clinical conditions relevant to older adults including myocardial infarction, stroke, and circulatory shock, as well as surgical interventions such as organ transplants, and cardiopulmonary bypass [1–4]. If the ischemic conditions are not relieved in a timely manner, hypoxic tissue may undergo necrosis. However, even restoring blood flow and oxygen delivery to the hypoxic tissue can lead to tissue damage, known as *reperfusion injury*. The severity of the reperfusion injury depends on the duration of ischemia and although the exact mechanisms have not been fully elucidated they involve decreased ATP levels, calcium loading, inflammation, accumulation of hydrogen ions, and oxidative stress [5–8].

Perhaps the most studied form of ischemia/reperfusion injury involves damage to the heart during myocardial infarction, an understanding of which is vital given that cardiovascular disease is the leading cause of morbidity and mortality in older adults in industrialized countries [9]. Myocardial infarction results from thrombotic occlusion of a coronary artery, causing the tissue normally perfused by the artery to become ischemic. This lack of oxygen disrupts mitochondrial metabolism, as the electron carriers in the electron transport chain are unable to hand off their electrons to oxygen, forcing these electron carriers to remain in reduced state. Furthermore, xanthine oxidoreductase, normally present in its reductase form, is converted to its oxidase form during ischemia. When oxygen is restored to tissue during reperfusion (e.g., as a result of angioplasty), the excess of electron carriers in reduced state react with this oxygen to form ROS, with the formation of ROS being exacerbated by xanthine oxidase, so that the surrounding tissues are presented with a massive oxidative insult. In particular ubiquinone, the mobile electron carrier which transports electrons from complexes I and II to complex III and cytochrome *c,* is a major contributor of electrons for the formation of ROS during ischemia/reperfusion. These free radicals can damage lipids, proteins, and nucleic acids, with the extent of lipid peroxidation being markedly increased by ischemia/reperfusion [10, 11]. Manifestations of postischemic reperfusion injury to myocardium can emerge over hours, days, and sometimes weeks after reperfusion and may present the greatest threat to survival for the patient [12]. Importantly, the extent of lipid peroxidation in the myocardium inversely relates to the antioxidant capacity of the myocardial tissue [11, 13] and negatively correlates with the postischemic functional recovery of the myocardium [11]. Three levels of ischemia/reperfusion injury have been described in the myocardium: level 1 occurs when the ischemic period is less than 5 min and is associated mainly with cardiac arrhythmias. Level 2 occurs when the ischemic period extends from 5 to 20 min and is referred to as myocardial stunning. This level is associated with reversible dysfunction without cell death. Level 3 occurs when the ischemic period is greater than 20 min and is the most severe, resulting in irreversible myocardial cell death from both necrosis and apoptosis. Both forms of cell death are regulated by the mitochondria through opening of the mitochondrial permeability transition pore (mPTP) and the mitochondrial apoptosis channel (mAC). The mPTP and mAC are opened in response to ischemia/reperfusion by the combination of elevated levels of oxidative stress and calcium. Opening of the mPTP causes depolarization of the mitochondrial membrane potential, mitochondrial swelling due to disruption of ion balance followed by osmotic uptake of water, and the release of proapoptotic factors (Fig. 14.1).

Although the literature on ischemia/reperfusion injury tends to focus on myocardial infarctions, ischemia/reperfusion is also relevant to other organs and clinical conditions. For example, organ transplantation by necessity involves ischemia starting at the recovery of organs from the donor until their reperfusion in the recipient. Reperfusion injury to the graft is a major concern in these procedures. The oxidative stress generated during the reperfusion upregulates proinflammatory molecules which compounds and perpetuates the early reperfusion injury, leading to cellular apoptosis/necrosis, and initiation of rejection episodes [14]. Extensive tissue ischemia/reperfusion also occurs during both the rupture and surgical repair of an abdominal aortic aneurysm. Surgical repair of a ruptured abdominal aortic aneurysm is associated with far higher mortality rates than elective repair of an abdominal aortic aneurysm. While in both cases the surgery requires lower torso ischemia induced by aortic clamp, ruptured aneurysm includes hemorrhagic shock in the interval prior to surgical repair [15]. Not surprisingly, patients with ruptured abdominal aortic aneurysm have elevated levels of lipid peroxidation pre-surgery compared to patients who undergo elective repair [10, 16]. Additionally, these patients are hit with a second wave of increased lipid peroxidation 3–5 days post-surgery [10].

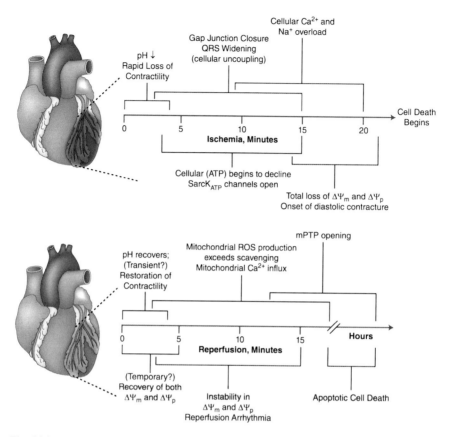

Fig. 14.1 Pathophysiological changes in rodent cardiac tissue during ischemia (*top*) and reperfusion (*bottom*). Postulated mechanisms involved in exercise-preconditioning noted in *red* font. Abbreviations: $\Delta\Psi p$ sarcolemmal membrane potential, $\Delta\Psi m$ mitochondrial membrane potential, *sarcKATP* sarcolemmal ATP-sensitive potassium channels, *PTP* permeability transition pore

In summary, prolonged tissue or organ ischemia followed by reperfusion results in an oxidative insult which can lead to organ injury and dysfunction [17]. The severity of ischemia/reperfusion injury depends on the degree and duration of ischemia as well as the sensitivity of the tissue to the lack of oxygen, the degree of inflammatory response, and the ability to neutralize the increase in ROS during reperfusion. Although the increased generation of ROS during both ischemia and reperfusion comes primarily from mitochondria, other sources also contribute including nicotinamide adenine dinucleotide phosphate (NADPH) oxidase and xanthine oxidase.

14.3 Effects of Aging on the Response to Ischemia/Reperfusion

One particular hallmark of aging is decreased stress resilience (i.e., the ability to return to homeostasis after a perturbation). Greater susceptibility to myocardial ischemia/reperfusion injury with aging has been shown in variety of animal models including mice, sheep, rabbits, and rats, [18–21] as well as in human myocardium [22, 23]. This age-associated increase in vulnerability to ischemia/reperfusion injury, including loss of contractile function, is associated with increased oxidative stress and greater calcium accumulation with negative consequences for ion-dependent enzymes, intracellular signaling, and ATP production [18, 24]. The combination of ROS and calcium results in opening of the mPTP, initiating necrosis and apoptosis. Inhibition of mPTP opening reduces ischemia/reperfusion injury in young but not in old, suggesting that the regulation of mPTP opening changes with age [25]. In addition, aging is associated with lower threshold of ROS to induce mPTP opening [25]. A recent study has provided evidence that apoptosis induced by ischemia/reperfusion is significantly greater in older patients, as compared to young [26]. These clinical results were confirmed in an animal study where myocardial apoptosis was measured directly, including caspase-3 activity [26]. Other age-related molecular changes in cardiomyocytes include DNA mutations, increased formation of ROS, protein oxidation, telomere shortening, and advanced glycation end products [27–30] Together, these changes result in a decline in function and decreased tolerance to stress [31].

Oxidative stress appears to play a key role in the increased susceptibility of older individuals to ischemia/reperfusion. In young adult myocardium, ischemia/reperfusion results in decreased complex II activity in the electron transport chain, secondary to high ROS release early in reperfusion. The oxidative stress-induced inhibition of complex II at restoration of blood flow, thus, has a protective effect on the recovery of cardiac metabolic efficiency by reducing subsequent mitochondrial ROS production. In middle-aged and older myocardium; however, complex II activity is maintained in response to ischemia/reperfusion. This leads to higher ROS production, as compared to young, and a delayed recovery of coronary flow during reperfusion [32]. A pharmacological inhibition of complex II has also been shown to protect the myocardium from ischemia/reperfusion injury [33, 34].

The role of oxidative stress in age-associated ischemia/reperfusion injury is further supported by data from intervention studies that have shown an improvement in post-ischemic myocardial function and reduced oxidative stress in response to antioxidant treatments [4, 11, 23, 35]. Furthermore, transgenic mice that overexpress manganese superoxide dismutase (MnSOD), a mitochondrial antioxidant enzyme, are more resistant to ischemia/reperfusion injury [36]. This area is covered in more detail in the section below on "Interventions to reduce ischemia/reperfusion injury."

The negative impact of age on the ischemia/reperfusion response is not limited to myocardial tissue. For example, the severity of hepatic ischemia/reperfusion injury, as commonly seen with liver transplants, is also associated with advancing age [37, 38]. A recent study found that among transplant patients, when the liver donor was older, the ischemia/reperfusion injury was greater and the graft was more apt to fail [39]. Age was also an independent risk factor for the transplant recipient in terms of graft injury and patient survival [39].

14.4 Effects of Gender on the Response to Ischemia/Reperfusion

Clearly, age is an independent risk factor for the impact of ischemia/reperfusion injury. What about gender? Although premenopausal women are protected against Cardiovascular Disease (CVD) compared to age-matched men, CVD increases within 10 years after the onset of menopause to levels matching or exceeding male incidence rates. CVD is the leading cause of death in women, accounting for approximately 40 % of total mortality. The marked difference in CVD occurrence before and after the menopause suggests that estrogen may have a cardioprotective effect. However, most of the studies showing beneficial effects of estrogen replacement on the risk for CVD have been observational; in contrast, randomized clinical trials have failed to detect such an effect. This discrepancy has been explained by the "timing hypothesis" whereby estrogen replacement has differential effects depending on the amount of time elapsed since the onset of menopause. Thus, the cardioprotective effects of estrogen would only be expected if initiated prior to the development of advanced atherosclerotic lesions and accumulation of mitochondrial mutations [40, 41]. Comorbidities, such as type 2 diabetes, in the presence of ischemic heart disease, generally worsen the outcome of a myocardial infarct and erase the gender-associated protection of premenopausal women. In fact, diabetic women have significantly higher mortality after myocardial infarction than diabetic men [42, 43].

The literature on gender differences and ischemia/reperfusion, in particular with aging, is still quite sparse and is noticeably an area which warrants more research. Increased resistance of the female heart to ischemia/reperfusion has been found in dogs, mice, rats, and rabbits [44–46]. Increasing estrogen levels in male rabbits increases the tolerance to ischemia/reperfusion, suggesting that estrogen may play the central role in these gender-related differences [47]. The effects of estrogen occur through activation of Estrogen Receptors (ERs) which act as a ligand-gated transcription factors (genomic effects). However, ERs are also found on the plasma membrane, where they can elicit rapid effects via activation of nongenomic signaling pathways [42]. Studies using isoform-specific ER modulators have shown that all ER isoforms including ER-α, ER-β, and G protein-coupled ER (GPR30) are involved in providing protection against ischemia/reperfusion injury using both genomic and nongenomic mechanisms [48–53]. Specifically, the activation of GPR30 has been shown to exert cardioprotective effect by inhibiting the mPTP opening associated with reperfusion injury and this effect occurs independent of gender [54, 55]. In addition to the effects of estrogen, other gender differences in heart function may contribute as well. For example, female rat hearts exhibit lower calcium uptake rates into mitochondria and better ability to maintain mitochondrial membrane potential, even at higher concentrations of calcium, as compared to male myocardium. Thus, male hearts are more susceptible to calcium overload which may exacerbate reperfusion injury. Moreover, female rats have lower cardiac mitochondrial content which makes them more efficient and there is lower generation of ROS [56, 57].

In a cross-sectional comparison of postmenopausal women that were on various forms of Estrogen Replacement Therapy (ERT) with an age-matched group of women that were not on estrogen replacement, the ERT group had significantly greater tolerance to forearm ischemia/reperfusion as indicated by lower increases in oxidative stress in response to the acute challenge [13]. This was, however, a small pilot study and the results should be verified in an intervention-based study with larger number of subjects.

14.5 Ischemic Preconditioning: Endogenous Protection Against Myocardial Ischemia

Brief, repeated episodes of myocardial ischemia and reperfusion, termed *ischemic preconditioning*, are a powerful endogenous protective mechanism against prolonged myocardial ischemia that was first described by Murry et al. in 1986 [58]. Ischemic preconditioning is an adaptive response to ischemia where the tissue adapts to brief occurrences of sublethal ischemic stress and becomes more resistant to a subsequent, prolonged ischemia resulting in diminished reperfusion injury. The preconditioning has two phases; an early phase which is manifested within minutes after the ischemic stress and lasts 2–3 h (classical preconditioning), and a late phase which has a slower onset and lasts up to 72 h (second window of protection) [59]. Both phases are associated with decreased ischemia/reperfusion injury and improved postischemic myocardial performance [60]. The beneficial adaptations to ischemic preconditioning are not limited to cardiac tissue and have been observed in other organs and tissue, in particular skeletal muscle, small intestine, and neuronal tissue [61–66].

Much of the data on preconditioning comes from experimentally induced short episodes of ischemia/reperfusion using direct occlusion by clamping coronary arteries. Nevertheless, there are clinical equivalents of ischemic preconditioning such as preinfarction angina, exercise, and various pharmacological preconditioning mimetics [67–69]. For example, preinfarction angina is associated with reduced left ventricular impairment, smaller infarct size, and improved short-term prognosis as well as in-hospital outcome, after an acute myocardial infarct [70–73]. These effects, however, were only seen in adult, nonelderly patients and did not extend to older adults. Indeed, both experimental and clinical data show that in general, the effects of preconditioning are lost or reduced with aging [18, 74, 75]. An exception to this rule is in female rat hearts, where it has been shown that the effect of ischemic preconditioning is absent in young but is observed in older female animals. This may be due to the fact that the young female animals already have high tolerance for ischemia/reperfusion injury and so the preconditioning does not afford any additional benefit. In older animals, where the susceptibility to ischemia/reperfusion injury is increased, the effects of preconditioning confer an advantage in the female heart [76]. The mechanisms for these differences are not fully understood.

The effects of preconditioning may occur through a temporary increase in oxidative stress triggering pathways involved in the upregulation of antioxidant enzymes. Wu et al. [77] studied patients undergoing coronary artery bypass grafting and measured free radical generation in the coronary sinus blood using spin-trap spectroscopy. The patients who were randomized to ischemic preconditioning (two cycles of 2 min ischemia followed by 3 min reperfusion prior to cross clamping) had improved ventricular functioning at 1 and 6 h following aortic declamping, as compared to the controls [77]. The ischemic preconditioning protocol resulted in measureable increases in free radicals. Furthermore, preconditioning-induced increase in free radical generation correlated with the left ventricular stroke work index at both 1 and 6 h after declamping. These data suggest that there is a hormetic effect of ischemic preconditioning resulting in a better capacity to constrain a subsequent oxidative insult, such as that seen with a more prolonged ischemia/reperfusion.

Norepinephrine is thought to be another possible mediator of the effect of preconditioning, specifically the release of norepinephrine from intramyocardial adrenergic nerves. The reduced effect of preconditioning with aging may be due to decreased release of norepinephrine in response to a myocardial infarct [74]. This concept is supported by data showing that the beneficial effect of preconditioning observed in adult and food-restricted senescent rats was blocked by pharmacological depletion of norepinephrine stores [74].

Full discussion on ischemic preconditioning and the cellular mechanisms involved in the adaptive response is beyond the scope of this chapter, but the reader is referred to some excellent reviews on this subject matter (e.g., [78–80]).

14.6 Forearm Ischemia/Reperfusion and Oxidative Stress

The majority of experimental data on the effects of cardiac ischemia/reperfusion come from animal studies since experimental induction of ischemia in the coronary arteries of humans cannot be done safely. A model used to study systemic responses to mild or moderate ischemia/reperfusion in humans is forearm ischemia/reperfusion. In this model, the forearm vasculature is occluded by inflating a blood pressure cuff on the upper arm, generally to a pressure of 200 mm Hg, and after a brief period of ischemia the cuff is released and reperfusion occurs.

Longer duration forearm ischemia (20 min rather than 5 min) generates increased levels of oxidative stress, activation of neutrophils, and a decrease in brachial flow mediated dilation; suggesting that this model produces modest ischemia/reperfusion injury to the endothelium of the brachial artery [81, 82]. We found that further modification of this procedure to a protocol utilizing three 10 min periods of forearm arterial occlusion separated by 2 min periods where the cuff was deflated to allow reperfusion, was better tolerated by study participants and resulted in slightly greater increases in oxidative stress [13]. In this model, the extent of oxidative stress induced in each individual in response to this ischemia/reperfusion is quantified by measuring levels of plasma F_2-isoprostanes at baseline (pre-I/R) and at several time points after the final cuff release, up to 240 min post I/R (see Fig. 14.3). F_2-isoprostanes are prostaglandin-like compounds formed from free-radical catalyzed peroxidation of arachadonic acid through a noncyclooxygenase enzyme pathway [83]. This marker of oxidative stress has been found to be the preferred biomarker of lipid peroxidation in terms of both sensitivity and reliability [84]. Increase in F_2-isoprostane levels in response to forearm ischemia/reperfusion is thus an index of free radical generation and the overall response to the challenge indicates how well the individual can neutralize or constrain this oxidative insult. This model has been shown to elicit significant increases in F_2-isoprostane levels in both young and older adults [13, 85].

Rather than just measuring oxidative stress in a "static" manner, the forearm ischemia/reperfusion trial (I/R) requires dynamic responses to an acute challenge such as upregulating endogenous antioxidant defenses. The forearm ischemia/reperfusion model, therefore, serves as a proxy marker for the capacity to respond to ischemia/reperfusion in the myocardium and other tissues.

Similarly to the results from studies on myocardial ischemia, the forearm ischemia/reperfusion trial demonstrates an impairment with aging in the ability to constrain the oxidative stress generated during an acute challenge [13]. Interestingly, resting levels of F_2-isoprostanes did not differ between the age groups, demonstrating the value of an acute challenge to unveil impairments in these defensive pathways such as upregulation of antioxidant defenses and/or downregulation of ROS-producing pathways. The older group had more acute peroxidation as exhibited by significantly higher levels of F_2-isoprostane levels at 15 min post I/R in the older group compared to the young. This higher response was sustained throughout the entire study period (240 min) in the older group, while the young group returned to baseline levels by the end of the study period (Fig. 14.2) [13].

This alteration in the resistance to oxidative stress may have significant physiological implications, as older adults have increased morbidity and mortality in a variety of serious medical conditions associated with oxidative stress, including ischemic heart disease, traumatic injury, burns, and organ transplantation [27–30, 86] Furthermore, the levels of oxidative stress in hospitalized older patients correlate with the severity of illness and mortality [87]. Thus, identifying the mechanisms that

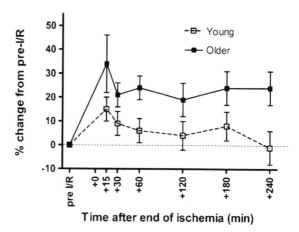

Fig. 14.2 Comparison of oxidative stress in young and older adults induced by forearm ischemia/reperfusion. Graph shows increases in plasma F_2-isoprostanes (percent change from baseline levels) in older adults (*filled squares*, $n = 20$) and young adults (*open squares*, $n = 20$). Significant effects of trial and age ($p < 0.05$). Reprinted from: Davies et al [13] with permission from Elsevier

underlie this reduced resistance as well interventions that effectively reverse age-related declines may be extremely helpful in preventing morbidity and mortality.

The age-related impairment in susceptibility to ischemia/reperfusion may occur gradually with advancing age or it may not manifest until late in life. A recent study comparing middle aged to young sedentary men and women found that the middle aged individuals exhibited greater and more sustained decrease in brachial flow mediated dilation, a measure of endothelial function, after forearm ischemia/reperfusion injury compared to the younger individuals [81]. Similar results have been seen in animal models including a progressive loss in the tolerance to ischemic insult as well as the ability to induce myocardial protection from ischemia/reperfusion injury through preconditioning [21, 88, 89]. Taken together, these data suggest that the decrement occurs more gradually with advancing age and underscore that potential interventions may be more successful if initiated earlier in life.

14.7 Interventions to Reduce Ischemia/Reperfusion Injury

Given the increased vulnerability of older adults to ischemia/reperfusion injury and oxidative damage, it is of great interest to find interventions that can improve tolerance to ischemia/reperfusion and reduce tissue reperfusion injury. The following sections will highlight research in areas that show promising effects on improving tolerance to ischemia/reperfusion including exercise, antioxidant supplementation, and dietary restriction.

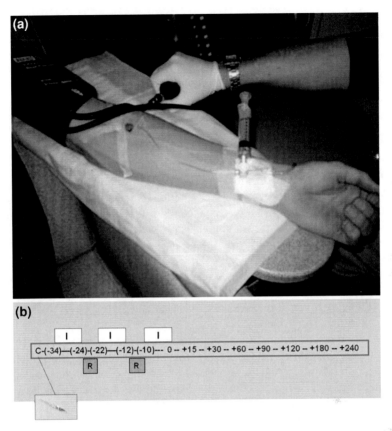

Fig. 14.3 Forearm Ischemia/Reperfusion (I/R) trial. Description of Trial: **a** An intravenous catheter is inserted in the antecubital vein and a baseline blood sample is collected (−34 min). The catheter is kept in situ with a slow saline drip throughout the trial in order to prevent clotting at the blood draw site. A blood pressure cuff is placed on the same arm, inflated to 200 mm Hg and kept inflated for 10 min, then released for 2 min. This inflation procedure is repeated twice more (total time = 34 min). **b** At the three time points of cuff inflation when blood flow is occluded, approximately 2 mL of heparin flush (10 kUSP/L) is injected into the blood sample tubing to further prevent any clotting at the site. A blood sample is drawn immediately after the final release of the blood pressure cuff and then at minutes 15, 30, 60, 120, 180, and 240 post final cuff deflation. Blood samples are centrifuged and the plasma is stored at −80 °C until analysis of F_2-isoprostanes

14.7.1 Exercise

Exercise, in particular endurance exercise, is perhaps the most practical and clinically relevant countermeasure against ischemia/reperfusion injury. Regular exercise or physical activity is well known to reduce risk for chronic diseases, including CVD, type 2 diabetes, hypertension, and certain cancers [90]. Acute exercise can be considered a type of preconditioning, especially if the intensity is high enough where the metabolic demand of the cardiac tissue is not fully met, inducing a temporary ischemia which is resolved when the exercise is halted. Regular exercise results in physiological adaptations in response to intermittent occurrences of acute exercise and the tissue becomes more resistant to subsequent bouts of exercise stress. This increased stress resistance with regular exercise does appear to translate to a general increased tolerance to ischemia/reperfusion injury.

14.7.1.1 Exercise-Induced Cardiac Preconditioning

The effect of ischemic preconditioning is reduced or lost with aging. For example; adult, but not elderly, patients with preinfarction angina (a clinical equivalent of ischemic preconditioning) have better outcomes after an acute myocardial infarct than patients who do not experience angina. However, a retrospective study found that among older patients (>65 years), high physical activity was associated with better outcome after an acute myocardial infarct [68]. Furthermore, when these patients were analyzed based on presence or absence of preinfarction angina, the effects of greater physical activity were only found in the group with angina [68]. These data suggest that regular exercise/physical activity preserves the preconditioning effect in aged individuals.

There is inherent difficulty in translating the beneficial effect seen with preconditioning in experimental settings to the real world of clinical practice because to do so would require knowledge of an impending ischemic event. In contrast, exercise can be carried out regularly without expectation of an impending ischemic event, is readily available and without harmful side-effects and can take numerous forms that best suit the patient's preferences [69]. Thus, exercise can be considered a sustainable form of preconditioning. Exercise-induced preconditioning appears to act through different mechanisms than ischemic preconditioning. Some potential candidates include adenosine receptors, opium receptors, transient ROS production, AMPK, and surges of inflammatory cytokines [69].

14.7.1.2 Exercise-Induced Increase in Tolerance to Ischemia/Reperfusion Injury

Epidemiological studies have shown that individuals that exercise regularly have a reduced risk of mortality during a myocardial ischemic event [91–94]. Moreover, studies employing exercise interventions show improved postischemic cardiac function in the exercise groups as compared to sedentary controls [95–98]. The protective effect afforded by exercise on ischemia/reperfusion related responses has been demonstrated both with short duration ischemia (5–20 min) as well as prolonged ischemia (20–60 min). Maximal benefits of exercise can be attained with a few days of consecutive exercise [97, 99, 100]; however, these beneficial effects are also quickly lost if regular exercise is not maintained [101].

There are multiple mechanisms that have been proposed and investigated as responsible for the exercise-induced increase in tolerance to ischemia/reperfusion injury. These include improved coronary collateral circulation, elevated heat shock proteins, elevated cyclooxygenase-2 (COX-2) activity, elevated endoplasmic reticulum stress proteins, sarcolemmal and mitochondrial ATP-sensitive potassium channels, and increased antioxidant capacity [102]. However, several studies show that the effects of exercise occur even when there are no changes in myocardial heat shock proteins, COX-2 activity, or endoplasmic reticulum stress proteins [99, 103–105]. Similarly, the time course of exercise-induced benefits against ischemia/reperfusion injury is much faster than the time it takes to attain coronary collateral circulation [102]. Thus, changes in antioxidant capacity may be the most plausible mechanism for exercise induced tolerance.

Exercise has been shown to increase antioxidant capacity, with increased MnSOD activity being the most consistent alteration [106–108]. For example, Starnes et al. [106] found that a single 60 min exercise bout increased stress response proteins (catalase and MnSOD activity, HSP70 content) and improved postischemic recovery of contractile function in old rats, 24 h after the exercise session. Importantly, exercise-induced changes in antioxidant activity are seen in both young and old [107]. The critical role of antioxidant enzymatic activity in exercise-induced cardioprotection has been shown by studies where blocking increases in MnSOD activity prevented the exercise-induced decrease in ischemia/reperfusion injury [101, 109, 110].

If exercise increases expression of antioxidant enzymes; then, we would anticipate that ischemia/reperfusion would induce less oxidative stress in physically active older adults than their sedentary counterparts. This appears to be the case, as we have recently shown that among older adults, physically fit individuals have a significantly lower F_2-isoprostane response to forearm ischemia/reperfusion than age-matched unfit controls [111]. The two groups did not differ in plasma concentration of small molecular weight antioxidants, nor could the reduced oxidative stress in the physically fit individuals be explained by other known effects of exercise training such as lower adiposity or higher HDL levels. Thus, exercise appears to significantly enhance the antioxidant capacity and oxidative resistance of older adults.

The exercise intensity required to invoke protection against ischemia/reperfusion injury in older adults is unclear. Wislöff et al. [112] compared high intensity exercise versus moderate intensity exercise in heart failure patients and found that the high intensity group had greater endothelial function, increased total antioxidant capacity, and decreased oxidative stress after a 12-week exercise intervention [112]. The effect on ischemia/reperfusion was not determined. There is some indication that an exercise intensity of approximately 70–75 % of maximal oxygen consumption ($VO_{2\ max}$) generally results in increased tolerance to ischemia/reperfusion [107], whereas the efficacy of low-intensity exercise (<60 % $VO_{2\ max}$) has been equivocal [110, 113]. Thus, future studies are need to resolve this important question.

14.7.2 Diet and Antioxidant Supplements

14.7.2.1 Flavonoids

In addition to exercise, another effective mechanism for inducing antioxidant enzymes and increasing resistance to ischemia/reperfusion injury may be through dietary consumption of flavonoids. While flavonoids are often touted as "antioxidants," their very poor bioavailability suggests that their concentration in plasma and tissue is too low for them to act directly as antioxidants. More likely, they evoke antioxidant effects because they are potent inducers of NF-E2-related factor (Nrf-2) activation [114, 115]. Nrf2 is a transcriptional factor that binds to the antioxidant response element of target genes and increases the transcription of a variety of proteins involved in antioxidant stress resistance and neutralization of ROS including catalase, MnSOD, and glutathione peroxidase. Thus, diets rich in flavonoids may be a relatively simple way to enhance resistance to ischemia/reperfusion damage in older adults. To test this notion, we examined the effect of consuming tart cherry juice on oxidative resistance using our forearm ischemia/reperfusion model, employing a double blind, placebo controlled, cross-over design in older men, and women [85]. Tart cherries have high levels of flavonoids, including polyphenols, and anthocyanins. The 2 week tart cherry juice intervention significantly reduced the F_2-isoprostane response to the forearm ischemia/reperfusion indicating an improved resistance to this oxidative insult [85].

14.7.2.2 Vitamins

While flavonoids likely act by inducing antioxidant enzymes, dietary vitamins such as tocopherols (vitamin E) and ascorbic acid (vitamin C) may increase antioxidant capacity and resistance to ischemia/reperfusion by acting directly as antioxidants, although with poor efficiency. Ascorbic acid was classified as a vitamin, because its deficiency leads to scurvy and not because of an essential role in protecting against oxidative stress. Nevertheless, animal studies suggest that vitamin C may be essential in protecting against oxidative stress when other arms of the antioxidant response are incapacitated. For example, when combined with selenium deficiency (which reduces glutathione peroxidase activity), vitamin C deficiency markedly increased levels of F_2-isoprostanes in guinea pigs [116]. Although similar studies have not been performed in humans for obvious ethical reasons, a longitudinal study of adult men found correlations between vitamin C intake and plasma F_2-isoprostane levels [117]. That megadoses of vitamin C may protect against oxidative damage during ischemia/reperfusion is suggested by a recent study in elderly patients undergoing bilateral total knee replacement, which requires tourniquet application and thus invokes ischemia/reperfusion injury. Patients who received vitamin C had lower levels of MDA and troponin I release compared to those who did not received vitamin C [118]. Further investigations of the effectiveness of vitamin C in surgical settings of ischemia/reperfusion involving the elderly are needed.

Like vitamin C, selenium deficiency to remove glutathione peroxidation activity unveils a role of vitamin E in maintaining oxidative balance, as dual vitamin E and selenium deficiency increased F_2-isoprostanes in rats [119]. However, the value of vitamin E supplementation in the absence of antioxidant enzyme deficiencies is less clear. For instance, vitamin E supplementation required extremely high doses (1,600 IU or higher) for prolonged periods (16 weeks) to significantly alter resting F_2-isoprostane levels in hypercholesterolemic women [120]. This previously unrecognized poor potency of vitamin E likely underlies the failure of clinical trials with vitamin E and CVD, as typical doses in these trials were 400 IU or lower. Whether high doses of vitamin E could protect against ischemia/reperfusion injury is unclear. A recent meta-analysis of nine clinical trials evaluating the efficacy of vitamin E supplementation on stroke, found a 10 % decrease in the risk of ischemic stroke from vitamin E supplementation, but a 22 % increase in risk for hemorrhagic stroke, leading the authors to conclude that indiscriminate use of vitamin E should be cautioned against [121].

One interesting question is whether there is an additive effect between exercise and dietary antioxidants or if dietary antioxidants might actually prevent the protective effect of exercise being exerted. This question was investigated by Hamilton et al. [122] using young rats that were either on control diet or an antioxidant diet, and within each diet group half were exercise trained and the other half were sedentary. The results showed that exercise and antioxidant supplementation independently decreased myocardial ischemia/reperfusion injury and that combining the two strategies did not enhance or worsen the effect of either treatment alone [122]. Whether combining both interventions might be more efficacious in humans is not known, but would be of particular interest to explore in older adults where the basal activities of antioxidant enzymes are lower than in young adults, so that inducers might have a more profound effect.

14.7.3 Dietary Restriction

Dietary Restriction (DR) is a term describing various interventions which either reduce caloric intake or limit certain nutrients such as protein or specific amino acids. It is well accepted that DR increases life span in many organisms and the limited data in humans show beneficial effects on markers of aging and age-related diseases.

The effects of DR on increasing tolerance to ischemia/reperfusion have been shown in both heart and brain [123–125]. The mechanisms underlying these beneficial effects of DR are thought to be a combination of reduced oxidative stress, upregulation of stress proteins, and attenuated inflammatory response. DR is associated with activation of Nrf2 resulting in increased cellular oxidative stress tolerance [126]. As previously described, the protective effect of ischemic preconditioning on ischemia/reperfusion injury is generally lost with aging. This age-associated loss of hormetic adaptation is preserved in older animals that have been calorically restricted [74]. Moreover, when DR and exercise were used in tandem in older animals, there was a synergistic effect and the benefits of ischemic preconditioning were restored to levels greater than that of either intervention alone [95].

Interestingly, the effects of DR are not limited to long-term intervention. Recent studies have shown that as little as 2–4 weeks of DR is sufficient to attenuate both reversible and irreversible ischemic myocardial damage in young and old rats [127]. Similarly, short-term DR improved survival and kidney function following renal ischemia/reperfusion injury in mice [128]. Even short-term fasting (1–3 days) elicits significant protection from ischemia/reperfusion in both kidney and liver [128, 129]. Surprisingly, the protection acquired after 3 day water only fasting was similar to that of 2–4 weeks of 30 % DR [128]. These results have great clinical relevance, because patients undergoing surgeries where ischemia/reperfusion is involved could potentially use "dietary preconditioning" prior to surgery, in order to improve their tolerance to ischemia/reperfusion and thereby minimize the impact of reperfusion injury. The reader is referred to two excellent reviews on the effect of caloric restriction and cardiovascular protection; one by Shinmura [130] and the other by Ungvari et al. [126].

14.8 Conclusions

The response to ischemia/reperfusion is a clinically relevant issue, especially in older individuals. The current literature illustrates the dire need for more translational studies both to identify vulnerable individuals and to identify appropriate interventions. Easily applied methodologies such as forearm ischemia/reperfusion, may be particularly useful in testing the efficacy of various potential interventions in humans. To date, the most consistent mechanism for reducing ischemia/reperfusion injury is through upregulation of endogenous antioxidant enzymes and stress proteins. Interventions that appear to act via this mechanism include exercise, short-term caloric restriction, and diets rich in flavonoids. In addition to general intervention studies, there is also a need to investigate the gender differences in oxidative stress responses to ischemia/reperfusion and in particular, the appropriateness of estrogen replacement for older women. Such recommendations await results from studies currently in progress. Hopefully, such studies will fill this obvious void in our knowledge of appropriate treatment strategies for older women.

References

1. McCord JM (1985) Oxygen-derived free radicals in postischemic tissue injury. N Engl J Med 312(3):159–163
2. Maxwell SRJ, Lip GYH (1997) Reperfusion injury: a review of the pathophysiology, clinical manifestations and therapeutic options. Int J Cardiol 58(2):95–117
3. Seal JB, Gewertz BL (2005) Vascular dysfunction in ischemia-reperfusion injury. Ann Vasc Surg 19:572–584

4. Zhang W-X, Zhou L-F, Zhang L et al (2011) Protective effects of glutamine preconditioning on ischemia-reperfusion injury in rats. Hepatobiliary Pancreat Dis Int 10(1):78–82
5. Downey JM (1990) Free radicals and thier involvement during long-term myocardial ischemia and reperfusion. Annu Rev Physiol 52:487–504
6. Hoffman JWJ, Gilbert TB, Poston RS, Silldorff EP (2004) Myocardial reperfusion injury: etiology, mechanisms, and therapies. J Extra Corpor Technol 36(4):391–411
7. Bolli R, Marbán E (1999) Molecular and cellular mechanisms of myocardial stunning. Physiol Rev 79(2):609–634
8. Piper HM, Meuter K, Schafer C (2003) Cellular mechanisms of ischemia-reperfusion injury. Ann Thorac Surg 75(2):S644–S648
9. Lakatta EG (2002) Age-associated cardiovascular changes in health: impact on cardiovascular disease in older persons. Heart Fail Rev 7(1):29–49
10. Lindsay TF, Luo XP, Lehotay DC et al (1999) Ruptured abdominal aortic aneurysm, a "two-hit" ischemia/reperfusion injury: evidence from an analysis of oxidative products. J Vasc Surg 30:219–228
11. Xia Z, Godin DV, Ansley DM (2003) Propofol enhances ischemic tolerance of middle-aged rat hearts: effects on 15-F2t-isoprostane formation and tissue antioxidant capacity. Cardiovasc Res 59(1):113–121
12. Webster KA (2009) Mitochondrial death channels. Am Sci 97:384–391
13. Davies SS, Traustadottir T, Stock AA et al (2009) Ischemia-reperfusion unveils impaired capacity of older adults to restrain oxidative insult. Free Radic Biol Med 47:1014–1018
14. Ozaki M (2002) New strategy for preventing ischemia/reperfusion-induced organ injury and promoting regeneration: a novel trial for improving transplant organ function by targeted regulation of cellular signals. Transplant Proc 34(7):2637–2639
15. Chen JC, Hildebrand HD, Salvian AJ, Hsiang YN, Taylor DC (1997) Progress in abdominal aortic aneurysm surgery: four decades of experience at a teaching center. Cardiovasc Surg 5(2):150–156
16. Aivatidi C, Vourliotakis G, Georgopoulos S, Sigala F, Bastounis E, Papalambros E (2011) Oxidative stress during abdominal aortic aneurysm repair—biomarkers and antioxidant's protective effect: a review. Eur Rev Med Pharmacol Sci 15(3):245–252
17. Ar'Rajab A, Dawidson I, Fabia R (1996) Reperfusion injury. New Horiz 4:224–234
18. Ataka K, Chen D, Levitsky S, Jimenez E, Feinberg H (1992) Effect of aging on intracellular Ca_2+, pHi, and contractility during ischemia and reperfusion. Circulation 86(suppl. II):II-371–II-376
19. Misare B, Krukencamp I, Levitsky S (1992) Age-dependent sensitivity to unprotected cardiac ischemia: the senescent myocardium. J Thorac Cardiovasc Surg 103:60–65
20. Tani M, Suganuma Y, Hasegawa H et al (1997) Decrease in ischemic tolerance with aging in isolated perfused fischer 344 rat hearts: relation to increases in intracellular na + after ischemia. J Mol Cell Cardiol 29(11):3081–3089
21. Willems L, Zatta A, Holmgren K, Ashton KJ, Headrick JP (2005) Age-related changes in ischemic tolerance in male and female mouse hearts. J Mol Cell Cardiol 38(2):245–256
22. Mariani J, Ou R, Bailey M et al (2000) Tolerance to ischemia and hypoxia is reduced in aged human myocardium. J Thorac Cardiovasc Surg 120:660–667
23. Rosenfeldt F, Pepe S, Linnane A et al (2002) The effects of ageing on the response to cardiac surgery: protective strategies for the ageing myocardium. Biogerontology 3(1):37–40
24. Shim YH (2010) Cardioprotection and ageing. Korean J Anesthesiol 58(3):223–230
25. Zhu J, Rebecchi MJ, Glass PSA, Brink PR, Liu L (2011) Cardioprotection of the aged rat heart by GSK-3β inhibitor is attenuated: age-related changes in mitochondrial permeability transition pore modulation. Am J Physiol: Heart Circ Physiol 300(3):H922–H930
26. Liu M, Zhang P, Chen M et al (2011) Aging might increase myocardial ischemia/reperfusion-induced apoptosis in humans and rats. AGE 34(3):621–632
27. Boengler K, Schulz R, Heusch G (2009) Loss of cardioprotection with ageing. Cardiovasc Res 83:247–261

28. Gutierrez C, Al-Faifi S, Chaparro C et al (2007) The effect of recipient's age on lung transplant outcome. Am J Transplant 7(5):1271–1277
29. Mosenthal AC, Lavery RF, Addis M et al (2002) Isolated traumatic brain injury: age is an independent predictor of mortality and early outcome. J Trauma 52(5):907–911
30. Pereira CT, Barrow RE, Sterns AM et al (2006) Age-dependent differences in survival after severe burns: a unicentric review of 1,674 patients and 179 autopsies over 15 years. J Am Coll Surg 202(3):536–548
31. Bernhard Laufer (2008) The aging cardiomyocyte: a mini-review. Gerontology 54(1):24–31
32. Mourmoura E, Leguen M, Dubouchaud H et al (2011) Middle age aggravates myocardial ischemia through surprising upholding of complex II activity, oxidative stress, and reduced coronary perfusion. AGE 33(3):321–336
33. Turan N, Csonka C, Csont Ts (2006) The role of peroxynitrite in chemical preconditioning with 3-nitropropionic acid in rat hearts. Cardiovasc Res 70(2):384–390
34. Wojtovich AP, Brookes PS (2008) The endogenous mitochondrial complex II inhibitor malonate regulates mitochondrial ATP-sensitive potassium channels: implications for ischemic preconditioning. Biochim Biophys Acta (BBA)-Bioenerg 1777(7–8):882–889
35. Besse S, Bulteau AL, Boucher F, Riou B, Swynghedauw B, De Leiris J (2006) Antioxidant treatment prevents cardiac protein oxidation after ischemia-reperfusion and improves myocardial function and coronary perfusion in senescent hearts. J Physiol Pharmacol 57(4):541–552
36. Chen Z, Siu B, Ho Y-S et al (1998) Overexpression of MnSOD protects against myocardial ischemia/reperfusion injury in transgenic mice. J Mol Cell Cardiol 30(11):2281–2289
37. Okaya T, Blanchard J, Schuster R et al (2005) Age-dependent responses to hepatic ischemia/reperfusion injury. Shock 24(5):421–427
38. Selzner M, Selzner N, Jochum W, Graf R, Clavien P-A (2007) Increased ischemic injury in old mouse liver: an atp-dependent mechanism. Liver Transplant 13(3):382–390
39. Selzner M, Kashfi A, Selzner N et al (2009) Recipient age affects long-term outcome and hepatitis C recurrence in old donor livers following transplantation. Liver Transplant 15(10):1288–1295
40. Mendelsohn ME, Karas RH (2005) Molecular and cellular basis of cardiovascular gender differences. Science 308(5728):1583–1587
41. Miller VM, Duckles SP (2008) Vascular actions of estrogens: functional implications. Pharmacol Rev 60(2):210–241
42. Sowers JR (1998) Diabetes mellitus and cardiovascular disease in women. Arch Intern Med 158(6):617–621
43. Natarajan S, Liao Y, Cao G, Lipsitz SR, McGee DL (2003) Sex differences in risk for coronary heart disease mortality associated with diabetes and established coronary heart disease. Arch Intern Med 163(14):1735–1740
44. Johnson MS, Moore RL, Brown DA (2006) Sex differences in myocardial infarct size are abolished by sarcolemmal KATP channel blockade in rat. Am J Physiol: Heart Circ Physiol 290(6):H2644–H2647
45. Ostadal B, Kolar F (2007) Cardiac adaptation to chronic high-altitude hypoxia: beneficial and adverse effects. Respir Physiol Neurobiol 158(2–3):224–236
46. Ostadal B, Prochazka J, Pelouch V, Urbanova D, Widimsky J (1984) Comparison of cardiopulmonary responses of male and female rats to intermittent high altitude hypoxia. Physiol Bohemoslov 33(2):129–138
47. Hale SL, Birnbaum Y, Kloner RA (1996) [beta]-Estradiol, but not [alpha]-estradiol, reduces myocardial necrosis in rabbits after ischemia and reperfusion. Am Heart J 132(2, Part 1):258–262
48. Lin J, Steenbergen C, Murphy E, Sun J (2009) Estrogen receptor-{beta} activation results in s-nitrosylation of proteins involved in cardioprotection. Circulation 120(3):245–254
49. Favre J, Gao J, Henry J-P et al (2010) Endothelial estrogen receptor {alpha} plays an essential role in the coronary and myocardial protective effects of estradiol in ischemia/reperfusion. Arterioscler Thromb Vasc Biol 30(12):2562–2567

50. Weil BR, Manukyan MC, Herrmann JL et al (2010) Signaling via GPR30 protects the myocardium from ischemia/reperfusion injury. Surgery 148(2):436–443
51. Deschamps AM, Murphy E, Sun J (2010) Estrogen receptor activation and cardioprotection in ischemia reperfusion injury. Trends Cardiovasc Med 20(3):73–78
52. Jeanes HL, Tabor C, Black D, Ederveen A, Gray GA (2008) Oestrogen-mediated cardioprotection following ischaemia and reperfusion is mimicked by an oestrogen receptor (ER)Î ± agonist and unaffected by an ERÎ² antagonist. J Endocrinol 197(3):493–501
53. Nikolic I, Liu D, Bell JA, Collins J, Steenbergen C, Murphy E (2007) Treatment with an estrogen receptor-beta-selective agonist is cardioprotective. J Mol Cell Cardiol 42(4):769–780
54. Bopassa JC, Eghbali M, Toro L, Stefani E (2010) A novel estrogen receptor GPER inhibits mitochondria permeability transition pore opening and protects the heart against ischemia-reperfusion injury. Am J Physiol: Heart Circ Physiol 298(1):H16–H23
55. Deschamps AM, Murphy E (2009) Activation of a novel estrogen receptor, GPER, is cardioprotective in male and female rats. Am J Physiol: Heart Circ Physiol 297(5): H1806–H1813
56. Ostadal B, Netuka I, Maly J, Besik J, Ostadalova I (2009) Gender differences in cardiac ischemic injury and protection–experimental aspects. Exp Biol Med 234(9):1011–1019
57. Lagranha CJ, Deschamps A, Aponte A, Steenbergen C, Murphy E (2010) Sex differences in the phosphorylation of mitochondrial proteins result in reduced production of reactive oxygen species and cardioprotection in females. Circ Res 106(11):1681–1691
58. Murry CE, Jennings RB, Reimer KA (1986) Preconditioning with ischemia: a delay of lethal cell injury in ischemic myocardium. Circulation 74(5):1124–1136
59. Bolli R (1996) The early and late phases of preconditioning against myocardial stunning and the essential role of oxyradicals in the late phase: an overview. Basic Res Cardiol 91(1): 57–63
60. Bolli R (2000) The late phase of preconditioning. Circ Res 87(11):972–983
61. Pan P-J, Chan R-C, Yang A-H, Chou C-L, Cheng Y-F, Chiu J-H (2008) Protective effects of preconditioned local somatothermal stimulation on neuromuscular plasticity against ischemia–reperfusion injury in rats. J Orthop Res 26(12):1670–1674
62. Sommer C (2009) Neuronal plasticity after ischemic preconditioning and TIA-like preconditioning ischemic periods. Acta Neuropathol 117(5):511–523
63. Saeki I, Matsuura T, Hayashida M, Taguchi T (2011) Ischemic preconditioning and remote ischemic preconditioning have protective effect against cold ischemia–reperfusion injury of rat small intestine. Pediatric Surgery International 27(8):857–862
64. Takeshita M, Tani T, Harada S et al (2011) Role of transcription factors in small intestinal ischemia-reperfusion injury and tolerance induced by ischemic preconditioning. Transpl Proc 42(9):3406–3413
65. Thaveau F, Zoll J, Rouyer O et al (2007) Ischemic preconditioning specifically restores complexes I and II activities of the mitochondrial respiratory chain in ischemic skeletal muscle. J Vasc Surg 46(3):541–547
66. Eberlin KR, McCormack MC, Nguyen JT, Tatlidede HS, Randolph MA, Austen WG (2008) Ischemic preconditioning of skeletal muscle mitigates remote injury and mortality. J Surg Res 148(1):24–30
67. Ahmed LA, Salem HA, Attia AS, Agha AM (2011) Pharmacological preconditioning with nicorandil and pioglitazone attenuates myocardial ischemia/reperfusion injury in rats. Eur J Pharmacol 663(1–3):51–58
68. Abete P, Ferrara N, Cacciatore F et al (2001) High level of physical activity preserves the cardioprotective effect of preinfarction angina in elderly patients. J Am Coll Cardiol 38(5):1357–1365
69. Frasier CR, Moore RL, Brown DA (2011) Exercise-induced cardiac preconditioning: how exercise protects your achy-breaky heart. J Appl Physiol 111(3):905–915
70. Muller DWM, Topol EJ, Califf RM et al (1990) Relationship between antecedent angina pectoris and short-term prognosis after thrombolytic therapy for acute myocardial infarction. Am Heart J 119(2, Part 1):224–231

71. Matsuda Y, Ogawa H, Moritani K et al (1984) Effects of the presence or absence of preceding angina pectoris on left ventricular function after acute myocardial infarction. Am Heart J 108(4, Part 1):955–958
72. Kloner RA, Shook T, Przyklenk K et al (1995) Coronary artery disease/myocardial infarction: previous angina alters in-hospital outcome in timi 4: a clinical correlate to preconditioning? Circulation 91(1):37–45
73. Ottani F, Galvani M, Ferrini D et al (1995) Coronary heart disease/myocardial infarction: prodromal angina limits infarct size: a role for ischemic preconditioning. Circulation 91(2):291–297
74. Abete P, Testa G, Ferrara N et al (2002) Cardioprotective effect of ischemic preconditioning is preserved in food-restricted senescent rats. Am J Physiol: Heart Circ Physiol 282(6):H1978–H1987
75. Abete P, Ferrara N, Cacciatore F et al (1997) Angina-induced protection against myocardial infarction in adult and elderly patients: a loss of preconditioning mechanism in the aging heart? J Am Coll Cardiol 30(4):947–954
76. Turcato S, Turnbull L, Wang GY et al (2006) Ischemic preconditioning depends on age and gender. Basic Res Cardiol 101(3):235–243
77. Wu Z-k, Tarkka MR, Eloranta J (2001) Effect of ischemic preconditioning on myocardial protection in coronary artery bypass graft patients*. Chest 119(4):1061–1068
78. Yang X, Cohen M, Downey J (2010) Mechanism of cardioprotection by early ischemic preconditioning. Cardiovasc Drugs Ther 24(3):225–234
79. Morris KC, Lin HW, Thompson JW, Perez-Pinzon MA (2011) Pathways for ischemic cytoprotection: Role of sirtuins in caloric restriction, resveratrol, and ischemic preconditioning. J Cereb Blood Flow Metab 31(4):1003–1019
80. Murphy E, Steenbergen C (2008) Mechanisms underlying acute protection from cardiac ischemia-reperfusion injury. Physiol Rev 88(2):581–609
81. DeVan AE, Umpierre D, Harrison ML et al (2011) Endothelial ischemia-reperfusion injury in humans: association with age and habitual exercise. Am J Physiol: Heart Circ Physiol 300(3):H813–H819
82. Kharbanda RK, Peters M, Walton B et al (2001) Ischemic preconditioning prevents endothelial injury and systemic neutrophil activation during ischemia-reperfusion in humans in vivo. Circulation 103(12):1624–1630
83. Morrow JD, Roberts LJ (1996) The isoprostanes : current knowledge and directions for future research. Biochem Pharmacol 51(1):1–9
84. Kadiiska MB, Gladen BC, Baird DD et al (2005) Biomarkers of oxidative stress study II. are oxidation products of lipids, proteins, and DNA markers of CCl4 poisoning? Free Radic Biol Med 38:698–710
85. Traustadóttir T, Davies SS, Stock AA et al (2009) Tart cherry juice decreases oxidative stress in healthy older men and women. J Nutr 139:1896–1900
86. Petcu EB, Sfredel V, Platt D, Herndon JG, Kessler C, Popa-Wagner A (2008) Cellular and molecular events underlying the dysregulated response of the aged brain to stroke: a mini-review. Gerontology 54(1):6–17
87. Powers JS, Roberts LJ 2nd, Tarvin E, Hongu N, Choi L, Buchowski M (2008) Oxidative stress and multiorgan failure in hospitalized elderly people. J Am Geriatr Soc 56(6): 1150–1152
88. Schulman D, Latchman DS, Yellon DM (2001) Effect of aging on the ability of preconditioning to protect rat hearts from ischemia-reperfusion injury. Am J Physiol: Heart Circ Physiol 281(4):H1630–H1636
89. Tani M, Honma Y, Hasegawa H, Tamaki K (2001) Direct activation of mitochondrial KATP channels mimics preconditioning but protein kinase C activation is less effective in middle-aged rat hearts. Cardiovasc Res 49(1):56–68
90. Booth FW, Gordon SE, Carlson CJ, Hamilton MT (2000) Waging war on modern chronic diseases: primary prevention through exercise biology. J Appl Physiol 88(2):774–787

91. Lakka TA, Venalainen JM, Rauramaa R, Salonen R, Tuomilehto J, Salonen JT (1994) Relation of leisure-time physical activity and cardiorespiratory fitness to the risk of acute myocardial infarction in men. N Engl J Med 330(22):1549–1554
92. Holtermann A, Mortensen OS, Burr H, Sögaard K, Gyntelberg F, Suadicani P (2010) Fitness, work, and leisure-time physical activity and ischaemic heart disease and all-cause mortality among men with pre-existing cardiovascular disease. Scand J Work Environ Health 36(5):366–372
93. Holtermann A, Mortensen OS, Burr H, Sögaard K, Gyntelberg F, Suadicani P (2010) Physical demands at work, physical fitness, and 30 year ischaemic heart disease and all-cause mortality in the copenhagen male study. Scand J Work Environ Health 36(5):357–365
94. Wislöff U, Nilsen TIL, Dröyvold WB, Mörkved S, Slördahl SA, Vatten LJ (2006) A single weekly bout of exercise may reduce cardiovascular mortality: how little pain for cardiac gain? 'the hunt study, norway'. Eur J Cardiovasc Prev Rehabil 13(5):798–804
95. Abete P, Testa G, Galizia G et al (2005) Tandem action of exercise training and food restriction completely preserves ischemic preconditioning in the aging heart. Exp Gerontol 40(1–2):43–50
96. Libonati JR, Gaughan JP, Hefner CA, Gow A, Paolone AM, Houser SR (1997) Reduced ischemia and reperfusion injury following exercise training. Med Sci Sports Exerc 29(4):509–516
97. Demirel HA, Powers SK, Zergeroglu MA et al (2001) Short-term exercise improves myocardial tolerance to in vivo ischemia-reperfusion in the rat. J Appl Physiol 91(5):2205–2212
98. Le Page C, Noirez P, Courty J, Riou B, Swynghedauw B, Besse S (2009) Exercise training improves functional post-ischemic recovery in senescent heart. Exp Gerontol 44(3):177–182
99. Hamilton KL, Powers SK, Sugiura T et al (2001) Short-term exercise training can improve myocardial tolerance to I/R without elevation in heat shock proteins. Am J Physiol: Heart Circ Physiol 281(3):H1346–H1352
100. Powers SK, Demirel HA, Vincent HK et al (1998) Exercise training improves myocardial tolerance to in vivo ischemia-reperfusion in the rat. AJP: Regul Integr Comp Physiol 275(5):R1468–R1477
101. Lennon SL, Quindry J, Hamilton KL et al (2004) Loss of exercise-induced cardioprotection after cessation of exercise. J Appl Physiol 96(4):1299–1305
102. Powers SK, Quindry JC, Kavazis AN (2008) Exercise-induced cardioprotection against myocardial ischemia-reperfusion injury. Free Radical Biol Med 44(2):193–201
103. Taylor RP, Harris MB, Starnes JW (1999) Acute exercise can improve cardioprotection without increasing heat shock protein content. Am J Physiol: Heart Circ Physiol 276(3):H1098–H1102
104. Murlasits Z, Lee Y, Powers SK (2007) Short-term exercise does not increase er stress protein expression in cardiac muscle. Med Sci Sports Exerc 39(9):1522–1528
105. Quindry J, French J, Hamilton K, Lee Y, Selsby J, Powers S (2010) Exercise does not increase cyclooxygenase-2 myocardial levels in young or senescent hearts. J Physiol Sci 60(3):181–186
106. Starnes JW, Taylor RP, Park Y (2003) Exercise improves postischemic function in aging hearts. Am J Physiol: Heart Circ Physiol 285(1):H347–H351
107. Quindry J, French J, Hamilton K, Lee Y, Mehta JL, Powers S (2005) Exercise training provides cardioprotection against ischemia-reperfusion induced apoptosis in young and old animals. Exp Gerontol 40(5):416–425
108. Siu PM, Bryner RW, Martyn JK, Alway SE (2004) Apoptotic adaptations from exercise training in skeletal and cardiac muscles. FASEB J 18(10):1150–1152
109. Yamashita N, Hoshida S, Otsu K, Asahi M, Kuzuya T, Hori M (1999) Exercise provides direct biphasic cardioprotection via manganese superoxide dismutase activation. J Exp Med 189(11):1699–1706
110. Lennon SL, Quindry JC, French JP, Kim S, Mehta JL, Powers SK (2004) Exercise and myocardial tolerance to ischaemia-reperfusion. Acta Physiol Scand 182(2):161–169

111. Traustadóttir T, Davies S, Su Y et al (2011) Oxidative stress in older adults: effects of physical fitness. AGE :33(E-pub ahead of print)
112. Wislöff U, Stöylen A, Loennechen JP et al (2007) Superior cardiovascular effect of aerobic interval training versus moderate continuous training in heart failure patients. a randomized study. Cirulation 115:3086–3094
113. Starnes JW, Taylor RP, Ciccolo JT (2005) Habitual low-intensity exercise does not protect against myocardial dysfunction after ischemia in rats. J Cardiovasc Risk 12(2):169–174
114. Lee-Hilz YY, Boerboom A-MJF, Westphal AH, van Berkel WJH, Aarts JMMJG, Rietjens IMCM (2006) Pro-oxidant activity of flavonoids induces epre-mediated gene expression. Chem Res Toxicol 19(11):1499–1505
115. Ghanim H, Sia CL, Korzeniewski K et al (2011) A Resveratrol and Polyphenol Preparation Suppresses Oxidative and Inflammatory Stress Response to a High-Fat, High-Carbohydrate Meal. J Clin Endocrinol Metab 96(5):1409–1414
116. Hill KE, Motley AK, May JM, Burk RF (2009) Combined selenium and vitamin C deficiency causes cell death in guinea pig skeletal muscle. Nutr Res 29(3):213–219
117. Helmersson J, Arnlov J, Larsson A, Basu S (2009) Low dietary intake of β-carotene, α-tocopherol and ascorbic acid is associated with increased inflammatory and oxidative stress status in a swedish cohort. Br J Nutr 101:1775–1782
118. Lee JY, Kim CJ, Chung MY (2010) Effect of high-dose vitamin c on oxygen free radical production and myocardial enzyme after tourniquet ischaemia-reperfusion injury during bilateral total knee replacement. J Int Med Res 38:1519–1529
119. Awad JA, Morrow JD, Hill KE, Roberts LJ, Burk RF (1994) Detection and localization of lipid peroxidation in selenium- and vitamin E-deficient rats using F2-isoprostanes. J Nutr 124(6):810–816
120. Roberts LJ, Oates JA, Linton MF et al (2007) The relationship between dose of vitamin E and suppression of oxidative stress in humans. Free Radical Biol Med 43(10):1388–1393
121. Schurks M, Glynn RJ, Rist PM, Tzourio C, Kurth T (2010) Effects of vitamin E on stroke subtypes: meta-analysis of randomised controlled trials. BMJ 341:c5702. doi:10.1136/bmj.c5702
122. Hamilton KL, Staib JL, Phillips T, Hess A, Lennon SL, Powers SK (2003) Exercise, antioxidants, and HSP72: protection against myocardial ischemia/reperfusion. Free Radical Biol Med 34(7):800–809
123. Yu ZF, Mattson MP (1999) Dietary restriction and 2-deoxyglucose administration reduce focal ischemic brain damage and improve behavioral outcome: evidence for a preconditioning mechanism. J Neurosci Res 57(6):830–839
124. Chandrasekar B, Nelson JF, Colston JT, Freeman GL (2001) Calorie restriction attenuates inflammatory responses to myocardial ischemia-reperfusion injury. Am J Physiol: Heart Circ Physiol 280(5):H2094–H2102
125. Ahmet I, Wan R, Mattson MP, Lakatta EG, Talan M (2005) Cardioprotection by intermittent fasting in rats. Circulation 112(20):3115–3121
126. Ungvari Z, Parrado-Fernandez C, Csiszar A, de Cabo R (2008) Mechanisms underlying caloric restriction and lifespan regulation: implications for vascular aging. Circ Res 102(5):519–528
127. Shinmura K, Tamaki K, Bolli R (2005) Short-term caloric restriction improves ischemic tolerance independent of opening of ATP-sensitive K + channels in both young and aged hearts. J Mol Cell Cardiol 39(2):285–296
128. Mitchell JR, Verweij M, Brand K et al (2010) Short-term dietary restriction and fasting precondition against ischemia reperfusion injury in mice. Aging Cell 9(1):40–53
129. Sumimoto R, Southard JH, Belzer FO (1993) Livers from fasted rats acquire resistance to warm and cold ischemia injury. Transplantations 55(4):728–732
130. Shinmura K (2011) Cardiovascular protection afforded by caloric restriction: essential role of nitric oxide synthase. Geriatr Gerontol Int 11(2):143–156
131. Lawler JM, Kwak H-B, Kim J-H, Suk M-H (2009) Exercise training inducibility of MnSOD protein expression and activity is retained while reducing prooxidant signaling in the heart of senescent rats. Am J Physiol 296:R1496–R1502

Chapter 15
Relationship of Oxidative Stress with Cardiovascular Disease

Richard E. White, Scott A. Barman, Shu Zhu and Guichun Han

Abstract More women die from complications related to cardiovascular disease (CVD) each year than men, yet dysfunction of the heart and blood vessels is still often considered to be primarily a "male" health issue. Emerging data indicate that oxidative stress is an important etiological factor for CVD in women, and it is apparent that female hormones, like estrogen, exert powerful influences on oxidative balance. This chapter will present recent findings and current concepts concerning oxidative stress and cardiovascular function in women. Prominent sources of oxidants in the heart and vasculature will be discussed (e.g., NADPH oxidase, xanthine oxidase (XO), mitochondria, and uncoupled NOS), as well as the effect of estrogen on activity and expression of these proteins in the context of normal hormonal levels and exogenous estrogen replacement therapy. We will also discuss three prominent CVDs that exhibit a rather marked—and at times, surprising—sexual dimorphism in their epidemiology, and consider the ability of estrogen to influence the development and progression of these pathophysiological

R. E. White (✉) · S. Zhu
Department of Basic Science, Georgia Campus—Philadelphia College of Osteopathic Medicine, 625 Old Peachtree Road, Suwanee, GA 30024, USA
e-mail: richardwh@pcom.edu

S. A. Barman
Department of Pharmacology & Toxicology, Medical College of Georgia, Georgia Health Sciences University, Augusta, GA 30912, USA
e-mail: rwhite@georgiahealth.edu

S. Zhu
e-mail: shuzh@pcom.edu

G. Han
Michael E. DeBakey Institute, College of Veterinary Medicine, Department of Veterinary Physiology and Pharmacology, Texas A&M University, Room#332 VMA, MS 4,466, Texas A&M University, College Station, TX 77845, USA
e-mail: ghan@cvm.tamu.edu

states in terms of cellular/molecular mechanisms. The overall goal of the chapter is to provide the reader with a rather comprehensive overview of how oxidative stress impacts women's cardiovascular health, and to review the potential role of estrogen as both a preventive and causative factor in CVD among women.

Keywords Oxidative balance · Cardiovascular disease · Sources of oxidants · NADPH oxidase · Xanthine oxidase · Mitochondria · Uncoupled NOS

15.1 Introduction

Cardiovascular disease (CVD) continues to be the leading cause of death in women, with the most recent statistics indicating that approximately half of all female deaths are due to dysfunction of the heart and blood vessels [1]. Despite the persistence of CVD as the primary cause of female mortality for many years, surveys reveal that women continue to view breast cancer as their most significant health threat [2]. What is more surprising is that only about 1 in 5 physicians are aware that more women die of CVD than men [3]. Such lack of appreciation for the frequency and lethality of CVD in women likely contributes to the fact that female patients often receive less aggressive care for cardiovascular disorders when seeking medical assistance (e.g., fewer admissions to coronary care units, less diagnostic cardiac catheterizations) [4].

A primary reason why CVD is often considered to be more of a "male disease" is the fact that during child-bearing years a woman exhibits a comparatively low risk for CVD compared to males [5]; however, the risk of CVD disease in postmenopausal women meets or exceeds that experienced by men. Such findings have led to the hypothesis that the female hormone estrogen (i.e., 17β-estradiol, E2) exerts a protective effect on cardiovascular function. Indeed, among the beneficial effects of E2 on the cardiovascular system are reduced atherosclerosis, preservation of endothelial cell function, inhibition of proliferation and migration of vascular smooth muscle (VSM) cells, increased NO production, decreased low-density lipoprotein (LDL), higher high-density lipoprotein (HDL), vasodilation (endothelium-dependent and -independent), lowering of blood pressure, stimulation of angiogenesis, prevention of ischemia, and a potential antioxidant effect [6–10]. Such findings from both experimental and clinical studies led to the concept that E2 might indeed by a key hormone that preserved and protected cardiovascular function throughout the aging process.

In contrast, large clinical trials, notably the Women's Health Initiative (WHI) [11], have indicated that E2 supplementation during the later menopausal years is actually detrimental to a woman's cardiovascular health. Although the reasons for this seemingly contradictory outcome are still not apparent, these findings have done much to curtail the prescription of hormone replacement therapy (HRT) for women. More recent and complete analysis of WHI data indicates that, in reality, younger menopausal women (i.e., ages 50–59) did indeed exhibit a beneficial

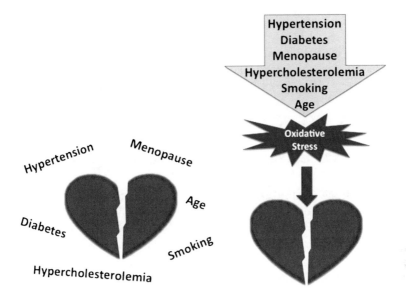

Fig. 15.1 Oxidative stress contributing to heart ischemia and dysfunction is caused by several pathophysiological factors such as diabetes, hypertension, menopause, aging, smoking, and hyperlipidemia

effect of E2 on cardiovascular endpoints [12]. So it would seem that the aging process brings with it a transition in how a woman's body responds to E2. In other words, there is increasing evidence that as a woman ages her cardiovascular system begins to respond differently to natural signals, such as hormones like E2. Thus, understanding how age and disease alter such normal responsiveness may unlock great therapeutic potential, particularly in terms of treating and/or preventing CVD in women—who can now expect to spend nearly one-third of life living in their postmenopausal years.

One significant change in a woman's physiology that is now receiving increased attention is the fact that oxidative stress increases with advancing age. Premenopausal women generally exhibit far less oxidative stress compared to men of similar age [13, 14]. With age, however, comes an increasing imbalance in oxidative state, such that elderly women exhibit even higher indices of oxidative stress than do elderly men [15]. Moreover, the rise in age-related oxidative stress in women parallels their steady increase in CVD [16]. As illustrated in Fig. 15.1, it is important to note that elevated oxidative stress is a common link associated with many of the major cardiovascular risk factors experienced by women, such as hypertension, hypercholesterolemia, diabetes, increased homocysteine, and cigarette smoking [16, 17]. Understanding the mechanisms underlying increased oxidative stress (i.e., increased production of reactive oxygen species (ROS) vs. decreased antioxidant defense mechanisms) in pathological states or in normal aging is an important challenge facing biomedical research that seeks to better the health of an increasingly aging U.S. population. A therapeutic means of rebalancing oxidant production with antioxidant defenses has the potential to reduce the incidence of CVD significantly in both women and men. For example, it is unclear how restoration of E2 via HRT might influence oxidative stress in aging women, as studies suggest HRT may [18] or may not [19] be protective against oxidative stress.

15.2 Sources of Oxidants in the Cardiovascular System

There are multiple mechanisms whereby ROS can be generated in the heart and vasculature. Details of ROS production have been delineated elsewhere in this volume. The present chapter will briefly outline major sites of ROS production in the cardiovascular system, and will then focus primarily on how these mechanisms are influenced by E2, with a view towards how ROS production and antioxidant defenses influence cardiovascular function and CVD in women.

Nicontinamide adenine dinucleotide phosphate (NADPH) oxidase. NADPH oxidase is a membrane-associated enzyme complex that is comprised of seven "NOX" isoforms (i.e., NOX1, NOX2/gp91phox, NOX3, NOX4, NOX5, DUOX1, and DUOX2). All isoforms are transmembrane proteins consisting of multiple subunits located in both the cytosol and cellular membrane. These proteins catalyze the one-electron reduction of oxygen using NADH or NADPH as the electron donor. Probably, the most well-known example of NADPH oxidase activity is the "oxidative/respiratory burst" of phagocytes in response to pathogens. In the cardiovascular system, NADPH oxidase also generates ROS from virtually all vascular cells—endothelial (NOX4, NOX2, NOX5) and VSM cells (NOX1, NOX4, NOX5), fibroblasts (NOX4) of the heart and vascular adventitia, and cardiomyocytes (NOX2 and NOX4). There appears to be a constituitive, albeit low level, production of ROS from NADPH oxidase, but this activity can be increased markedly by endogenous substances. Angiotensin II (acting via the AT_1 receptor) increases superoxide production from NADPH oxidase in VSM cells and adventitial fibroblasts [20]. Thrombin, platelet-derived growth factor, and various cytokines also induce NADPH oxidase activity [21]. Tumor necrosis factor alpha (TNF-α) increases VSM NADPH oxidase activity over 24 h via increased transcription of p22phox, an important subunit of NADPH oxidase [22]. In addition to pharmacological stimulation, ROS production from endothelial cell NADPH oxidase is also enhanced by a physiological stimulus: shear stress within the vessel [23]. Thus, a variety of physiological and physical stimuli can modulate the activity of NADPH oxidase in the cardiovascular system.

Xanthine oxidase. Xanthine oxidase (XO) is another enzymatic mechanism of ROS generation in the cardiovascular system; however, it has received far less attention as a potential target to manage oxidative stress [24]. XO and xanthine dehydrogenase are encoded by the same gene, and are the two inter-convertible forms of xanthine oxidoreductase, a molybdenum iron-sulfur flavin hydrolase.

The primary role of these enzymes is to oxidize hypoxanthine to xanthine, and xanthine into uric acid. Two molecules of hydrogen peroxide and superoxide are generated for every molecule of xanthine that is reduced. Although XO has a fairly wide distribution, expression in the cardiovascular system seems to be localized mainly to endothelial cells [25]. However, xanthine oxidoreductase has also been detected in myocardial cells, where it appears to be localized to the sarcoplasmic reticulum [26]. In addition to endothelium-bound XO, circulating XO can bind to glycosaminoglycans on the endothelium surface, and this form of circulating/depositing XO may have a greater pathological influence on endothelial cell function than XO expressed in endothelial cells [27]. XO activity is regulated at both the transcriptional and posttranslational levels. Proinflammatory cytokines and hypoxia enhance transcription of XO, and elevated XO levels are associated with CVD (e.g., coronary artery disease, heart failure, hypertension) [28]. Further, immunoblot studies have demonstrated that expression of XO protein is elevated in myocardial tissue from patients with heart failure due to cardiomyopathy, and that inhibition of XO activity by allopurinol improved myocardial performance significantly in failing hearts [29]. Similarly, vascular stress in patients with congestive heart failure is associated with higher levels of vascular XO activity [30]. Hypercholesterolemia elevates XO activity in young, asymptomatic individuals suggesting that activation of vascular XO is an early contributing factor to endothelial dysfunction [31]. In addition, upregulation of XO expression and superoxide production is noted arteries from aged animals compared to controls [32]. Clearly, there is increasing evidence that ROS generated from XO may be a contributing factor in cardiovascular pathophysiology.

Mitochondrial respiration. The role of mitochondria extends beyond simply generating cellular energy via ATP production. Mitochondria also regulate the homeostasis of NO, ROS, and calcium in the heart and vasculature [33]. In fact, mitochondria are the primary intracellular source of ROS as they generate oxidative-reduction reactions and consume oxygen [34], and recent studies indicate that scavenging mitochondrial ROS can improve the function of vascular endothelium and reduce hypertension [35]. Mitochondrial function involves the transfer of electrons from NADH or flavoprotein-linked dehydrogenases resulting, ultimately, in ATP production (with water generated as a by-product). However, during electron transport roughly 2 % of electrons escape the electron transport chain complexes I (NADH-ubiquinone oxidoreductase) and III (succinate-ubiquinone oxidoreductase) to reduce molecular oxygen and form reactive oxygen radicals (e.g., superoxide). In addition to electron leakage, mitochondrial-derived ROS may be released by opening of the mitochondrial permeability transition pore (PTP). Activity of the mitochondrial ATP-sensitive potassium (K_{ATP}) channel, which is opened by NO, attenuates the release of ROS via the PTP [34], making this mitochondrial channel a potential therapeutic target to reduce oxidative stress. Much like NADPH oxidase, ROS production from mitochondria can be stimulated by angiotensin II [36], epidermal growth factor [37], transforming growth factor-β [38], and TNF-α [39].

Angiotensin II, a well-known stimulator of NADPH oxidase activity, has also been shown to increase production of mitochondrial-derived ROS in vascular endothelial cells, leading to an increase in cellular superoxide levels and reduced NO bioavailability; however, this stimulatory effect of angiotensin II is dependent upon NADPH oxidase activity, as pharmacological or molecular attenuation of NADPH oxidase impairs angiotensin II-induced generation of mitochondrial superoxide [40]. Further, overexpression of the mitochondrial antioxidant enzyme thioredoxin 2 makes mice resistant to angiotensin II-induced hypertension and endothelial dysfunction [41]. Interestingly, mitochondrial-derived superoxide has recently been shown to stimulate extramitochondrial NADPH oxidase activity, strongly suggesting that a powerful positive feedback loop for ROS production exists in the vascular wall, and involves an interplay between NADPH oxidase and mitochondrial ROS generation [35]. These findings suggest the therapeutic potential for mitochondrial-directed antioxidant agents as a treatment for hypertensive patients which exhibit increased NADPH oxidase activity [42–44].

Uncoupled nitric oxide synthase. Nitric oxide synthases are a family of multifunctional enzymes with similar, yet distinct isoforms. For example, the three NOS isoforms (NOS1/neuronal (n)NOS; NOS2/inducible (i)NOS; NOS3/endothelial (e)NOS) are products of different genes bearing only 51–57 % homology among the human isoforms. Both nNOS and eNOS are expressed constituitively, and their activity is stimulated by increases in intracellular calcium levels. On the other hand, iNOS is fully active at basal levels of calcium, so it is considered to be the "calcium-independent" NOS isoform whose activity is regulated primarily by level of expression. Moreover, NOS isoforms exhibit different localization, regulation, catalytic properties, and sensitivity to inhibitors. Proinflammatory cytokines or endotoxin induces expression of iNOS in a variety of tissues. While eNOS is expressed primarily in vascular endothelial cells, nNOS is found predominately in neurons and muscle cells (including skeletal, cardiac, and smooth muscle) [45].

The designation "nitric oxide synthase" is somewhat misleading insofar as NOS can produce either NO or superoxide anion depending on the immediate biochemical environment of the enzyme. Under most physiological conditions NO is the primary product of NOS activity, but NOS can become "uncoupled" to produce superoxide under some pathological states. In the uncoupled state (i.e., in the absence of the essential cofactor tetrahydrobiopterin (BH_4) or the substrate L-arginine) electrons, which normally flow from the reductase domain of one subunit to the oxygenase domain of the other subunit, are diverted to molecular oxygen rather than to L-arginine, resulting in production of superoxide rather than NO [46]. Compared to dismutation via superoxide dismutase, superoxide combines 3–4 times more rapidly with NO to generate peroxynitrite, which can activate downstream signaling pathways leading to cellular injury [47]. Once formed, peroxynitrite can also reduce BH_4 to BH_2, leading to BH_4 depletion and further uncoupling of NOS (with greater superoxide production). Thus, NOS has the capacity to enhance oxidative stress both directly (via superoxide) and indirectly (via peroxynitrite from the combination of NO with superoxide).

Because a diversity of cell types can express NOS, these isoforms constitute a potentially significant source of ROS in the cardiovascular and other systems. Although all NOS isoforms have the capacity to catalyze superoxide formation, it is uncoupled eNOS that is generally considered to be the major contributor to cardiovascular dysfunction [48, 49]. Endothelial dysfunction is associated with major CVD risk factors such as hypertension, vascular occlusive disease, hypercholesterolemia, hyperglycemia, diabetes, obesity, inflammation, and smoking [50]. Nonetheless, it is in reality the nNOS isoform that has the greater propensity to produce superoxide [51], but we are only now beginning to understand the important role of nNOS in cardiovascular regulation. For example, superoxide generated from uncoupled nNOS can promote abnormal coronary artery reactivity [52] and impaired cardiac relaxation, diastolic stiffness, and remodeling [53].

15.3 Female Hormones, Oxidative Stress, and Cardiovascular Function

Although there is increasing evidence that as a woman ages her risk for CVD increases in parallel with increased oxidative stress [15, 16], a direct etiological connection between ROS and CVD in women has not been firmly established. Despite the fact that antioxidant therapy has been successful in ameliorating CVD in some animal models, reversal of human CVD by antioxidant supplementation has not proven to be a consistently effective therapeutic measure [42, 54]. Nonetheless, there is experimental evidence for a gender disparity regarding oxidative stress. For example, females with neurodegenerative disease exhibit higher levels of oxidative stress compared to similarly affected males, suggesting a higher susceptibility to oxidative injury for women [55]. In addition, a higher oxidative stress status was observed in elderly women compared to men of similar age, and this stress was a strong independent risk factor for coronary artery disease in these women [56]. Further, data from the WHI indicated that antioxidant vitamin supplementation reduced the risk of major cardiovascular events for women ≥ 65 years of age, but had little protective effect in younger women [57]. Such studies suggest that the oxidant status of a woman changes with age (and possibly with changing hormonal status), and such increased oxidative stress may predispose a women for increased risk of CVD and other maladies.

The extent to which female hormones influence oxidant status remains somewhat unclear, as highlighted in a recent review describing both anti- and pro-oxidative actions of estrogens [58]. Because of their chemical structure, estrogens have the potential to function as natural antioxidants. Most estrogens (e.g., E2) possess a phenolic hydroxyl group on the A ring of the steroid molecule, and this moiety can quench excess ROS independent of E2 receptor (ER) activation [59]. The caveat associated with this direct antioxidant capability, however, is that in vitro studies indicate that significant ROS scavenging by E2 occurs mainly at supraphysiological (i.e., µM) concentrations, whereas the more physiological range of plasma E2 concentration varies between approximately 0.1 and 10 nM. Therefore, the physiological relevance of a direct antioxidant capacity of E2 is somewhat questionable. In contrast to this inherent chemical property, estrogens can also modulate the activity of enzymes and antioxidants that influence oxidant status.

Estrogen and lower oxidative stress in the cardiovascular system. Evidence indicates that estrogens inhibit the activity of NADPH oxidase in cardiovascular cells. Cerebral arteries from male rats generate twice as much NADPH-derived superoxide compared to arteries from females, which also exhibit lower expression of NOX1, and NOX4 catalytic subunits [60]. Interestingly, ovariectomy abolished this gender advantage for female animals, suggesting that E2 attenuates NADPH oxidase activity. Other studies have also demonstrated E2-induced downregulation of NADPH oxidase subunit expression. In vitro application of E2 to human vascular endothelial cells decreases expression of NOX2 and p22phox, and also suppresses ROS production by NADPH oxidase [61]. E2-induced suppression of NADPH oxidase activity is also reported in rat VSM cells [62]. Similarly, E2 treatment inhibits oxidative stress in a mouse model of congestive heart failure, and this reduction was associated with a significant attenuation of myocardial NADPH oxidase activity and ROS production [63]. Other studies have indicated that E2 impedes the development of myocardial hypertrophy by inhibiting expression of NADPH oxidase [64]. Thus, there appears to be consistent experimental evidence that estrogens exert an inhibitory influence upon ROS generated from NADPH oxidase in both cardiac and vascular cells. Interestingly, inhibition of NADPH oxidase activity lowered blood pressure [54] and oxidative stress [65] in male hypertensive animals, but had no effect on females. These studies are consistent with the idea that NADPH oxidase activity is lower in females, possibly because of a tonic inhibitory influence exerted by E2 on NADPH oxidase.

There is also evidence that E2 may modulate the activity of XO, but findings regarding the effect of E2 on XO have not been consistent. For example, E2 prevents the hypoxia-induced increase in XO activity in pulmonary artery endothelial cells [66]. On the other hand, E2 was able to elevate ROS production in human endothelial cells—a response which was completely inhibited by a combination of allopurinol and rotenone, a mitochondrial inhibitor [67]. Consistent with this study was a previous report of higher XO activity in females than males [68], suggesting a stimulatory effect of E2 on XO activity. In contrast, male rats are reported to exhibit 59 % greater hepatic XO activity compared to females [69], which is consistent with gout/hyperuricemia traditionally being considered primarily a "male" disease [70]. Thus, our understanding of the overall effect of E2 on XO activity, and its potential impact of CVD in women, is far from complete. More studies are clearly required to better understand the potential effects of E2 on XO, and how this could impact the function of the cardiovascular and other systems.

E2 can also influence mitochondrial function, and can either suppress or stimulate mitochondrial ROS production depending on cell type. Felty et al. [67] have identified mitochondria as a major source of E2-induced ROS production in breast cancer and neuroblastoma cells, with physiological levels of E2 producing a rapid (15-min) generation of ROS. In contrast, physiological E2 concentrations have been shown to suppress mitochondrial superoxide production in brain and PC-12 cells, a cell culture model of neurons [71]. In cerebral blood vessels E2 can reduce ROS production from mitochondria. This antioxidant effect is apparently mediated via an ERα-dependent signaling mechanism which may increase the efficiency of oxidative phosphorylation due to increased expression of mitochondrial cytochrome c, and thereby increase the efficiency of electron transport with less ROS generation [72]. Another study indicates that chronic E2 treatment enhances the capacity for mitochondrial oxidative phosphorylation in cerebral vessels, probably by enhancing the activity of complex IV and citrate synthase [73]. In addition, E2 may also lower ROS levels by upregulating mitochondrial MnSOD levels in the vasculature [74] and in other cell types [75]. ERβ has been detected in cardiac mitochondria [76] indicating that E2 may also influence the function of mitochondria in cardiac myocytes. Thus, there appear to be multiple mechanisms whereby estrogens can limit the level of ROS produced by mitochondrial sources in the cardiovascular system, i.e., inhibition of ROS production or increased superoxide dismutation via mitochondrial MnSOD.

Given the ability of E2 to inhibit ROS generation from both NADPH oxidase and from mitochondrial sources in the cardiovascular system, one would expect that a woman's level of oxidative stress would increase with age; especially after menopause when E2 levels decline and its tonic suppression of enzymatic ROS generation subsides. As noted above, women do experience an age-related increase in oxidative stress [16], which would be consistent with an attenuated inhibitory influence of E2 upon NADPH oxidase and mitochondria in postmenopausal women. However, this model is somewhat at odds with results from the WHI trial which demonstrated that restoration of E2 via HRT *increased* the risk of CVD in older postmenopausal women [11]. Thus, the actions of E2 on NADPH oxidase activity or mitochondrial ROS production cannot account for the observed deleterious effects of E2 on older postmenopausal women. At present, it remains unclear why and how aging alters how a woman's physiology responds to E2, and especially why a hormone which suppresses ROS formation would produce deleterious effects on cardiovascular function in postmenopausal women.

A well-known mechanism whereby E2 can promote healthy cardiovascular function is stimulation of NO production. Production of NO within the cardiovascular system promotes a variety of beneficial effects, such as vasodilation, attenuated release of endothelin-1 and thromboxane A2, inhibition of platelet aggregation, inhibition of VSM proliferation, and stimulation of endothelial cell proliferation/angiogenesis [77]. Interestingly, there is evidence for gender differences in cardiovascular NO production. For example, arteries from female animals produce more NO than those from males [78], and NO measurements in females are higher when E2 is elevated during the menstrual cycle [79] or during pregnancy [80]. Such studies strongly suggest that E2 enhances NOS activity in the cardiovascular system. The primary NOS isoform targets of E2 action in the vasculature are likely eNOS and nNOS, although there is some evidence that E2 can also upregulate iNOS in the myocardium [81, 82]. E2 is known to stimulate the activity of eNOS [83, 84] or nNOS

[85, 86] by a transduction mechanism involving heat shock protein 90 and PI3 kinase—Akt phosphorylation. This stimulation of vascular NOS activity appears to be mediated primarily via the ERα in both endothelium and VSM [87, 88]. Thus, E2 can increase NO bioavailability in the cardiovascular system, help quench superoxide production, and promote normal function.

Estrogen-stimulated ROS production in the cardiovascular system. Overall, E2 appears to exert a predominately antioxidant influence on the cardiovascular system and evidence suggests this protective effect is mediated by inhibition of NADPH oxidase activity, suppression of mitrochondrial ROS generation, and stimulation of NOS activity. Of course, the "double-edged sword" of NOS activity is that the enzyme can become uncoupled from NO production, and instead generate superoxide anion. Such "uncoupling" may be fairly limited under most physiological conditions; however, an enhanced tendency for uncoupling of NOS activity has been reported in various pathological states (e.g., atherosclerosis, diabetes, endothelial dysfunction). Furthermore, uncoupling also may commonly occur during the normal aging process [9, 51, 89, 90].

Studies indicate that as a woman ages her serum L-arginine levels decrease, which can lead to a decline in NO bioavailability [91]. This decline in L-arginine may be related to the fact that vascular arginase activity is upregulated with aging [92]. Because L-arginine is the substrate for NO production, significant decreases in levels of this critical amino acid could attenuate NO production, and enhance the probability of NOS uncoupling with age. In addition to the possibility of substrate depletion, there is evidence for a concomitant decline in BH_4 levels with age in women [93]. BH_4 levels in the resistance vasculature of aged mice were ≤ 50 % of those measured in arteries of young mice, and this age-associated cardiovascular BH_4 deficiency was correlated with a markedly lower expression of GTP cyclohydrolase 1 (the rate-limiting enzyme in BH_4 synthesis) in the aged mice [89]. As mentioned earlier, BH_4 is an essential cofactor that maintains NOS activity "coupled" to NO production. In light of these observed age-related declines in factors critical for "normal" NOS function (i.e., NO production), it would seem very likely that as a woman ages there may be a gradual and increasing tendency for uncoupling of NOS activity, which would thereby predispose her to enhanced oxidative stress. Because E2 is a powerful NOS agonist in the cardiovascular and other systems, it follows that E2 would be more likely to stimulate superoxide production from mostly uncoupled NOS in older postmenopausal women, thus leading to deleterious consequences.

This hypothesis raises the intriguing possibility that menopause—instead of being a "disease" that must be treated—could actually be a highly favorable physiological adaptation in elderly women; i.e., menopause would naturally reduce a woman's level of E2 at the transition time in her life when E2 would begin to do more harm than good (i.e., produce more ROS than NO), thus contributing to a number of pathological consequences. Support for this hypothesis is derived from experiments detecting E2-stimulated superoxide production in blood vessels: E2 stimulates superoxide production from uncoupled nNOS expressed in coronary artery smooth muscle [52]. In this study, the normal NO-dependent vasodilatory effect of E2 on coronary arteries was transformed into a vasoconstrictor response due to E2-

stimulated superoxide production. Thus, there is direct experimental evidence demonstrating that E2 can function as a prooxidant hormone in cardiovascular tissues under certain conditions (i.e., uncoupled NOS), but whether this prooxidant effect of E2 occurs as a function of normal aging, and to what extent, has yet to be determined.

Estrogen: timing is everything. Because E2 inhibits the activity of NADPH oxidase and lowers ROS production from mitochondria, uncoupled NOS would seem to be positioned as a likely mechanism whereby E2 could increase oxidant stress in the cardiovascular system of elderly women. Based upon this model, one would predict that E2 supplementation in older postmenopausal women (i.e., women with decreased levels of L-arginine and/or BH_4, and thus a higher tendency toward NOS uncoupling) would actually enhance oxidative stress and possibly even promote cardiovascular dysfunction. Interestingly, this hypothesis is very consistent with findings of the WHI trial which reported an increased risk of coronary heart disease (CHD) and stroke in older women (most of whom were already a decade past the onset of menopause) receiving HRT [11]. In contrast to the deleterious effects of HRT in older menopausal subjects, E2 supplementation reduced the risk of CVD when given to "younger" women who were closer to the onset of menopause—thus giving rise to the so-called "Timing Hypothesis" of HRT [94, 95]. More recent analysis of data obtained from younger menopausal women has also reported a cardiovascular benefit when HRT is initiated early in menopause [96]. Thus, it appears that age (i.e., age-related changes in biochemistry) alter how a woman's cardiovascular system responds to E2—which is cardioprotective when given closer to the onset of menopause, but is more likely to promote CVD when initiated years after the onset of menopause [95]. Although the biological basis for the Timing Hypothesis is yet unknown (and is probably multifactorial), age-dependent uncoupling of NOS activity in the heart and blood vessels (and other organs) could be a contributing factor to this altered responsiveness to E2. In other words, age may gradually transform E2 from a protective, antioxidant hormone into a potentially harmful, prooxidative agent. Regardless, what seems clear is that there is a greater potential for E2 to promote cardiovascular health in women when given earlier in life, but this beneficial effect is gradually abolished with age.

15.4 Oxidative Stress and Gender Differences in CVD

Atherosclerosis and endothelial dysfunction. Although the number of deaths due to coronary atherosclerotic heart disease has decreased steadily in men over recent years, mortality from atherosclerotic disease in women has remained relatively unchanged [97]. Clearly, our understanding of how atherosclerosis impacts women's health is lagging, and this lack of knowledge is also true regarding the mechanisms whereby female hormones affect atherosclerotic disease. The pathogenesis of atherosclerosis will not be reviewed here. Instead, we will focus more on the impact of oxidative stress on atherogenesis, and discuss the current ideas about how gender and female hormones are believed to influence atherosclerotic disease. The impact of oxidative stress on endothelial dysfunction, an important early warning sign of atherosclerosis, will be considered in light of current concepts and potential therapeutic interventions.

A major etiological factor for the development and progression of atherosclerosis is oxidative stress, and ROS contribute to atherogenesis via multiple mechanisms in the blood vessel wall. For example, ROS are known to oxidize LDLs, modulate adhesion molecules and chemotactic factor expression, stimulate VSM proliferation and migration, increase endothelial cell apoptosis, and enhance matrix metalloproteinase activity with vascular remodeling and plaque rupture [98, 99]. Association of atherosclerosis with increased ROS levels has been indicated in a long-term model of atherosclerosis: superoxide generation in diseased aorta was threefold higher compared to control vessels, and increased ROS production was measured throughout the vessel wall in atherosclerosis [100]. As noted above, ROS stimulates the proliferation of VSM cells with subsequent migration from the medial into the intimal layer of blood vessels, and such vascular remodeling is attenuated in mice deficient for either p47phox (i.e., NADPH oxidase activity) or eNOS [99, 101]. These and other studies in a variety of models have clearly established an association of oxidative stress with atherosclerotic disease of the vasculature. Disappointingly, however, clinical trials with antioxidant supplements have seldom demonstrated a clear protective effect on atherosclerotic disease, but such agents may have some efficacy as a preventative measure [99].

Much less is known regarding oxidative stress and the development of atherosclerosis specifically in the female cardiovascular system. In general, females during their child-bearing years exhibit more favorable plasma lipoprotein profiles than men. Younger women have lower LDL and higher HDL levels compared to men of the same age; however, as women age LDL increases and HDL decreases in parallel with an increased risk of CVD [4]. These changes in plasma lipoproteins have been associated with hormonal status, as a well-known effect of E2 (certainly in premenopausal women) is to lower LDL and raise HDL; however, it has been estimated that only 25 % of the atheroprotective effect of E2 is related to changes in plasma lipoproteins, whereas about 75 % was more likely due to direct effects on arteries and inhibition of oxidized LDL [102]. For example, the spontaneous atherosclerosis that develops in apolipoprotein E-deficient (apoE-/-) mice is attenuated by E2, which reduces atherosclerosis formation and intimal thickening, and concomitantly reduces vascular superoxide production [103]. Young, ovariectomized mice undergoing E2 supplementation exhibited decreased superoxide production and NADPH oxidase expression in aorta, along with upregulation of Cu/ZnSOD, and MnSOD [103]. Despite the protective and antioxidant effects described for E2 in younger women and animals, it appears to be less efficacious in lowering LDL particles in older, postmenopausal women [104]. For example, E2-containing oral contraceptive therapy is associated with reduced severity of coronary artery disease in both nonhuman primates (i.e., cynomolgus monkeys) [97] and premenopausal women [105], but E2 supplementation began after the onset of menopause may be far less effective in limiting atherosclerotic progression [6]. Thus, again, there appears to be a time-/age-dependency of how E2 influences cardiovascular health in females—basically, the younger, the better.

Consistent with the "Timing Hypothesis" are studies indicating that E2 produces beneficial cardiovascular effects when administered soon after endogenous E2 deficiency (i.e., before significant atherosclerosis has progressed), but is ineffective or even deleterious if administered after established plaque formation [97]. Studies in cynomolgus monkeys have demonstrated that when E2 replacement is initiated immediately after surgical menopause the progression of coronary artery atherosclerosis is inhibited 70 % [106, 107]. Further, E2 can inhibit fatty streak development in arteries, but has limited effect on already established lesions [108]. These studies are consistent with data from the Nurse's Health Study (over 70,000 postmenopaual women) indicating that women beginning HRT at 50–59 years of age experienced approximately a 50 % reduction in relative risk for a major coronary event, whereas women who were $\geq$60 years of age at the time of HRT initiation had no risk reduction [109]. At present, it is unclear why E2 is able to help retard the development of atherosclerotic disease, but is much less effective at reversing an ongoing process.

One mechanism that may contribute to altered E2 responses in diseased vessels is the fact that E2 receptor expression is altered by atherosclerosis. Studies have shown that the ability of E2 to prevent lipid deposition on the vascular wall requires expression of ERα in the vascular endothelium [110]. Further, ER expression has been detected in human coronary artery smooth muscle cells in nearly 3 out of 4 normal arteries, but only about one-third of atherosclerotic arteries expressed significant levels of ER [111]. Thus, the effect of E2 on arteries may be increasingly diminished as the expression pattern of ER proteins, especially ERα, is altered by atherosclerosis. Such downregulation of ER expression would very likely abolish a protective influence of E2 on the extent of atherosclerotic lesions. If verified, this concept raises the possibility that development of novel therapeutic means of stimulating the ERα transduction pathway downstream from the receptor might hold substantial promise as a way to restore the beneficial effects of E2 to diseased vessels (and possibly help reverse disease progression). Previous studies have demonstrated that an important target of ERα-mediated E2 action is NOS expressed in both endothelium and VSM [83, 85, 88].

Endothelial dysfunction is often considered to be the initial insult common to atherosclerosis, restenosis, and hypertensive vascular remodeling, and can result from hypercholesterolemia, oxidative stress, diabetes, and other pathological processes. A decline in NO bioavailability is a common factor in these maladies, and is believed to originate primarily from attenuated eNOS activity. Decreased production of NO from the endothelium leads to the enhanced oxidative stress that precedes the development of overt atherosclerosis [112]. In fact, all major risk factors for atherosclerotic vascular disease have been associated with decreased NO bioavailability and endothelial dysfunction.

There is increasing evidence that NOS uncoupling in atherosclerosis coincides with decreased levels of BH_4 and L-arginine. Atherosclerosis increases the activity of endothelial arginase II, which enhances degradation of the substrate required for NO production [113]. In addition, studies of rabbit aorta indicate that atherosclerosis depletes BH_4 levels by nearly 30-fold [114], whereas oxidized LDLs reduce cellular BH_4 levels and increase superoxide production in human aortic endothelial cells [112]. Loss of vascular L-arginine and BH_4 would increase the likelihood of NOS uncoupling, thus making superoxide the primary product of eNOS (and nNOS) activity instead of NO and thereby worsening oxidative stress/atherosclerosis (i.e., possibly generating a localized positive feedback mechanism for atherosclerotic disease). If atherosclerosis converts NOS activity from a protective into a prooxidative influence, one would expect that upregulating activity of "uncoupled" NOS would worsen atherosclerosis. Interestingly, such a scenario has been demonstrated in the atherosclerotic apo$^{E-/-}$ mouse model: overexpression of eNOS, instead of increasing NO levels, generated more superoxide and accelerated atherosclerotic lesion progression compared to mice with normal levels of eNOS [115]. In this same study, the deleterious effects of NOS activity were reversed by "re-coupling" NOS function with exogenous BH_4, and this restoration of normal NOS function reduced the size of lesions. Similarly, treating human endothelial cells or hypercholesterolemic human subjects with exogenous BH_4 restored normal endothelial function [112], whereas BH_4-deficiency increased arterial superoxide content and exaggerated neointimal formation following vascular injury [116]. These studies strongly implicate uncoupled NOS activity within the vascular wall as an important oxidant producing, atherosclerotic mechanism, and further suggest that re-coupling of NOS might be a means of retarding or possibly even reversing atherosclerosis.

In summary, there is increasing evidence from both animal and human studies that an important early marker of atherosclerosis—dysfunction of the vascular endothelium with increased ROS production and lowered NO bioavailability—is associated with depletion of L-arginine and BH_4, with subsequent uncoupling of NOS activity. Such studies propose a strong argument for the idea that uncoupled NOS helps tips the delicate oxidative balance more toward oxidant stress, and thereby contributes to the pathogenesis of atherosclerosis and endothelial dysfunction. This model is consistent with clinical findings measuring elevated levels of oxidative stress, oxidized LDL, and BH_2 (i.e., oxidized BH_4) in patients after acute myocardial infarction [117]. Nonetheless, other factors besides L-arginine/BH_4 depletion are also implicated in elevating ROS early in atherosclerosis. For example, age-induced impairment of mouse carotid artery relaxation was not associated with changes in either BH_4 or GTPCH, but was still associated with increased superoxide formation [118]. Further, there is also evidence for upregulation of GTPCH and BH_4 levels in aorta from apo$^{E-/-}$ mice [119], although this response could be compensatory to enhanced oxidative stress. As should be expected, there appears to be several potential links between ROS and atherosclerosis; however, the ability of E2 to stimulate NOS activity in both endothelial and VSM cells would suggest that uncoupled NOS might have greater atherosclerotic potential in women compared to men.

Diabetes and oxidative stress in women. Diabetes has been called "the epidemic of the twenty first century" [120], and cardiovascular complications account for 80 % of mortality in diabetes. Despite the fact that premenopausal women suffer less CVD than men of similar age, it is women who experience more severe (and premature) CVD in diabetes. Compared to diabetic men, women with diabetes have a four- to sixfold greater risk of CHD, a poorer prognosis after myocardial infarction, a higher risk of death from CVD, a greater atherogenic dyslipidemia, atherosclerotic plaques which are more prone to rupture, a greater incidence of hypertension, a higher rate of congestive heart failure, more symptoms associated with hyperglycemia, and a greater incidence of depression [121]. Diabetes produces particularly disparate gender outcomes for CHD, with risk of fatal CHD 50 % higher in diabetic women than in men [122], and diabetes more than doubles mortality rate of CHD in women compared to men (90 vs. 40 %, respectively) [123]. In addition, women with diabetes exhibit a tenfold increase in CHD-related deaths compared to non diabetic women [124, 125]. Obviously, diabetes promotes significant coronary artery pathology/dysfunction; yet, the reasons why women suffer greater cardiovascular risk in diabetes than men—particularly in coronary arteries—are not known. What is apparent, however, is that diabetes blunts the cardiovascular benefit of female gender, and in fact, transforms this normal gender advantage into a detriment. The contribution of gender-related hormones (e.g., estrogens, androgens) to this curious phenomenon is still unknown.

Diabetic CVD has much in common with the aging process, as both are characterized by a higher incidence of oxidative stress, atherosclerosis, hypertension, and CHD. As discussed above, oxidative stress is a common etiological factor among cardiovascular disorders, and is also increased significantly in diabetes via a number of mechanisms. For example, there is evidence that hyperglycemia increases superoxide levels in aortic endothelial cells by processes involving uncoupled NOS [126] and mitochondria [127]. In fact, superoxide production is enhanced throughout the arterial wall in diabetes [128]. In addition to superoxide, hyperglycemia elevates hydrogen peroxide production in human aortic smooth muscle cells [129]. Other factors that would also elevate oxidative stress in diabetes include a reduction in antioxidant enzymes (e.g., Cu–Zn SOD) and an inhibition of antioxidant enzymatic activity due to enhanced glycation [130]. Thus, diabetes can produce an imbalance in oxidative state by multiple mechanisms, all contributing to enhanced oxidative stress and cardiovascular dysfunction/accelerated aging. Moreover, this diabetes-induced oxidative imbalance appears to be worse in women than in men [121].

Although diabetes increases oxidative stress in both men and women, a definite sexual dimorphism exists with regard to CVD. The reasons for this gender disparity are not apparent, but we and others have hypothesized that greater oxidative stress could be a contributing factor to the higher level of CVD observed in diabetic women [9, 121]. For example, lipid peroxidation is increased in women with diabetes compared to healthy females, but this same indicator of oxidative stress did not increase in diabetic men [131]. As stated above, there are various means for increasing ROS in both diabetic men and women; and E2 exerts an important regulatory influence on these mechanisms—particularly in women. Such sex hormone specific effects would likely help promote gender differences in CVD among diabetics. For example, hyperglycemia can stimulate activity of NADPH oxidase [132]; however, E2s are known to downregulate activity of this enzyme in cerebral arteries [60], VSM [62], and human endothelial cells [61]. Thus, it seems unlikely that NADPH oxidase activity would be elevated in diabetic women compared to diabetic males. There is evidence for enhanced oxidant production from XO [133] and mitochondrial sources [127] in diabetes, but E2 has been shown to decrease the potential for mitochondrial oxidant production [71] while increasing the ROS-reducing ability of mitochondrial MnSOD [74].

It is known that diabetes tends to reduce E2 in women. For example, E2 levels decline by approximately 30 % in premenopausal women with type 1 diabetes [134]. Because E2 tends to inhibit ROS production from two prominent sources (i.e., NADPH oxidase and mitochondrion), the normal "braking influence" of E2 on oxidant production could be attenuated in diabetic women as E2 levels decline, thus contributing to greater oxidative stress and subsequent CVD. On the other hand, we have already discussed that E2 is a powerful stimulator of NOS activity in the cardiovascular system. Normally, E2 stimulated NO production would help quench superoxide and lower oxidative stress, and a decrease in E2 levels should lessen NOS stimulation in diabetic women. In diabetes, however, there is strong evidence that NOS exhibits an increased tendency for uncoupling, thereby making uncoupled NOS a potential source of superoxide in the disease state. Even the lower levels of E2 measured in diabetes would still stimulate the activity of uncoupled NOS, and enhance production of ROS in diabetic women (but not in healthy women of the same age). Further, the influence of this prooxidative effect of E2 would be much less in diabetic men, who, obviously, have far lower levels of circulating E2. Thus, it is plausible that E2 stimulation of uncoupled NOS could contribute to the increased risk of CVD observed in diabetic women compared to healthy women or diabetic males.

Several lines of evidence support uncoupled NOS as a source of oxidant production in diabetes. In addition to studies cited above [126], endothelial cells also exhibit a reduced expression of GTPCH [135] and substantially increased activity of arginase [136] under diabetic conditions. Decreased levels of L-arginine, coupled with diabetes-stimulated increases in vascular asymmetric dimethylarginine [137], could certainly promote NOS uncoupling in diabetes, as would BH_4 depletion. Importantly, restoration of BH_4 levels by overexpression of GTPCH in endothelial cells reversed diabetes-stimulated superoxide production and helped restore normal endothelium function [138]. Thus, there is strong evidence that diabetes enhances eNOS uncoupling to promote ROS production. These findings in endothelial cells are highly consistent with our work in coronary artery smooth muscle. We found that the potent vasodilatory response to E2 can be converted into a powerful vasoconstriction due to E2-stimulated superoxide production from uncoupled nNOS [52]. These studies unmasked a prooxidative effect of E2 on vascular tissue that completely reversed how arteries normally respond to E2. We extended these findings by observing that although E2 relaxed coronary arteries from healthy pigs under normal conditions, it constricted similar vessels from diabetic pigs via stimulation of superoxide production [9]. Such an abnormal vasoconstrictor response due to E2-stimulated ROS production would be expected if NOS expressed in these "diabetic" vessels existed primarily in the uncoupled state. In other words, E2 enhances NO production in normal arteries, but generates ROS in arteries from diabetic animals. In these studies, we found that tempol (an SOD mimetic) completely prevented E2-induced contraction of coronary arteries from diabetic animals, thus indicating this abnormal vasoconstrictor response was mediated by E2-stimulated superoxide production. Taken together, these studies of endothelium and VSM lend strong support to the idea that uncoupled NOS (i.e., both eNOS and nNOS) are potential sources of ROS production in the diabetic cardiovascular system. Moreover, the fact that E2 is a powerful agonist of uncoupled NOS suggests a novel mechanism that could contribute to greater oxidative stress in women with diabetes compared to normal women and diabetic men. Elevated oxidative stress in diabetic women would be expected to promote CVD, and could contribute to the observed gender disparity in CHD and other cardiovascular dysfunction.

Pulmonary hypertension (PAH). There are several features of the pulmonary circulation that make it rather unique in comparison to other circulatory systems. For example, the pulmonary vasculature is the only circulation that receives 100 % of cardiac output. Further, although hypoxia tends to induce vasodilation in the systemic circulation, the pulmonary circulation responds with vasoconstriction to shunt perfusion to more well-ventilated areas of the lungs, and thereby promote efficient gas exchange. Although CVD and hypertension are far more prevalent in older adults, it is younger adults who are more commonly afflicted with PAH, a group of diseases characterized by high pulmonary artery pressure and pulmonary vascular resistance [139]. Interestingly, and in contrast to what is generally true regarding CVD, there is a female preponderance in the incidence of idiopathic PAH, with women being diagnosed 2–4 times more often than men [140]. At present, however, there is no clear explanation as to why more women suffer from idiopathic PAH compared to their male counterparts. In contrast, there appears to be much less sexual dimorphism in the incidence of hypoxic PAH.

Formally, PAH is defined as having a sustained mean pulmonary arterial pressure exceeding 25 mmHg at rest or more than 30 mmHg during exercise [141]. Progressive elevation of this pressure leads to remodeling of pulmonary arteries, abnormal vascular function, right ventricular overload, failure, and, ultimately, death [142, 143]. Increased pulmonary arterial resistance results from enhanced vasoconstriction and proliferation of VSM and endothelial cells, and therapies often include substances which induce relaxation of pulmonary artery smooth muscle. Although the etiology of PAH is not completely understood, we and others have suggested that abnormal function or expression of potassium channels in pulmonary arterial VSM could contribute to excessive vasoconstriction in PAH [144–146]. In addition, it is likely that impaired release of vasodilatory substances (e.g., NO, prostacyclin) or increased production of vasoconstrictors (e.g., endothelin-1, thromboxane) from pulmonary vascular endothelium may contribute to increased pulmonary vascular resistance. Therefore, it appears quite likely that dysfunction of both VSM and endothelial cells in the pulmonary circulation contributes to the development of PAH. In summary, evidence indicates that PAH develops due to an imbalance among vasoconstrictive, proinflammatory, mitogenic, and thrombogenic factors versus vasodilatory, antimitogenic, and anticoagulant mechanisms [141].

There is increasing evidence that oxidative stress plays a major role in the pathogenesis of PAH by enhancing pulmonary arterial resistance via multiple mechanisms [143]. Oxidative stress is enhanced in the failing right ventricle (but not in the corresponding left ventricle) in a monocrotaline model of PAH, whereas ROS scavenging can improve right ventricular systolic function [147]. Other studies employing a surgical model of persistent PAH of the newborn (PPHN) indicate increased expression and activity of the NADPH oxidase complex and uncoupled eNOS in pulmonary arteries [148]. Using a similar model, NO-induced relaxation of pulmonary artery rings was attenuated in vessels from PPHN animals, and this depressed relaxation was improved by either tiron (superoxide scavenger) or L-NAME (NOS inhibitor) [149]. In contrast, relaxation of arteries from control animals was unaffected by these agents, implicating superoxide derived from uncoupled NOS as an important mechanism contributing to vasoconstriction of pulmonary arteries in PAH. Supporting this idea are studies demonstrating that mice deficient in BH_4 (hph-1 mice) are prone to developing PAH even under normoxic conditions, but increasing BH_4 synthesis (i.e., promoting "re-coupling" of NOS activity) in endothelium prevented the development of hypoxia-induced PAH [150]. In addition to NADPH oxidase and uncoupled NOS as sources of ROS in PAH, XO activity, and superoxide production were enhanced in a model of hypoxia-induced PAH in neonatal rats, whereas allopurinol prevented the pulmonary vascular remodeling associated with PAH [151]. In summary, these finding from cardiac and vascular tissue studied in multiple experimental models of PAH strongly indicate that oxidant stress can play an etiological role in the development of PAH, and that there are multiple potential sources of ROS that contribute to this pathogenesis.

At present, it is unclear why more women suffer from idiopathic PAH than men. As summarized in a recent review [140], it is speculated that this gender disparity could be related to the influence of sex hormones, increased tendency toward autoimmune disorders, or possibly environmental factors. To date, most research has considered a differential effect of gonadal steroids on the pulmonary circulation: E2s could promote PAH while androgens would promote more normal vascular resistance. However, results have indicated that E2 produces an acute relaxation effect on pulmonary arteries and can limit hypoxic-induced pulmonary vasoconstriction by stimulating NO production [152, 153]. Furthermore, female rodents experience less severe hypoxic PAH compared to males, who are less resistant to hypoxic PAH [154, 155]. These findings suggest a protective effect of E2 against the development of hypoxic PAH, and raise the question that testosterone might instead promote development of this condition. In contrast, however, testosterone relaxes isolated human pulmonary arteries and induces vasodilation in isolated perfused human lungs (albeit at supraphysiological concentrations) [156]. Therefore, evidence indicates that both E2 and testosterone exert a vasodilatory effect on pulmonary arteries, making it appear less likely that direct vascular effects of sex hormones alone could account for the female predilection for idiopathic PAH. Nonetheless, there are conflicting studies on how exogenous estrogens influence the development of PAH. HRT reduced the risk of PAH in postmenopausal women with systemic sclerosis—an effect which was independent of autoantibodies [157]. Yet, on the other hand, there is a report of HRT being associated with onset of PAH in a menopausal woman genetically predisposed to familial PAH [158], a correlation between postmenopausal HRT and incidence of PAH [159], and increased risk of idiopathic PAH in women using oral contraceptives [160, 161]. Clearly, more research must be done to identify the mechanism(s) responsible for the gender disparity associated with idiopathic PAH.

PAH is associated with increased oxidative stress in both the heart [147] and lungs [162], and there is increasing evidence suggesting a link between E2, oxidative stress, and pulmonary vasoconstriction. For example, isolated perfused lungs from female rats exhibit a greater vasoconstrictor response to stimulation of thromboxane receptors than do lungs from males, and this hyper responsiveness was enhanced by E2 [163]. These studies suggested that E2 could intensify constrictor influences exerted by arachidonic acid metabolites in the pulmonary circulation. More recent studies have demonstrated that arachidonic acid produces a greater endothelium-dependent contraction in rabbit pulmonary arteries from females compared to males, and that this is contraction mediated by the lipoxygenase metabolite 15-hydroxyeicosatetraenoic acid (15-HETE) [139]. Furthermore, treating these pulmonary arteries with 1 μM E2 increased production of 15-HETE and expression of 15-lipoxygenase. Interestingly, other studies have demonstrated that PAH, hypoxia, and oxidative stress are associated with increased production of 15-HETE and related compounds [162, 164]. In addition to these functional studies, 15-HETE has also been found to attenuate apoptosis of pulmonary VSM cells, which could promote PAH by stimulating vessel remodeling and medial hypertrophy [165]. Thus, there is increasing evidence that estrogens can enhance vasoconstrictor influences in pulmonary arteries by stimulating synthesis of vasoactive prostanoids—which can also generate ROS as byproducts of cyclooxygenase and lipoxygenase activity. Conversely, arachidonic acid metabolites can stimulate ROS production [166], setting up a potential positive feedback mechanisms for ROS generation in the pulmonary vasculature.

Taken together, these findings suggest a potential synergy among E2, ROS, and arachidonic acid metabolism in pulmonary arteries that could contribute to the higher incidence of idiopathic PAH in women. In support of this idea are studies indicating that E2 can directly stimulate activity of the cyclooxygenase 2 (COX-2) isoform in the mice vasculature, probably via simulation of ERα [167, 168]. Furthermore, we have demonstrated that indomethacin inhibits E2-induced contraction of pig coronary arteries due to ROS generation, and thereby reverses abnormal vascular reactivity [52]—an effect which could be mediated via hydrogen peroxide-induced stimulation of arachidonic acid metabolism [169]. Although further studies are certainly required to shed further light on the gender disparity associated with idiopathic PAH, a greater understanding of how E2 and ROS impact arachidonic acid metabolism may hold much promise for novel therapeutic approaches to treating this devastating disease that afflicts more women than men.

15.5 Conclusion

As is usually the case in biomedical research, experimental findings regarding the role of oxidative stress and CVD in women have led to even more unanswered questions. This fascinating (and even frustrating) problem is illustrated by our evolving attitudes regarding how female hormones influence a woman's cardiovascular health. Prior to 2002, HRT was a highly prescribed regimen to help prevent or reverse age-related declines in cardiovascular function. After 2002, however, HRT was more rarely employed because of fears associated with increased risk of cardiovascular and other diseases. More recent evidence and retrospective analysis has now led to the concept that HRT is a beneficial, and possibly even protective, option for women during early stages of menopause (i.e., 10 years or less from onset), but should be increasingly avoided as a woman ages beyond that time frame. Because E2 exerts an important regulatory influence upon oxidant balance, it appears likely that oxidative stress may be a key player in influencing how a woman's cardiovascular system responds to HRT. Evidence suggests that during her child-bearing years through early menopause E2 functions primarily as an anti-oxidant hormone to help protect cardiovascular function. In contrast, it appears that E2 may gradually exert an increasingly more prooxidant influence during the postmenopausal years, and could promote dysfunction of the cardiovascular and other systems. The fact that cardiovascular dysfunction in diabetes and PAH is more severe for women than men, coupled with the emerging idea that oxidative stress plays an etiological role in these diseases, supports the concept that perturbations in oxidative balance can have profound consequences for women's health. Future studies will certainly help clarify the complicated interplay among gonadal steroids, oxidative stress, cardiovascular function, and gender. In so doing, such studies will help enhance both the quantity and quality of life for women of all ages.

Acknowledgments The authors wish to recognize funding sources that have helped support their research: The National Heart, Lung, and Blood Institute (HL07389, HL68026, White and Barman) and the American Heart Association (Scientist Development Grant, Han; 055149B, Barman).

References

1. Roger VL, Go AS, Lloyd-Jones DM et al (2011) Heart disease and stroke statistics–2011 update: a report from the american heart association. Circulation 123(4):e18–e209
2. Research SfWsH (2005) What Diseases Do Women Fear Most?: International Communications Research
3. Mosca L, Linfante AH, Benjamin EJ et al (2005) National study of physician awareness and adherence to cardiovascular disease prevention guidelines. Circulation 111(4):499–510
4. Sweitzer NK, Douglas PS (2005) Cardiovascular disease in women. In: Zipes DP, Libby P, Bonow RO, Braunwald E (eds) Heart disease: a textbook of cardiovascular medicine. Elsevier Saunders, Philadelphia, pp 1951–1964
5. Barrett-Connor E (1997) Sex differences in coronary heart disease. why are women so superior? The 1995 ancel keys lecture. Circulation 95(1):252–264
6. Vitale C, Mendelsohn ME, Rosano GM (2009) Gender differences in the cardiovascular effect of sex hormones. Nat Rev Cardiol 6(8):532–542
7. Mendelsohn ME, Karas RH (1999) The protective effects of estrogen on the cardiovascular system. N Engl J Med 340(23):1801–1811
8. White RE (2002) Estrogen and vascular function. Vascul Pharmacol 38:73–80
9. White RE, Gerrity R, Barman SA, Han G (2010) Estrogen and oxidative stress: a novel mechanism that may increase the risk for cardiovascular disease in women. Steroids 75(11):788–793
10. Lenfant F, Tremollieres F, Gourdy P, Arnal JF (2011) Timing of the vascular actions of estrogens in experimental and human studies: why protective early, and not when delayed? Maturitas 68(2):165–173
11. Rossouw JE, Anderson GL, Prentice RL et al (2002) Risks and benefits of estrogen plus progestin in healthy postmenopausal women: principal results from the women's health initiative randomized controlled trial. JAMA 288(3):321–333
12. Manson JE, Allison MA, Rossouw JE et al (2007) Estrogen therapy and coronary-artery calcification. N Engl J Med 356(25):2591–2602
13. Baker L, Meldrum KK, Wang M et al (2003) The role of estrogen in cardiovascular disease. J Surg Res 115(2):325–344
14. Rossouw JE (2002) Hormones, genetic factors, and gender differences in cardiovascular disease. Cardiovasc Res 53(3):550–557
15. Agrawal A, Lourenco EV, Gupta S, La Cava A (2008) Gender-based differences in leptinemia in healthy aging, non-obese individuals associate with increased marker of oxidative stress. Int J Clin Exp Med 1(4):305–309
16. Vassalle C, Mercuri A, Maffei S (2009) Oxidative status and cardiovascular risk in women: keeping pink at heart. World J Cardiol 1(1):26–30
17. Fearon IM, Faux SP (2009) Oxidative stress and cardiovascular disease: novel tools give (free) radical insight. J Mol Cell Cardiol 47(3):372–381
18. Unfer TC, Conterato GM, da Silva JC, Duarte MM, Emanuelli T (2006) Influence of hormone replacement therapy on blood antioxidant enzymes in menopausal women. Clin Chim Acta 369(1):73–77
19. Maffei S, Mercuri A, Prontera C, Zucchelli GC, Vassalle C (2006) Vasoactive biomarkers and oxidative stress in healthy recently postmenopausal women treated with hormone replacement therapy. Climacteric 9(6):452–458
20. Li JM, Wheatcroft S, Fan LM, Kearney MT, Shah AM (2004) Opposing roles of p47phox in basal versus angiotensin II-stimulated alterations in vascular o2- production, vascular tone, and mitogen-activated protein kinase activation. Circulation 109(10):1307–1313
21. Elahi MM, Kong YX, Matata BM (2009) Oxidative stress as a mediator of cardiovascular disease. Oxid Med Cell Longev 2(5):259–269
22. Manea A, Manea SA, Gafencu AV, Raicu M, Simionescu M (2008) AP-1-dependent transcriptional regulation of NADPH oxidase in human aortic smooth muscle cells: role of p22phox subunit. Arterioscler Thromb Vasc Biol 28(5):878–885

23. Duerrschmidt N, Stielow C, Muller G, Pagano PJ, Morawietz H (2006) NO-mediated regulation of NAD(P)H oxidase by laminar shear stress in human endothelial cells. J Physiol. 576(Pt 2):557–567
24. George J, Struthers AD (2009) Role of urate, xanthine oxidase and the effects of allopurinol in vascular oxidative stress. Vasc Health Risk Manag 5(1):265–272
25. Nees S, Gerbes AL, Gerlach E, Staubesand J (1981) Isolation, identification, and continuous culture of coronary endothelial cells from guinea pig hearts. Eur J Cell Biol 24(2):287–297
26. Khan SA, Lee K, Minhas KM et al (2004) Neuronal nitric oxide synthase negatively regulates xanthine oxidoreductase inhibition of cardiac excitation-contraction coupling. Proc Natl Acad Sci U S A 101(45):15944–15948
27. Panus PC, Wright SA, Chumley PH, Radi R, Freeman BA (1992) The contribution of vascular endothelial xanthine dehydrogenase/oxidase to oxygen-mediated cell injury. Arch Biochem Biophys 294(2):695–702
28. Weseler AR, Bast A (2010) Oxidative stress and vascular function: implications for pharmacologic treatments. Curr Hypertens Rep 12(3):154–161
29. Cappola TP, Kass DA, Nelson GS et al (2001) Allopurinol improves myocardial efficiency in patients with idiopathic dilated cardiomyopathy. Circulation 104(20):2407–2411
30. Landmesser U, Spiekermann S, Dikalov S et al (2002) Vascular oxidative stress and endothelial dysfunction in patients with chronic heart failure: role of xanthine-oxidase and extracellular superoxide dismutase. Circulation 106(24):3073–3078
31. Spiekermann S, Landmesser U, Dikalov S et al (2003) Electron spin resonance characterization of vascular xanthine and NAD(P)H oxidase activity in patients with coronary artery disease: relation to endothelium-dependent vasodilation. Circulation 107(10):1383–1389
32. Jacobson A, Yan C, Gao Q et al (2007) Aging enhances pressure-induced arterial superoxide formation. Am J Physiol Heart Circ Physiol 293(3):H1344–H1350
33. Davidson SM (2010) Endothelial mitochondria and heart disease. Cardiovasc Res 88(1):58–66
34. Chang JC, Kou SJ, Lin WT, Liu CS (2010) Regulatory role of mitochondria in oxidative stress and atherosclerosis. World J Cardiol 2(6):150–159
35. Dikalova AE, Bikineyeva AT, Budzyn K et al (2010) Therapeutic targeting of mitochondrial superoxide in hypertension. Circ Res 107(1):106–116
36. Kimura S, Zhang GX, Nishiyama A et al (2005) Role of NAD(P)H oxidase- and mitochondria-derived reactive oxygen species in cardioprotection of ischemic reperfusion injury by angiotensin II. Hypertension 45(5):860–866
37. Krieg T, Cui L, Qin Q, Cohen MV, Downey JM (2004) Mitochondrial ROS generation following acetylcholine-induced EGF receptor transactivation requires metalloproteinase cleavage of proHB-EGF. J Mol Cell Cardiol 36(3):435–443
38. Herrera B, Alvarez AM, Sanchez A et al (2001) Reactive oxygen species (ROS) mediates the mitochondrial-dependent apoptosis induced by transforming growth factor (beta) in fetal hepatocytes. FASEB J 15(3):741–751
39. Chen KH, Reece LM, Leary JF (1999) Mitochondrial glutathione modulates TNF-alpha-induced endothelial cell dysfunction. Free Radic Biol Med 27(1–2):100–109
40. Doughan AK, Harrison DG, Dikalov SI (2008) Molecular mechanisms of angiotensin II-mediated mitochondrial dysfunction: linking mitochondrial oxidative damage and vascular endothelial dysfunction. Circ Res 102(4):488–496
41. Widder JD, Fraccarollo D, Galuppo P et al (2009) Attenuation of angiotensin II-induced vascular dysfunction and hypertension by overexpression of Thioredoxin 2. Hypertension 54(2):338–344
42. O'Connor PM, Gutterman DD (2010) Resurrecting hope for antioxidant treatment of cardiovascular disease: focus on mitochondria. Circ Res 107(1):9–11
43. Moreno MU, San Jose G, Fortuno A, Beloqui O, Diez J, Zalba G (2006) The C242T CYBA polymorphism of NADPH oxidase is associated with essential hypertension. J Hypertens 24(7):1299–1306

44. Graham D, Huynh NN, Hamilton CA et al (2009) Mitochondria-targeted antioxidant MitoQ10 improves endothelial function and attenuates cardiac hypertrophy. Hypertension 54(2):322–328
45. Zhou L, Zhu DY (2009) Neuronal nitric oxide synthase: structure, subcellular localization, regulation, and clinical implications. Nitric Oxide 20(4):223–230
46. Schulz E, Jansen T, Wenzel P, Daiber A, Munzel T (2008) Nitric oxide, tetrahydrobiopterin, oxidative stress, and endothelial dysfunction in hypertension. Antioxid Redox Signal 10(6):1115–1126
47. Ungvari Z, Gupte SA, Recchia FA, Batkai S, Pacher P (2005) Role of oxidative-nitrosative stress and downstream pathways in various forms of cardiomyopathy and heart failure. Curr Vasc Pharmacol 3(3):221–229
48. Briones AM, Touyz RM (2010) Oxidative stress and hypertension: current concepts. Curr Hypertens Rep 12(2):135–142
49. Pepine CJ (2009) The impact of nitric oxide in cardiovascular medicine: untapped potential utility. Am J Med 122(5 Suppl):S10–S15
50. Seals DR, Jablonski KL, Donato AJ (2011) Aging and vascular endothelial function in humans. Clin Sci (Lond) 120(9):357–375
51. Alderton WK, Cooper CE, Knowles RG (2001) Nitric oxide synthases: structure, function and inhibition. Biochem J 357(Pt 3):593–615
52. White RE, Han G, Dimitropoulou C et al (2005) Estrogen-induced contraction of coronary arteries is mediated by superoxide generated in vascular smooth muscle. Am J Physiol Heart Circ Physiol 289(4):H1468–H1475
53. Jessup JA, Zhang L, Chen AF et al (2011) Neuronal nitric oxide synthase inhibition improves diastolic function and reduces oxidative stress in ovariectomized mRen2. Lewis rats. Menopause
54. Sartori-Valinotti JC, Iliescu R, Fortepiani LA, Yanes LL, Reckelhoff JF (2007) Sex differences in oxidative stress and the impact on blood pressure control and cardiovascular disease. Clin Exp Pharmacol Physiol 34(9):938–945
55. Schuessel K, Leutner S, Cairns NJ, Muller WE, Eckert A (2004) Impact of gender on upregulation of antioxidant defence mechanisms in Alzheimer's disease brain. J Neural Transm 111(9):1167–1182
56. Vassalle C, Maffei S, Boni C, Zucchelli GC (2008) Gender-related differences in oxidative stress levels among elderly patients with coronary artery disease. Fertil Steril 89(3):608–613
57. Lee IM, Cook NR, Gaziano JM et al (2005) Vitamin E in the primary prevention of cardiovascular disease and cancer: the Women's Health Study: a randomized controlled trial. JAMA 294(1):56–65
58. Kumar S, Lata K, Mukhopadhyay S, Mukherjee TK (2010) Role of estrogen receptors in pro-oxidative and anti-oxidative actions of estrogens: a perspective. Biochim Biophys Acta 1800(10):1127–1135
59. Green PS, Gordon K, Simpkins JW (1997) Phenolic A ring requirement for the neuroprotective effects of steroids. J Steroid Biochem Mol Biol 63(4–6):229–235
60. Miller AA, Drummond GR, Mast AE, Schmidt HH, Sobey CG (2007) Effect of gender on NADPH-oxidase activity, expression, and function in the cerebral circulation: role of estrogen. Stroke 38(7):2142–2149
61. Wagner AH, Schroeter MR, Hecker M (2001) 17beta-estradiol inhibition of NADPH oxidase expression in human endothelial cells. FASEB J 15(12):2121–2130
62. Laufs U, Adam O, Strehlow K et al (2003) Down-regulation of Rac-1 GTPase by Estrogen. J Biol Chem 278(8):5956–5962
63. Satoh M, Matter CM, Ogita H et al (2007) Inhibition of apoptosis-regulated signaling kinase-1 and prevention of congestive heart failure by estrogen. Circulation 115(25):3197–3204
64. Xu Y, Armstrong SJ, Arenas IA, Pehowich DJ, Davidge ST (2004) Cardioprotection by chronic estrogen or superoxide dismutase mimetic treatment in the aged female rat. Am J Physiol Heart Circ Physiol 287(1):H165–H171

65. Dantas AP, Franco Mdo C, Silva-Antonialli MM et al (2004) Gender differences in superoxide generation in microvessels of hypertensive rats: role of NAD(P)H-oxidase. Cardiovasc Res 61(1):22–29
66. Budhiraja R, Kayyali US, Karamsetty M et al (2003) Estrogen modulates xanthine dehydrogenase/xanthine oxidase activity by a receptor-independent mechanism. Antioxid Redox Signal 5(6):705–711
67. Felty Q (2006) DNA Estrogen-induced synthesis in vascular endothelial cells is mediated by ROS signaling. BMC Cardiovasc Disord 6:16
68. Relling MV, Lin JS, Ayers GD, Evans WE (1992) Racial and gender differences in N-acetyltransferase, xanthine oxidase, and CYP1A2 activities. Clin Pharmacol Ther 52(6):643–658
69. Levinson DJ, Chalker D (1980) Rat hepatic xanthine oxidase activity: age and sex specific differences. Arthritis Rheum 23(1):77–82
70. Bhole V, de Vera M, Rahman MM, Krishnan E, Choi H (2010) Epidemiology of gout in women: Fifty-two-year followup of a prospective cohort. Arthritis Rheum 62(4):1069–1076
71. Razmara A, Duckles SP, Krause DN, Procaccio V (2007) Estrogen suppresses brain mitochondrial oxidative stress in female and male rats. Brain Res 1176:71–81
72. Stirone C, Duckles SP, Krause DN, Procaccio V (2005) Estrogen increases mitochondrial efficiency and reduces oxidative stress in cerebral blood vessels. Mol Pharmacol 68(4):959–965
73. Duckles SP, Krause DN, Stirone C, Procaccio V (2006) Estrogen and mitochondria: a new paradigm for vascular protection? Mol Interv 6(1):26–35
74. Faraci FM, Didion SP (2004) Vascular protection: superoxide dismutase isoforms in the vessel wall. Arterioscler Thromb Vasc Biol 24(8):1367–1373
75. Yager JD, Chen JQ (2007) Mitochondrial estrogen receptors–new insights into specific functions. Trends Endocrinol Metab 18(3):89–91
76. Yang SH, Liu R, Perez EJ et al (2004) Mitochondrial localization of estrogen receptor beta. Proc Natl Acad Sci U S A 101(12):4130–4135
77. Frishman WH, Helisch A, Naseer N, Lyons J, Hays RM (2003) Nitric oxide donor drugs in the treatment of cardiovascular disease. In: Frishman WH, Sonnenblick EH, Sica DA (eds) Cardiovascular Pharmacotherapeutics. McGraw-Hill, New York, pp 565–588
78. Hayashi T, Fukuto JM, Ignarro LJ, Chaudhuri G (1992) Basal release of nitric oxide from aortic rings is greater in female rabbits than in male rabbits: implications for atherosclerosis. Proc Natl Acad Sci U S A 89(23):11259–11263
79. Kharitonov SA, Logan-Sinclair RB, Busset CM, Shinebourne EA (1994) Peak expiratory nitric oxide differences in men and women: relation to the menstrual cycle. Br Heart J 72(3):243–245
80. Conrad KP, Joffe GM, Kruszyna H et al (1993) Identification of increased nitric oxide biosynthesis during pregnancy in rats. FASEB J 7(6):566–571
81. Nuedling S, Kahlert S, Loebbert K et al (1999) 17 Beta-estradiol stimulates expression of endothelial and inducible NO synthase in rat myocardium in vitro and in vivo. Cardiovasc Res 43(3):666–674
82. Fraser H, Davidge ST, Clanachan AS (2000) Activation of Ca(2+)-independent nitric oxide synthase by 17beta-estradiol in post-ischemic rat heart. Cardiovasc Res 46(1):111–118
83. Haynes MP, Sinha D, Russell KS et al (2000) Membrane estrogen receptor engagement activates endothelial nitric oxide synthase via the PI3-kinase-Akt pathway in human endothelial cells. Circ Res 87(8):677–682
84. Bucci M, Roviezzo F, Cicala C, Pinto A, Cirino G (2002) 17-beta-oestradiol-induced vasorelaxation in vitro is mediated by eNOS through hsp90 and akt/pkb dependent mechanism. Br J Pharmacol 135(7):1695–1700
85. Han G, Ma H, Chintala R et al (2007) Nongenomic, endothelium-independent effects of estrogen on human coronary smooth muscle are mediated by type I (neuronal) NOS and PI3-kinase-Akt signaling. Am J Physiol Heart Circ Physiol 293(1):H314–H321

86. Han G, Ma H, Chintala R, Fulton DJ, Barman SA, White RE (2009) Essential role of the 90-kilodalton heat shock protein in mediating nongenomic estrogen signaling in coronary artery smooth muscle. J Pharmacol Exp Ther 329(3):850–855
87. Tan E, Gurjar MV, Sharma RV, Bhalla RC (1999) Estrogen receptor-alpha gene transfer into bovine aortic endothelial cells induces eNOS gene expression and inhibits cell migration. Cardiovasc Res 43(3):788–797
88. Han G, Yu X, Lu L et al (2006) Estrogen receptor alpha mediates acute potassium channel stimulation in human coronary artery smooth muscle cells. J Pharmacol Exp Ther 316(3):1025–1030
89. Yang YM, Huang A, Kaley G, Sun D (2009) eNOS uncoupling and endothelial dysfunction in aged vessels. Am J Physiol Heart Circ Physiol 297(5):H1829–H1836
90. Kuzkaya N, Weissmann N, Harrison DG, Dikalov S (2005) Interactions of peroxynitrite with uric acid in the presence of ascorbate and thiols: implications for uncoupling endothelial nitric oxide synthase. Biochem Pharmacol 70(3):343–354
91. Reckelhoff JF, Kellum JA, Blanchard EJ, Bacon EE, Wesley AJ, Kruckeberg WC (1994) Changes in nitric oxide precursor, L-arginine, and metabolites, nitrate and nitrite, with aging. Life Sci 55(24):1895–1902
92. Berkowitz DE, White R, Li D et al (2003) Arginase reciprocally regulates nitric oxide synthase activity and contributes to endothelial dysfunction in aging blood vessels. Circulation 108(16):2000–2006
93. Chen EY, Kallwitz E, Leff SE et al (2000) Age-related decreases in GTP-cyclohydrolase-I immunoreactive neurons in the monkey and human substantia nigra. J Comp Neurol 426(4):534–548
94. Manson JE, Hsia J, Johnson KC et al (2003) Estrogen plus progestin and the risk of coronary heart disease. N Engl J Med 349(6):523–534
95. Taylor HS, Manson JE (2011) Update in hormone therapy use in menopause. J Clin Endocrinol Metab 96(2):255–264
96. Salpeter SR, Cheng J, Thabane L, Buckley NS, Salpeter EE (2009) Bayesian meta-analysis of hormone therapy and mortality in younger postmenopausal women. Am J Med. Nov 122(11):1016–1022 e1011
97. Clarkson TB, Mehaffey MH (2009) Coronary heart disease of females: lessons learned from nonhuman primates. Am J Primatol 71(9):785–793
98. Szasz T, Thakali K, Fink GD, Watts SW (2007) A comparison of arteries and veins in oxidative stress: producers, destroyers, function, and disease. Exp Biol Med (Maywood) 232(1):27–37
99. Papaharalambus CA, Griendling KK (2007) Basic mechanisms of oxidative stress and reactive oxygen species in cardiovascular injury. Trends Cardiovasc Med 17(2):48–54
100. Miller FJ Jr, Gutterman DD, Rios CD, Heistad DD, Davidson BL (1998) Superoxide production in vascular smooth muscle contributes to oxidative stress and impaired relaxation in atherosclerosis. Circ Res 82(12):1298–1305
101. Castier Y, Brandes RP, Leseche G, Tedgui A, Lehoux S (2005) p47phox-dependent NADPH oxidase regulates flow-induced vascular remodeling. Circ Res 97(6):533–540
102. Clarkson TB (2007) Estrogen effects on arteries vary with stage of reproductive life and extent of subclinical atherosclerosis progression. Menopause 14(3 Pt 1):373–384
103. Wing LY, Chen YC, Shih YY, Cheng JC, Lin YJ, Jiang MJ (2009) Effects of oral estrogen on aortic ROS-generating and -scavenging enzymes and atherosclerosis in apoE-deficient mice. Exp Biol Med (Maywood) 234(9):1037–1046
104. Hsia J, Otvos JD, Rossouw JE et al (2008) Lipoprotein particle concentrations may explain the absence of coronary protection in the women's health initiative hormone trials. Arterioscler Thromb Vasc Biol 28(9):1666–1671
105. Merz CN, Johnson BD, Berga S, Braunstein G, Reis SE, Bittner V (2006) Past oral contraceptive use and angiographic coronary artery disease in postmenopausal women: data from the National Heart, Lung, and Blood Institute-sponsored Women's Ischemia Syndrome Evaluation. Fertil Steril 85(5):1425–1431

106. Clarkson TB, Anthony MS, Klein KP (1996) Hormone replacement therapy and coronary artery atherosclerosis: the monkey model. Br J Obstet Gynaecol 103(Suppl 13):53–57 discussion 57–58
107. Clarkson TB (2002) The new conundrum: do estrogens have any cardiovascular benefits? Int J Fertil Womens Med 47(2):61–68
108. Rosenfeld ME, Kauser K, Martin-McNulty B, Polinsky P, Schwartz SM, Rubanyi GM (2002) Estrogen inhibits the initiation of fatty streaks throughout the vasculature but does not inhibit intra-plaque hemorrhage and the progression of established lesions in apolipoprotein E deficient mice. Atherosclerosis 164:251–259
109. Grodstein F, Manson JE, Stampfer MJ (2006) Hormone therapy and coronary heart disease: the role of time since menopause and age at hormone initiation. J Womens Health (Larchmt) 15(1):35–44
110. Billon-Gales A, Fontaine C, Douin-Echinard V et al (2009) Endothelial estrogen receptor-alpha plays a crucial role in the atheroprotective action of 17beta-estradiol in low-density lipoprotein receptor-deficient mice. Circulation 120(25):2567–2576
111. Losordo DW, Kearney M, Kim EA, Jekanowski J, Isner JM (1994) Variable expression of the estrogen receptor in normal and atherosclerotic coronary arteries of premenopausal women. Circulation 89(4):1501–1510
112. Cosentino F, Hurlimann D (2008) Delli Gatti C, et al. Chronic treatment with tetrahydrobiopterin reverses endothelial dysfunction and oxidative stress in hypercholesterolaemia. Heart 94(4):487–492
113. Ryoo S, Gupta G, Benjo A et al (2008) Endothelial arginase II: a novel target for the treatment of atherosclerosis. Circ Res 102(8):923–932
114. Vasquez-Vivar J, Duquaine D, Whitsett J, Kalyanaraman B, Rajagopalan S (2002) Altered tetrahydrobiopterin metabolism in atherosclerosis: implications for use of oxidized tetrahydrobiopterin analogues and thiol antioxidants. Arterioscler Thromb Vasc Biol 22(10):1655–1661
115. Ozaki M, Kawashima S, Yamashita T et al (2002) Overexpression of endothelial nitric oxide synthase accelerates atherosclerotic lesion formation in apoE-deficient mice. J Clin Invest 110(3):331–340
116. Wang CH, Li SH, Weisel RD et al (2005) Tetrahydrobiopterin deficiency exaggerates intimal hyperplasia after vascular injury. Am J Physiol Regul Integr Comp Physiol 289(2):R299–R304
117. Yada T, Kaji S, Akasaka T et al (2007) Changes of asymmetric dimethylarginine, nitric oxide, tetrahydrobiopterin, and oxidative stress in patients with acute myocardial infarction by medical treatments. Clin Hemorheol Microcirc 37(3):269–276
118. Blackwell KA, Sorenson JP, Richardson DM et al (2004) Mechanisms of aging-induced impairment of endothelium-dependent relaxation: role of tetrahydrobiopterin. Am J Physiol Heart Circ Physiol 287(6):H2448–H2453
119. d'Uscio LV, Katusic ZS (2006) Increased vascular biosynthesis of tetrahydrobiopterin in apolipoprotein E-deficient mice. Am J Physiol Heart Circ Physiol 290(6):H2466–H2471
120. Amos AF, McCarty DJ, Zimmet P (1997) The rising global burden of diabetes and its complications: estimates and projections to the year 2010. Diabet Med 14(Suppl 5):S1–S85
121. Legato MJ, Gelzer A, Goland R et al (2006) Gender-specific care of the patient with diabetes: review and recommendations. Gend Med 3(2):131–158
122. Huxley R, Barzi F, Woodward M (2006) Excess risk of fatal coronary heart disease associated with diabetes in men and women: meta-analysis of 37 prospective cohort studies. BMJ 332(7533):73–78
123. Zuanetti G, Latini R, Maggioni AP, Santoro L, Franzosi MG (1993) Influence of diabetes on mortality in acute myocardial infarction: data from the GISSI-2 study. J Am Coll Cardiol 22(7):1788–1794
124. Juutilainen A, Kortelainen S, Lehto S, Ronnemaa T, Pyorala K, Laakso M (2004) Gender difference in the impact of type 2 diabetes on coronary heart disease risk. Diabetes Care 27(12):2898–2904

125. Sowers JR (2004) Diabetes in the elderly and in women: cardiovascular risks. Cardiol Clin 22(4):541–551 vi
126. Cosentino F, Hishikawa K, Katusic ZS, Luscher TF (1997) High glucose increases nitric oxide synthase expression and superoxide anion generation in human aortic endothelial cells. Circulation 96(1):25–28
127. Nishikawa T, Edelstein D, Du XL et al (2000) Normalizing mitochondrial superoxide production blocks three pathways of hyperglycaemic damage. Nature 404(6779):787–790
128. Lund DD, Faraci FM, Miller FJ Jr, Heistad DD (2000) Gene transfer of endothelial nitric oxide synthase improves relaxation of carotid arteries from diabetic rabbits. Circulation 101(9):1027–1033
129. Fukumoto H, Naito Z, Asano G, Aramaki T (1998) Immunohistochemical and morphometric evaluations of coronary atherosclerotic plaques associated with myocardial infarction and diabetes mellitus. J Atheroscler Thromb 5(1):29–35
130. Rains JL, Jain SK (2011) Oxidative stress, insulin signaling, and diabetes. Free Radic Biol Med 50(5):567–575
131. Evans RW, Orchard TJ (1994) Oxidized lipids in insulin-dependent diabetes mellitus: a sex-diabetes interaction? Metabolism 43(9):1196–1200
132. Inoguchi T, Li P, Umeda F et al (2000) High glucose level and free fatty acid stimulate reactive oxygen species production through protein kinase C–dependent activation of NAD(P)H oxidase in cultured vascular cells. Diabetes 49(11):1939–1945
133. Desco MC, Asensi M, Marquez R et al (2002) Xanthine oxidase is involved in free radical production in type 1 diabetes: protection by allopurinol. Diabetes 51(4):1118–1124
134. Ahmed B, Bairey Merz CN, Johnson BD et al (2008) Diabetes mellitus, hypothalamic hypoestrogenemia, and coronary artery disease in premenopausal women (from the National Heart, Lung, and Blood Institute sponsored WISE study). Am J Cardiol 102(2):150–154
135. Wenzel P, Daiber A, Oelze M et al (2008) Mechanisms underlying recoupling of eNOS by HMG-CoA reductase inhibition in a rat model of streptozotocin-induced diabetes mellitus. Atherosclerosis 198(1):65–76
136. Romero MJ, Platt DH, Tawfik HE et al (2008) Diabetes-induced coronary vascular dysfunction involves increased arginase activity. Circ Res 102(1):95–102
137. Fard A, Tuck CH, Donis JA et al (2000) Acute elevations of plasma asymmetric dimethylarginine and impaired endothelial function in response to a high-fat meal in patients with type 2 diabetes. Arterioscler Thromb Vasc Biol 20(9):2039–2044
138. Cai S, Khoo J, Channon KM (2005) Augmented BH4 by gene transfer restores nitric oxide synthase function in hyperglycemic human endothelial cells. Cardiovasc Res 65(4):823–831
139. Pfister SL (2011) Role of lipoxygenase metabolites of arachidonic Acid in enhanced pulmonary artery contractions of female rabbits. Hypertension 57(4):825–832
140. Pugh ME, Hemnes AR (2010) Development of pulmonary arterial hypertension in women: interplay of sex hormones and pulmonary vascular disease. Womens Health (Lond Engl) 6(2):285–296
141. Wu SC, Caravita S, Lisi E et al (2009) Pulmonary arterial hypertension. Intern Emerg Med 4(6):459–470
142. Rabinovitch M (2008) Molecular pathogenesis of pulmonary arterial hypertension. J Clin Invest 118(7):2372–2379
143. Perez-Vizcaino F, Cogolludo A, Moreno L (2010) Reactive oxygen species signaling in pulmonary vascular smooth muscle. Respir Physiol Neurobiol 174(3):212–220
144. Archer SL, Huang J, Henry T, Peterson D, Weir EK (1993) A redox-based O2 sensor in rat pulmonary vasculature. Circ Res 73(6):1100–1112
145. Barman SA, Zhu S, White RE (2005) Hypoxia modulates cyclic AMP activation of BkCa channels in rat pulmonary arterial smooth muscle. Lung 183(5):353–361
146. Barman SA, Zhu S, White RE (2004) Protein kinase C inhibits BKCa channel activity in pulmonary arterial smooth muscle. Am J Physiol Lung Cell Mol Physiol 286(1):L149–L155

147. Redout EM, van der Toorn A, Zuidwijk MJ et al (2010) Antioxidant treatment attenuates pulmonary arterial hypertension-induced heart failure. Am J Physiol Heart Circ Physiol 298(3):H1038–H1047
148. Grobe AC, Wells SM, Benavidez E et al (2006) Increased oxidative stress in lambs with increased pulmonary blood flow and pulmonary hypertension: role of NADPH oxidase and endothelial NO synthase. Am J Physiol Lung Cell Mol Physiol 290(6):L1069–L1077
149. Konduri GG, Bakhutashvili I, Eis A, Pritchard K Jr (2007) Oxidant stress from uncoupled nitric oxide synthase impairs vasodilation in fetal lambs with persistent pulmonary hypertension. Am J Physiol Heart Circ Physiol 292(4):H1812–H1820
150. Khoo JP, Zhao L, Alp NJ et al (2005) Pivotal role for endothelial tetrahydrobiopterin in pulmonary hypertension. Circulation 111(16):2126–2133
151. Jankov RP, Kantores C, Pan J, Belik J (2008) Contribution of xanthine oxidase-derived superoxide to chronic hypoxic pulmonary hypertension in neonatal rats. Am J Physiol Lung Cell Mol Physiol 294(2):L233–L245
152. Lahm T, Patel KM, Crisostomo PR et al (2007) Endogenous estrogen attenuates pulmonary artery vasoreactivity and acute hypoxic pulmonary vasoconstriction: the effects of sex and menstrual cycle. Am J Physiol Endocrinol Metab 293(3):E865–E871
153. Lahm T, Crisostomo PR, Markel TA et al (2008) Selective estrogen receptor-alpha and estrogen receptor-beta agonists rapidly decrease pulmonary artery vasoconstriction by a nitric oxide-dependent mechanism. Am J Physiol Regul Integr Comp Physiol 295(5):R1486–R1493
154. Rabinovitch M, Gamble WJ, Miettinen OS, Reid L (1981) Age and sex influence on pulmonary hypertension of chronic hypoxia and on recovery. Am J Physiol 240(1):H62–H72
155. Stupfel M, Pesce VH, Gourlet V, Bouley G, Elabed A, Lemercerre C (1984) Sex-related factors in acute hypoxia survival in one strain of mice. Aviat Space Environ Med 55(2):136–140
156. Smith AM, Bennett RT, Jones TH, Cowen ME, Channer KS, Jones RD (2008) Characterization of the vasodilatory action of testosterone in the human pulmonary circulation. Vasc Health Risk Manag 4(6):1459–1466
157. Beretta L, Caronni M, Origgi L, Ponti A, Santaniello A, Scorza R (2006) Hormone replacement therapy may prevent the development of isolated pulmonary hypertension in patients with systemic sclerosis and limited cutaneous involvement. Scand J Rheumatol 35(6):468–471
158. Morse JH, Horn EM, Barst RJ (1999) Hormone replacement therapy: a possible risk factor in carriers of familial primary pulmonary hypertension. Chest 116(3):847
159. Sweeney L, Voelkel NF (2009) Estrogen exposure, obesity and thyroid disease in women with severe pulmonary hypertension. Eur J Med Res 14(10):433–442
160. Lahm T, Crisostomo PR, Markel TA et al (2008) The effects of estrogen on pulmonary artery vasoreactivity and hypoxic pulmonary vasoconstriction: potential new clinical implications for an old hormone. Crit Care Med 36(7):2174–2183
161. Kleiger RE, Boxer M, Ingham RE, Harrison DC (1976) Pulmonary hypertension in patients using oral contraceptives. A report of six cases. Chest 69(2):143–147
162. Bowers R, Cool C, Murphy RC et al (2004) Oxidative stress in severe pulmonary hypertension. Am J Respir Crit Care Med 169(6):764–769
163. Farhat MY, Ramwell PW (1992) Estradiol potentiates the vasopressor response of the isolated perfused rat lung to the thromboxane mimic U-46619. J Pharmacol Exp Ther 261(2):686–691
164. Zhu D, Medhora M, Campbell WB, Spitzbarth N, Baker JE, Jacobs ER (2003) Chronic hypoxia activates lung 15-lipoxygenase, which catalyzes production of 15-HETE and enhances constriction in neonatal rabbit pulmonary arteries. Circ Res 92(9):992–1000
165. Zhang L, Ma J, Li Y et al (2010) 15-Hydroxyeicosatetraenoic acid (15-HETE) protects pulmonary artery smooth muscle cells against apoptosis via HSP90. Life Sci 87(7–8):223–231

166. Cho KJ, Seo JM, Kim JH (2011) Bioactive lipoxygenase metabolites stimulation of NADPH oxidases and reactive oxygen species. Mol Cells 32(1):1–5
167. Funk CD, FitzGerald GA (2007) COX-2 inhibitors and cardiovascular risk. J Cardiovasc Pharmacol 50(5):470–479
168. Egan KM, Lawson JA, Fries S et al (2004) COX-2-derived prostacyclin confers atheroprotection on female mice. Science 306(5703):1954–1957
169. Barlow RS, White RE (1998) Hydrogen peroxide relaxes porcine coronary arteries by stimulating BKCa channel activity. Am J Physiol 275(4 Pt 2):H1283–H1289

Chapter 16
Female Infertility and Free Radicals: Potential Role in Endometriosis and Adhesions

Zeynep Alpay Savasan

Abstract Oxidative stress is widely implicated in various forms of reproduction failure. Postoperative adhesions and endometriosis are the two common conditions associated with female infertility with several similarities in their development and clinical behavior. The pathophysiology of both conditions involves peritoneal surface inflammation, tissue healing, and fibrogenesis, in which alterations in oxidative metabolism appear to be operational. Enhanced free radical generation and/or decreased scavenging have been shown in the endometriosis and postoperative adhesion tissues. Interventions targeting oxidative metabolism for the prevention and treatment of either condition have given mixed results. Studies utilizing contemporary methods may shed further light on the involvement of oxidative stress leading to development of more targeted therapies in these conditions.

Keywords Female infertility · Free radicals · Oxidative stress · Endometriosis · Postoperative adhesion · Reproduction failure · Tissue healing

Postoperative adhesions and endometriosis are common conditions associated with female infertility. Both are progressive in nature and they have a tendency to recur after treatment. The proposed developmental mechanisms of these conditions are hypothetical and have limitations in helping to establish successful prevention and/or treatment approaches. In recent years, free radicals have been shown to play a role in the development of postoperative adhesions and endometriosis. This review summarizes the possible contribution of free radical metabolism to the development of these two conditions.

Z. Alpay Savasan (✉)
Department of Obstetrics and Gynecology, Hutzel Women's Hospital, Wayne State University, 3990 John R, 4BRUSH, Detroit, MI 48201, USA
e-mail: zalpay@med.wayne.edu; zeynepalpay@yahoo.com

16.1 Free Radicals and Oxidative Stress

A free radical is an unstable and reactive molecule, which has one or more unpaired electrons. These molecules are essential for normal cellular function and immune defense in particular. They act as secondary messengers in signaling pathways and can help in eliminating microorganisms. Their interaction with macromolecules may result in damage within the cell through oxidation, and may lead to mutations or apoptosis. However, cellular antioxidant homeostatic systems are capable of protecting the cell. The balance between the free radical activity and the antioxidant systems help to maintain the normal cellular function. The excess production of free radicals or the lack of or defective antioxidant defense mechanisms may lead to a state called "oxidative stress". The oxidative stress as a result of imbalance in the free radical system has been implicated in the pathophysiology of various processes, such as aging, cataract, rheumatologic, neurologic, and cardiovascular diseases [1–12].

There are various types of free radicals, which are generated by different sources and enzyme systems (Table 16.1). Superoxide anions, hydrogen peroxide, hydroxyl radicals are some of the most common free radicals and are generated from molecular oxygen (Fig. 16.1). Xanthine oxidase generates superoxide, heme oxygenase, cyclooxygenase, and lipooxygenase generate hydroxyl and peroxyl radicals (lipid hydroxyperoxidases) (Fig. 16.1). The scavenger and antioxidant molecules are either endogenous enzymes and molecules or exogenous compounds (Table 16.1).

Nitric oxide is one of the important free radicals with a regulatory role. It is synthesized by endothelial cells and regulates multiple cellular functions, such as smooth muscle tone, neurotransmission, cell proliferation, apoptosis and macrophage-mediated cytotoxicity. Nitric oxide is synthesized by a group of enzymes called nitric oxide synthases (NOS). There are three forms of NOS: endothelial NOS (eNOS), inducible NOS (iNOS), and neuronal NOS (nNOS). Nitric oxide is synthesized during conversion of arginine to citrulline (Fig. 16.2). Nitric oxide synthases catalyze this step and require tetrohydrobiopterin (H4B) and nicotinamide adenine dinucleotide dihydro phosphate (NADPH) as cofactors. This process can be inhibited by two methylated arginine analogues; asymmetric dimethyl arginine (ADMA); and monomethyl arginine (L-NMMA) both of which are competitive inhibitors of NOS (Fig. 16.2).

16.2 Endometriosis

Endometriosis is a progressive chronic inflammatory disease of female genital tract. The diagnosis is based on the identification of ectopic endometrial tissue outside the uterus. Women with endometriosis commonly experience pelvic pain, dysmenorrhea, dyspareunia, and infertility. Chronic, incapacitating pelvic pain and

16 Female Infertility and Free Radicals

Table 16.1 Free radicals and scavenger antioxidants

Free radicals	Scavenger antioxidants
•Singlet oxygen	•Superoxide dismutase
•Superoxide radical	•Catalase
•Hydroxyl radical	•Glutathione peroxidase
•Alkoxyl radical	•Beta-carotene
•Peroxyl radical	•Vitamin A, E, C
•Hydrogen peroxide	•Selenium
•Lipid peroxides	

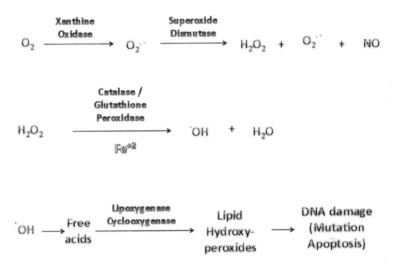

Fig. 16.1 Common free radicals generated from molecular oxygen

Fig. 16.2 Nitric oxide synthesized during conversion of arginine to citrulline

infertility impact the physical and mental health status, and quality of life negatively and substantially [13]. There is a good correlation between the stage of the disease and the painful symptoms. On the contrary, the link between endometriosis and infertility is not clear, but the clinical association is well-accepted [14]. The accumulating evidence has shown that infertility in endometriosis is multifactorial and could be associated with free radical metabolism.

16.3 Pathogenesis of Endometriosis

Similar to normal endometrial tissue, endometriotic implants are hormone-dependent and they regress in postmenopausal years. Hormone suppression by contraceptive medications and establishing a pseudopregnancy state help the regression of the lesions. In some instances, the disease has been shown to occur in distant sites (abdominal wall, previous scar, genitourinary, gastrointestinal systems, and thorax) [15–18]. Moreover, there are case reports of male patients with urogenital endometriosis after high-dose estrogen exposure [19–25].

There are several pathogenetic mechanisms proposed to explain the development of this disease. Sampson's theory is one of the earliest proposed mechanisms, which suggests the peritoneal implantation after menstrual dissemination of endometrial tissue [26]. These implants can behave similar to eutopic endometrium in each menstrual cycle. They proliferate with hormonal stimulation and bleed with progesterone withdrawal. It has been shown that the incidence of endometriosis is higher in women who have short cycles with longer menstrual flow [27]. Similarly, patients with mullerian anomalies with outflow tract obstruction have increased incidence of the disease [28]. Celomic metaplasia is another theory suggesting the induction of metaplasia in the celomic epithelium which gives rise to peritoneum, ovarian epithelium (the most common sites), and pleura. Furthermore, menstrual dissemination of endometrial tissue into the venous circulation may result in metastatic or embolic endometriosis [29].

Development and survival of endometriosis depend on the new blood vessel formation. The current literature highlights the fact that neovascularization of endometriotic lesions is not only driven by angiogenesis, but also involves de novo formation of microvessels from circulating endothelial progenitor cells which is also called postnatal vasculogenesis [30]. This is also a characteristic of various pathogenic processes, such as tumor growth and atherosclerosis, and typically comprises the activation, mobilization, and recruitment of bone marrow-derived endothelial progenitor cells to the sites of tissue hypoxia. Review of the recent studies indicates that up to 37 % of the microvascular endothelium of ectopic endometrial tissue originates from endothelial progenitor cells [30]. Accordingly, blockade of endothelial progenitor cells recruitment effectively inhibits the formation of microvascular networks in developing endometriotic lesions, indicating that vasculogenesis represents an integral part of the pathogenesis of endometriosis [30].

Additional suggested contribution to the pathogenesis of endometriosis is the increased adhesive and proliferative potential of endometriotic cells in response to specific extracellular matrix components. The stronger adhesion potential to extracellular matrix proteins is mediated partly through integrins. Integrins which are a class of cell adhesion molecules act as a bridge between cells and extracellular matrix, such as collagen, fibronectin, or laminin. The alteration of integrin expression and enhanced metalloproteinase activity have a role in the development of endometriosis not only by increasing adhesive and proliferative properties of endometriotic cells [31] but also by reducing the cellular apoptosis and hence increasing the survival in the peritoneal cavity [32].

Immunologic response is altered in women with endometriosis. The lymphocyte proliferation and T-cell-mediated cytotoxity are decreased, but macrophage activity is increased in women with endometriosis [33]. Natural killer (NK) cell activity is also defective in peripheral blood and peritoneal fluid toward both the autologous and heterologous endometrium [34]. Not only the innate immune system, but also adaptive immunity is also altered in endometriosis. There is an increased antibody production due to an alteration in B-lymphocyte function [35]. Moreover, the volume of peritoneal fluid in women with endometriosis is increased and contains increased amounts of inflammatory cells, angiogenic factors, cytokines, prostaglandins, and growth factors [36–38]. The patterns of the immunologic response may differ with the stage of the disease providing evidence for the immunologic component of the disease progression [38].

16.4 Role of Free Radicals in Endometriosis

Endometriosis is a multifactorial disease and the inflammatory process is the main component implicated in the pathogenesis. Free radicals and oxidative stress are closely related to various inflammatory conditions. The pro-inflammatory mediators, such as inflammatory cells, cytokines, prostaglandins, growth factors, and angiogenic factors are increased in the peritoneal fluid of patients with endometriosis [36–38]. Furthermore, the number of peritoneal macrophages and their phagocytic activity are significantly increased in this disease [39–42]. Development of several inflammatory conditions, such as arthritis, cardiovascular and neurologic diseases, inflammatory lung and bowel diseases has been associated with free radical metabolism and oxidative stress [5–9, 43–47]. Nitric oxide production and iNOS expression are increased in peritoneal macrophages of patients with endometriosis [48], similar to other inflammatory diseases (synovial macrophages in arthritis, or alveolar macrophages in inflammatory lung conditions) [6, 9, 45, 49]. Peritoneal content of nitrite and nitrate is also increased in endometriosis [48]. The peritoneal macrophages of patients with endometriosis have higher NO and iNOS production than the normal patients after LPS, interferon-α, and interferon-γ stimulation [48]. Moreover, lipid peroxide concentration, which is a strong chemotaxin for monocytes and inducer of cytokine secretion is increased both in the peritoneal fluid and the eutopic and ectopic endometrial tissues of patients with endometriosis [50–53].

Refluxed menstrual debris includes erythrocytes, endometrial cells, apoptotic cells, and inflammatory cells, which are the proposed potent oxidative stress inducers. Free iron and heme, if not chelated, have an important role in the free radical production and are deposited more commonly in the close proximity of the endometriotic lesions [54]. Severe hemolysis occurring during the development of endometriosis results in high levels of free heme and iron which can oxidatively modify lipids and proteins, leading to cell and DNA damage, and subsequently fibrosis development [55]. Moreover, oxidative stress contributes to mesothelial cell damage and enhances the adhesion of the ectopic endometrial cells [56].

There is a link between free radical metabolism and angiogenesis. Alteration of endometrial eNOS gene expression may be an inducer of angiogenesis [57]. Oxidative stress generated after incubation of endometriotic implants with minimally oxidized low density lipoprotein, increases the vascular endothelial growth factor (VEGF) production in an in vitro study [58]. This is mediated by the glycoprotein called glycodelin, which is increased in oxidative stress [58]. Glycodelin concentrations are elevated in the peritoneal fluid of patients with endometriosis and this elevation is associated with the severity of menstrual pain [59]. The role of angiogenetic factors in the invasion and progression of the endometriotic implants probably through the positive feedback of local oxidative stress have been demonstrated as potential therapeutic targets both in in vivo and in vitro studies [60–64].

Oxidative stress related ovulation and implantation abnormalities may result in infertility in patients with endometriosis. Free radicals are elevated in the peritoneal and fallopian tube fluid of infertile patients with endometriosis. This can have an impact on the ovulation, fertilization and even gamete, and embryo transportation and implantation [57, 65]. Increased NO in the peritoneal fluid of patients with endometriosis may decrease the oocyte fertilization rate and embryo development [66]. In a mouse model, when sodium nitroprusside (NO donor) and peritoneal fluid samples from human patients with endometriosis are used in in vitro fertilization culture medium, oocyte fertilization, and embryo development are adversely affected in a dose-dependent manner [66]. Furthermore, NO and iNOS expressions are increased in the eutopic and ectopic endometrium and the elevated NO activity in the endometrium may affect the receptivity of endometrium and reduce fecundity [67]. Alpha (v) beta (3) integrin, which is a uterine receptivity marker, and eNOS are expressed in endometrial glandular epithelium. They show similar pattern of expression during the normal menstrual cycle. However, the expression of the receptivity marker integrin is reduced in the endometrial glandular epithelium during the mid-luteal phase whereas eNOS expression is increased in patients with endometriosis [68]. Elevated NO production and decreased alpha (v) beta (3) integrin in the endometrium may contribute to impaired embryo implantation and development seen in the infertility patients with endometriosis [69].

The oxidative stress environment induces the secretion of pro-inflammatory mediators. Moreover, pro-inflammatory mediators such as tumor necrosis factor (TNF)-α can impair the antioxidant defense systems. Patients with unexplained recurrent first trimester miscarriages have higher levels of NO, malondialdehyde (an oxidative stress marker), and TNF-α, but lower levels of antioxidant mediator production when compared to control patients [70]. TNF-α also plays an important role in endometrial proliferation and normal menstruation. Elevated TNF-α concentrations have been reported in the peritoneal fluid of patients with endometriosis. There is also a positive correlation between TNF-α production with the progression of the disease [37, 38]. Furthermore, sperm quality is shown to decrease after TNF-α treatment in a dose and time-dependent manner suggesting the hampering effect of elevated TNF-α on the sperm quality in the peritoneal fluid of patients with endometriosis [71].

Local defense mechanisms are also altered in endometriosis. Endometrial manganese and copper/zinc superoxide dismutase expressions are increased in women with endometriosis during menstruation [72]. There is an aberrant expression of glutathione peroxidase and xanthine oxidase in both ectopic and eutopic endometrium of patients with endometriosis [73, 74]. Furthermore, peritoneal fluid from patients with endometriosis has significantly decreased levels of antioxidant Vitamin E and superoxide dismutase [51–53].

Collectively, the development of endometriosis initially depends on the attachment and angiogenesis, which lead to cell growth and survival. The additional factors contributing to the persistence of endometriotic lesions probably include hormonal imbalance, genetic predisposition, altered immune surveillance, and oxidative stress. The mediators of inflammatory response, oxidative stress, and fibrosis are likely involved in the symptoms [55].

16.5 Postoperative Adhesions

Postoperative adhesions occur commonly (50–90 %) after abdominal/pelvic surgeries [75, 76]. Pelvic adhesions are well-known to cause infertility (15–40 %) [77, 78], small bowel obstruction (60–70 %) [78], and chronic pelvic pain (17–24 %) [78–82]. Furthermore, presence of adhesions prolongs the surgery time, increases the risk of iatrogenic injuries and makes future surgeries difficult or even sometimes impossible [83, 84]. Adhesions can recur and progress after repetitive and even corrective surgeries. The most important risk factor for adhesion development is the type of surgery and the extent of peritoneal damage. Surgeries of the colon and rectum have higher risk of adhesion development and related complications than the surgeries of small bowel, appendix, or gallbladder [85]. The highest incidence for adhesion-related complication is reported after total colectomy with ileal-pouch-anal anastomosis as 19.3 %. Other high risk surgeries include gynecologic surgeries and open colectomy with a risk of 11.1 and 9.5 %; respectively. Additional possible risk factors for adhesion development are age younger than 60 years, previous laparotomy within 5 years, peritonitis, multiple laparotomies, emergency surgeries, omental resection, and penetrating abdominal trauma especially gunshot wounds [86–89]. In general, open surgeries with exception of appendectomy, have a higher risk for development of adhesions than laparoscopic interventions [90].

16.6 Pathogenesis of Postoperative Adhesions

Adhesion is a non-anatomical fibrinous tissue extending between anatomical spaces. An insult to a normal tissue is the triggering factor in the development of adhesion. This insult is usually a direct surgical trauma to the peritoneal surface, but it could also be an infection, inflammation, or hypoxia-mediated injury.

Normal peritoneum is lined by a mesothelial cell layer. Under this layer there are fibroblasts, macrophages, and extracellular matrix (collagen, glycoprotein). This layer also includes blood vessels and lymphatics. An insult to the area triggers a local release of histamine and kinins, which results in an increase in the capillary leak of fibrin-rich exudate and covers the denuded tissue. Fibrin gel matrix forms in this exudate and produces fibrinous bands between the injured tissues. In the meantime, fibrinolytic activity starts to lyse and remodel the fibrinous bands. If the remodeling is defective or if there is an imbalance between fibrin deposition and degradation process, adhesions develop. The proliferating fibroblasts invade the formed fibrinous bands and deposit collagen to form the adhesion tissue. If the process continues, angiogenesis takes over and the adhesion tissue becomes vascularized (Fig. 16.3).

Hypoxia and ischemic insult have been proposed as the strongest contributory factors in the postoperative adhesion development. The abdominal peritoneum has a protective fibrinolytic activity which is essential in the normal healing process. Mesothelial cell layer and the peritoneal fibroblasts have two enzyme activities; tissue plasminogen activator (tPA) which is the tissue modeling enzyme that degrades fibrin into soluble degradation products and plasminogen activator inhibitor type-1 (PAI-1) which is the regulatory enzyme of tPA activity [91, 92]. The two enzyme systems have a delicate balance for the normal healing process. This balance has been shown to be defective in the adhesion tissue. Adhesion fibroblasts have higher PAI-1 activity than the normal peritoneal fibroblasts [91]. The hypoxic environment increases PAI-1 activity in both adhesion and normal peritoneal fibroblasts supporting the adverse effect of ischemia on the protective fibrinolytic system. There is a decrease in tPA activity during surgery, suggesting the depression of peritoneal fibrinolysis [93, 94]. The balance between two systems is indicated by the ratio between tPA:PAI-1. This ratio is significantly higher in normal peritoneal fibroblasts than the adhesion tissue fibroblasts [91]. The hypoxic environment has been shown to be the main inducer of this imbalance by decreasing this ratio in normal and adhesion fibroblasts (90 vs 98 %; respectively) [91]. Furthermore, these markers have been proposed as preoperative predictors for the postoperative adhesion development. There is a strong correlation between preoperative plasma concentrations of PAI-1 and the extent of postoperative adhesion development [95]. The modulators of this system have been studied extensively to prevent the postoperative adhesion development [96–100].

Extracellular matrix regulators are the other important factors involved in the normal healing process. Matrix metalloproteinases (MMP) are the proteolytic enzymes of the extracellular matrix and their regulators are tissue inhibitors of

Pathophysiology of Adhesion Development

Insult to peritoneal surface
↓
Fibrin-rich exudate
↓
Fibrin polymerization & interaction with fibronectin (fibrin gel matrix)
↓
Fibrinous bands between injured areas
↓
Fibrinolysis – remodeling
↓
Imbalance between fibrin deposition and fibrinolysis
↓
Persistence of bands
↓
Invasion of bands by proliferating fibroblasts
↓
Angiogenesis due to prolonged ischemia
↓
Adhesions

Fig. 16.3 Pathophysiology of adhesion development

metalloproteinases (TIMP). Similar to fibrinolytic system, there is a balance between these two counteracting enzymes for a normal extracellular tissue modeling and healing. Hypoxia can adversely affect this balance resulting an inhibition in MMP and enhancement in TIMP activity leading to fibrosis [101, 102].

Inflammation is one of the main inducers of defective tissue healing and adhesion development [103]. The pro-inflammatory cytokines and mediators have been shown to be elevated in adhesion tissue. Transforming growth factor (TGF)-$\beta1$ and TGF-$\beta3$ levels are increased in adhesion fibroblasts [104]. Moreover, TGF-$\beta1$ expression increases under hypoxic conditions and results in elevation in both PAI-1 and plasminogen activator activity [105]. It is important to mention that amniotic membrane, which is a well-studied anti-inflammatory graft in many fields promotes re-epithelization and its use has been proposed as a potential adhesion preventing method [106, 107].

Normal peritoneal and adhesion fibroblasts and their characteristics have been studied extensively [108–113] to understand the pathophysiology of adhesion development. Accumulating evidence shows that they have different phenotypes and cellular responses [108–113]. Hypoxia treatment changes the cellular characteristics of normal peritoneal fibroblasts into the adhesion fibroblasts supporting the role of hypoxia in the cellular differentiation and adhesion development [109–112].

Alterations in immunologic system have been detected in adhesion tissue. Cell surface immunologic marker expressions and immune responses differ in normal peritoneal fibroblasts and adhesion tissue fibroblasts [108, 109, 113] Adhesion tissue fibroblasts are more susceptible to lymphokine-activated killer (LAK) cell elimination than the normal fibroblasts, suggesting a possible impaired natural immune response in adhesion development [108]. Hypoxia treatment also alters the in vitro immune response in normal peritoneal fibroblasts. LAK cell elimination of hypoxia-treated normal peritoneal fibroblasts is enhanced similar to adhesion fibroblasts [109]. Moreover, natural killer-T (NKT) cell deficient mice are resistant to adhesion development, whereas reconstitution of NKT cells results in the development of severe adhesions [114]. Supporting the role of immunological mechanisms in the pathogenesis of adhesion development, expression of NKG2D receptors which play a fundamental role in the NK cell elimination and cytotoxicity were found to be 20 % lower in adhesion fibroblasts compared to normal peritoneal fibroblasts [113].

16.7 Free Radicals in Postoperative Adhesion

Ischemia/reperfusion process has been implicated in pathophysiology of several diseases. Local free radical and oxidative stress generation after ischemia/reperfusion plays an important role in adhesion development supported by the overproduction of free radicals and lack of scavenger mediators [115]. The administration of free radical scavengers have been shown to decrease the adhesion formation [116]. Hypoxic insult (direct tissue trauma, damaged vascular supply, and hypoxic damage by pneumoperitoneum) and hyperoxic environment during surgery (room air exposure in laparotomy) have been proposed as the triggering factors for excessive free radical production [117].

The carbon dioxide (CO_2) gas used to generate pneumoperitoneum can result in mesothelial cell trauma by causing a hypoxic, hypercarbic, and acidotic environment [118, 119]. The antioxidant and scavenger levels significantly decrease by the pneumoperitoneum in a dose and time related manner [120]. It has been demonstrated that the shorter the time of exposure to the gas, the lower the incidence of adhesion development [121]. Similarly, helium gas can generate the same unfavorable environment with an increase in oxidative stress and a decrease in the scavengers [122]. Although, any approach to the abdominal cavity (laparotomy, laparoscopy, or minilaparotomy) can cause trauma and hypoxic damage and increase local inflammatory response, some investigators believe that laparoscopic surgery can prevent postoperative adhesion formation. The free radical production and inflammatory reaction have been shown to increase more after minilaparotomy and full laparatomy than laparoscopy. Laparoscopy has also been shown to preserve systemic immune function [123]. Furthermore, microlaparoscopy has been shown to result in a more favorable environment with shorter and lesser exposure of CO_2 gas than the conventional laparoscopy. The scavenger levels remain significantly higher resulting in lower incidence of postoperative adhesions [121]. Additionally, in animal experiments, administration of oxygen to the pneumoperitoneum environment (CO_2 or helium) or application of assisted ventilation and providing supplemental systemic oxygen helped to prevent adhesion development supporting the hypoxia-mediated mechanisms in the adhesion formation [124, 125].

Nitric oxide is an important counter-regulatory mediator of free radical metabolism in fibrosis formation. It inhibits collagen synthesis, proline hydroxylation, and elastase activity [126]. It also induces myofibroblast apoptosis, enhances fibrinolysis, collagenolysis, and TGF-β production [127]. Elevated NO levels inhibit free radical-mediated collagen synthesis through fibrinolysis [128]. Moreover, NO has been shown to decrease excessive collagen deposition [126, 129, 130]. Normal peritoneal fibroblasts contain higher levels of NO than the adhesion fibroblasts [131]. Low NO levels in adhesion fibroblasts are associated with increased extracellular matrix and collagen type I production with an increase in fibrosis. Furthermore, hypoxia treatment decreases NO secretion in normal fibroblast similar to adhesion phenotype [131]. Inducible-NOS gene and protein levels are affected after hypoxia treatment in normal fibroblasts promoting the adhesion phenotype and fibrosis. Moreover, administration of iNOS inhibitors induces fibrosis [112, 115–121, 124–126, 129, 130, 132–134].

The majority of the adhesion preventing agents has anti-inflammatory and antioxidant potentials. Neurokinin-1 receptor (Substance P) is a known leukocyte recruiter and free radical generator. Neurokinin-1 receptor antagonist administration has been shown to reduce adhesion development by reducing peritoneal tissue oxidative stress markers, such as 8-isoprostane, protein carbonyl, NADPH oxidase, myeloperoxidase mRNA, increasing antioxidant mediators, and inhibiting neutrophil recruitment in the peritoneal fluid of an animal model [135]. Similarly, intraabdominal methylene blue administration decreases the incidence of adhesion development by increasing antioxidant mediators and peritoneal fibrinolytic activity by inhibiting NADPH oxidase (98 %) and myeloperoxidase (78 %) activity [136]. Furthermore, intraperitoneal administration of melatonin decreases the extent of peritoneal adhesions by lowering the malondialdehyde and NO concentrations and increasing the reduced glutathione concentrations in the mice peritoneal tissues [137]. Additionally, there is a reduction in the adhesion formation after oral supplements of vitamin E in rats [138]. However, this effect of vitamin E was shown on adhesions developed after scraping of the caecum with mesh gauze, but the same effect was not seen on the adhesions developed after denuding the serous surface of the uterus [139]. The difference in this observed effect could be explained by a possible target tissue or dose-related generation of free radicals and antioxidants.

Intraabdominal influx of neutrophils after surgical peritoneal trauma plays an important role in postoperative adhesion formation. Preventing the intraabdominal influx of neutrophils in the early postoperative inflammatory reaction by antineutrophil serum and increasing antioxidant and free radical scavengers by superoxide dismutase, catalase, and mannitol, to scavenge the superoxide, hydrogen peroxide, and hydroxyl radicals, respectively have been shown to decrease postoperative adhesion development supporting inflammatory and oxidative stress-related nature of the process [140].

Several adhesion preventive techniques and materials have been accepted and used by many surgeons. There are conflicting results in the literature on the efficacy of these preventive measures in reducing the incidence of adhesion development [65, 141]. The inconsistent conclusions in the studies and failure of not establishing the optimal preventive modality could be explained by the limited understanding of the pathophysiology of the process. Determination of alterations in the defense and immunologic systems and healing processes of patients who have adhesions is critical to understanding the adhesion development. This will be a pivotal contribution to the improvement of the preventive measures and adhesion-related complications.

16.8 Conclusion

Enhanced free radical production and subsequent oxidative stress has been shown to have a central role in endometriosis and postoperative adhesion development. The similarities between these two conditions further support the involvement of common mechanistic pathways. Although the initiation process and triggering stimuli could be different, there appear to be shared factors contributing to the predisposition and persistence. It is possible that free radical metabolism could be the most important central player in these two complicated and debilitating diseases of the reproductive medicine. Further research involving free radical metabolism in the field of endometriosis and postoperative adhesions will potentially help to answer some of the fundamental questions.

References

1. Babizhayev MA (2011) Mitochondria induce oxidative stress, generation of reactive oxygen species and redox state unbalance of the eye lens leading to human cataract formation: disruption of redox lens organization by phospholipid hydroperoxides as a common basis for cataract disease. Cell Biochem Funct 29:183–206
2. Gemma C, Vila J, Bachstetter A, Bickford PC (2007) Oxidative stress and the aging brain: from theory to prevention. CRC Press, Boca Raton
3. Gredilla R (2010) DNA damage and base excision repair in mitochondria and their role in aging. J Aging Res 2011:257093 1–9
4. Khizhkin EA, Ilukha VA, Ilyina TN, Unzhakov AR, Vinogradova IA, Anisimov VN (2010) Antioxidant system and energy provision of the rat heart during aging depend on illumination regimen and are resistant to exogenous melatonin. Bull Exp Biol Med 149:354–358
5. Kim SK, Kim KS, Lee YS, Park SH, Choe JY (2010) Arterial stiffness and proinflammatory cytokines in fibromyalgia syndrome. Clin Exp Rheumatol 28:S71–S77

6. Kundu S, Bala A, Ghosh P, Mukhopadhyay D, Mitra A, Sarkar A, Bauri AK, Ghosh A, Chattopadhyay S, Chatterjee M (2011) Attenuation of oxidative stress by allylpyrocatechol in synovial cellular infiltrate of patients with rheumatoid arthritis. Free Radic Res 45: 518–526
7. Pizza V, Agresta A, D'Acunto CW, Festa M, Capasso A (2011) Neuroinflamm-Aging and neurodegenerative diseases: an overview. CNS Neurol Disord Drug Targets 10:621–634
8. Pizza V, Agresta A, D'Acunto CW, Festa M, Capasso A (2011) Neuroinflammation and ageing: current theories and an overview of the data. Rev Recent Clin Trials 6:189–203
9. Pizzolla A, Gelderman KA, Hultqvist M, Vestberg M, Gustafsson K, Mattsson R, Holmdahl R (2011) CD68-expressing cells can prime T cells and initiate autoimmune arthritis in the absence of reactive oxygen species. Eur J Immunol 41:403–412
10. Ristow M, Schmeisser S (2011) Extending life span by increasing oxidative stress. Free Radic Biol Med 51:327–336
11. Wojcik M, Burzynska-Pedziwiatr I, Wozniak LA (2010) A review of natural and synthetic antioxidants important for health and longevity. Curr Med Chem 17:3262–3288
12. Zhao HF, Li Q, Li Y (2011) Long-term ginsenoside administration prevents memory loss in aged female C57BL/6J mice by modulating the redox status and up-regulating the plasticity-related proteins in hippocampus. Neurosci 183:189–202
13. Fourquet J, Baez L, Figueroa M, Iriarte RI, Flores I (2011) Quantification of the impact of endometriosis symptoms on health-related quality of life and work productivity. Fertil Steril 96:107–112
14. de Ziegler D, Borghese B, Chapron C (2010) Endometriosis and infertility: pathophysiology and management. Lancet 376:730–738
15. Channabasavaiah AD, Joseph JV (2010) Thoracic endometriosis: revisiting the association between clinical presentation and thoracic pathology based on thoracoscopic findings in 110 patients. Med 89:183–188
16. Giannella L, La MA, Ternelli G, Menozzi G (2010) Rectus abdominis muscle endometriosis: case report and review of the literature. J Obstet Gynaecol Res 36:902–906
17. Leite GK, Carvalho LF, Korkes H, Guazzelli TF, Kenj G, Viana AT (2009) Scar endometrioma following obstetric surgical incisions: retrospective study on 33 cases and review of the literature. Sao Paulo Med J 127:270–277
18. Nissotakis C, Zouros E, Revelos K, Sakorafas GH (2010) Abdominal wall endometrioma: a case report and review of the literature. AORN J 91:730–742
19. Beckman EN, Pintado SO, Leonard GL, Sternberg WH (1985) Endometriosis of the prostate. Am J Surg Pathol 9:374–379
20. Giannarini G, Scott CA, Moro U, Grossetti B, Pomara G, Selli C (2006) Cystic endometriosis of the epididymis. Urology 68:203
21. Martin JD Jr, Hauck AE (1985) Endometriosis in the male. Am Surg 51:426–430
22. Oliker AJ, Harris AE (1971) Endometriosis of the bladder in a male patient. J Urol 106: 858–859
23. Ornstein MH, Kershaw DR (1985) Cysts of the seminal vesicle are Mullerian in origin. J R Soc Med 78:1050–1051
24. Pinkert TC, Catlow CE, Straus R (1979) Endometriosis of the urinary bladder in a man with prostatic carcinoma. Cancer 43:1562–1567
25. Schrodt GR, Alcorn MO, Ibanez J (1980) Endometriosis of the male urinary system: a case report. J Urol 124:722–723
26. Sampson J (1927) Peritoneal endometriosis due to menstrual dissemination of endometrial tissue into the peritoneal cavity. Am J Obstet Gynecol 14:422
27. Berube S, Marcoux S, Langevin M, Maheux R (1998) Fecundity of infertile women with minimal or mild endometriosis and women with unexplained infertility. The Canadian collaborative group on endometriosis. Fertil Steril 69:1034–1041
28. Olive DL, Henderson DY (1987) Endometriosis and mullerian anomalies. Obstet Gynecol 69:412–415

29. Sampson JA (1927) Metastatic or embolic endometriosis, due to the menstrual dissemination of endometrial tissue into the venous circulation. Am J Pathol 3:93–110
30. Laschke MW, Giebels C, Menger MD (2011) Vasculogenesis: a new piece of the endometriosis puzzle. Hum Reprod Update 17:628–636
31. Adachi M, Nasu K, Tsuno A, Yuge A, Kawano Y, Narahara H (2011) Attachment to extracellular matrices is enhanced in human endometriotic stromal cells: a possible mechanism underlying the pathogenesis of endometriosis. Eur J Obstet Gynecol Reprod Biol 155:85–88
32. Chrobak A, Gmyrek GB, Sozanski R, Sieradzka U, Paprocka M, Gabrys M, Jerzak M (2004) The influence of extracellular matrix proteins on T-cell proliferation and apoptosis in women with endometriosis or uterine leiomyoma. Am J Reprod Immunol 51:123–129
33. Gilmore SM, Aksel S, Hoff C, Peterson RD (1992) In vitro lymphocyte activity in women with endometriosis–an altered immune response? Fertil Steril 58:1148–1152
34. Oosterlynck DJ, Cornillie FJ, Waer M, Vandeputte M, Koninckx PR (1991) Women with endometriosis show a defect in natural killer activity resulting in a decreased cytotoxicity to autologous endometrium. Fertil Steril 56:45–51
35. Evers JL, Dunselman GA, Van der Linden PJ (1993) Markers for endometriosis. Baillieres Clin Obstet Gynaecol 7:715–739
36. Oosterlynck DJ, Meuleman C, Sobis H, Vandeputte M, Koninckx PR (1993) Angiogenic activity of peritoneal fluid from women with endometriosis. Fertil Steril 59:778–782
37. Bedaiwy MA, Falcone T, Sharma RK, Goldberg JM, Attaran M, Nelson DR, Agarwal A (2002) Prediction of endometriosis with serum and peritoneal fluid markers: a prospective controlled trial. Hum Reprod 17:426–431
38. Bedaiwy MA, Falcone T (2003) Peritoneal fluid environment in endometriosis. Clinicopathological implications. Minerva Ginecol 55:333–345
39. Tran LV, Tokushige N, Berbic M, Markham R, Fraser IS (2009) Macrophages and nerve fibres in peritoneal endometriosis. Hum Reprod 24:835–841
40. Na YJ, Lee DH, Kim SC, Joo JK, Wang JW, Jin JO, Kwak JY, Lee KS (2011) Effects of peritoneal fluid from endometriosis patients on the release of monocyte-specific chemokines by leukocytes. Arch Gynecol Obstet 283:1333–1341
41. Lousse JC, Van LA, Gonzalez-Ramos R, Defrere S, Renkin E, Donnez J (2008) Increased activation of nuclear factor-kappa B (NF-kappaB) in isolated peritoneal macrophages of patients with endometriosis. Fertil Steril 90:217–220
42. Na YJ, Jin JO, Lee MS, Song MG, Lee KS, Kwak JY (2008) Peritoneal fluid from endometriosis patients switches differentiation of monocytes from dendritic cells to macrophages. J Reprod Immunol 77:63–74
43. Sugamura K, Keaney JF Jr (2011) Reactive oxygen species in cardiovascular disease. Free Radic Biol Med 51:978–992
44. Patel RS, Al Mheid I, Morris AA, Ahmed Y, Kavtaradze N, Ali S, Dabhadkar K, Brigham K, Hooper WC, Alexander RW et al (2011) Oxidative stress is associated with impaired arterial elasticity. Atherosclerosis 218:90–95
45. Piechota A, Polanczyk A, Goraca A (2011) Protective effects of endothelin-A receptor antagonist BQ123 against LPS-induced oxidative stress in lungs. Pharmacol Rep 63:494–500
46. Whaley-Connell A, McCullough PA, Sowers JR (2011) The role of oxidative stress in the metabolic syndrome. Rev Cardiovasc Med 12:21–29
47. Kotliarov AA, Mosina LM, Chibisov SM, Selezneva NM, Shmyreva MV, Efremova EN, Surotkina SA (2011) Efficacy of nebivolol in patients with coronary heart disease and concomitant chronic obstructive pulmonary disease. Klin Med (Mosk) 89:44–48
48. Osborn BH, Haney AF, Misukonis MA, Weinberg JB (2002) Inducible nitric oxide synthase expression by peritoneal macrophages in endometriosis-associated infertility. Fertil Steril 77:46–51
49. Weinberg JB (1998) Nitric oxide production and nitric oxide synthase type 2 expression by human mononuclear phagocytes: a review. Mol Med 4:557–591
50. Murphy AA, Palinski W, Rankin S, Morales AJ, Parthasarathy S (1998) Macrophage scavenger receptor(s) and oxidatively modified proteins in endometriosis. Fertil Steril 69:1085–1091

51. Polak G, Koziol-Montewka M, Gogacz M, Blaszkowska I, Kotarski J (2001) Total antioxidant status of peritoneal fluid in infertile women. Eur J Obstet Gynecol Reprod Biol 94:261–263
52. Szczepanska M, Kozlik J, Skrzypczak J, Mikolajczyk M (2003) Oxidative stress may be a piece in the endometriosis puzzle. Fertil Steril 79:1288–1293
53. Vinatier D, Cosson M, Dufour P (2000) Is endometriosis an endometrial disease? Eur J Obstet Gynecol Reprod Biol 91:113–125
54. Van LA, Casanas-Roux F, Donnez J (2002) Iron overload in the peritoneal cavity of women with pelvic endometriosis. Fertil Steril 78:712–718
55. Kobayashi H, Yamada Y, Kanayama S, Furukawa N, Noguchi T, Haruta S, Yoshida S, Sakata M, Sado T, Oi H (2009) The role of iron in the pathogenesis of endometriosis. Gynecol Endocrinol 25:39–52
56. Arumugam K, Yip YC (1995) De novo formation of adhesions in endometriosis: the role of iron and free radical reactions. Fertil Steril 64:62–64
57. Gupta S, Agarwal A, Krajcir N, Alvarez JG (2006) Role of oxidative stress in endometriosis. Reprod Biomed Online 13:126–134
58. Park JK, Song M, Dominguez CE, Walter MF, Santanam N, Parthasarathy S, Murphy AA (2006) Glycodelin mediates the increase in vascular endothelial growth factor in response to oxidative stress in the endometrium. Am J Obstet Gynecol 195:1772–1777
59. Scholl B, Bersinger NA, Kuhn A, Mueller MD (2009) Correlation between symptoms of pain and peritoneal fluid inflammatory cytokine concentrations in endometriosis. Gynecol Endocrinol 25:701–706
60. Xu H, Becker CM, Lui WT, Chu CY, Davis TN, Kung AL, Birsner AE, D'Amato RJ, Wai Man GC, Wang CC (2011) Green tea epigallocatechin-3-gallate inhibits angiogenesis and suppresses vascular endothelial growth factor C/vascular endothelial growth factor receptor 2 expression and signaling in experimental endometriosis in vivo. Fertil Steril 96:1021–1028
61. Laschke MW, van Vorsterman Oijen AE, Scheuer C, Menger MD (2011) In vitro and in vivo evaluation of the anti-angiogenic actions of 4-hydroxybenzyl alcohol. Br J Pharmacol 163:835–844
62. Imesch P, Samartzis EP, Schneider M, Fink D, Fedier A (2011) Inhibition of transcription, expression, and secretion of the vascular epithelial growth factor in human epithelial endometriotic cells by romidepsin. Fertil Steril 95:1579–1583
63. Numao A, Hosono K, Suzuki T, Hayashi I, Uematsu S, Akira S, Ogino Y, Kawauchi H, Unno N, Majima M (2011) The inducible prostaglandin E synthase mPGES-1 regulates growth of endometrial tissues and angiogenesis in a mouse implantation model. Biomed Pharmacother 65:77–84
64. Katayama H, Katayama T, Uematsu K, Hiratsuka M, Kiyomura M, Shimizu Y, Sugita A, Ito M (2010) Effect of dienogest administration on angiogenesis and hemodynamics in a rat endometrial autograft model. Hum Reprod 25:2851–2858
65. Alpay Z, Saed GM, Diamond MP (2006) Female infertility and free radicals: potential role in adhesions and endometriosis. J Soc Gynecol Investig 13:390–398
66. Luo Q, Chen XJ, Ding GL, Dong MY, Huang HF (2010) Downregulative effects of nitric oxide on oocyte fertilization and embryo development: possible roles of nitric oxide in the pathogenesis of endometriosis-associated infertility. Cell Physiol Biochem 26:1023–1028
67. Wu MY, Chao KH, Yang JH, Lee TH, Yang YS, Ho HN (2003) Nitric oxide synthesis is increased in the endometrial tissue of women with endometriosis. Hum Reprod 18:2668–2671
68. Khorram O, Lessey BA (2002) Alterations in expression of endometrial endothelial nitric oxide synthase and alpha(v)beta(3) integrin in women with endometriosis. Fertil Steril 78:860–864
69. Khorram O (2002) Nitric oxide and its role in blastocyst implantation. Rev Endocr Metab Disord 3:145–149
70. El-Far M, El-Sayed IH, El-Motwally A, Hashem IA, Bakry N (2007) Tumor necrosis factor-alpha and oxidant status are essential participating factors in unexplained recurrent spontaneous abortions. Clin Chem Lab Med 45:879–883

71. Said TM, Agarwal A, Falcone T, Sharma RK, Bedaiwy MA, Li L (2005) Infliximab may reverse the toxic effects induced by tumor necrosis factor alpha in human spermatozoa: an in vitro model. Fertil Steril 83.1665–1673
72. Ota H, Igarashi S, Hatazawa J, Tanaka T (1999) Immunohistochemical assessment of superoxide dismutase expression in the endometrium in endometriosis and adenomyosis. Fertil Steril 72:129–134
73. Ota H, Igarashi S, Kato N, Tanaka T (2000) Aberrant expression of glutathione peroxidase in eutopic and ectopic endometrium in endometriosis and adenomyosis. Fertil Steril 74:313–318
74. Ota H, Igarashi S, Tanaka T (2001) Xanthine oxidase in eutopic and ectopic endometrium in endometriosis and adenomyosis. Fertil Steril 75:785–790
75. Beck DE, Cohen Z, Fleshman JW, Kaufman HS, van Goor H, Wolff BG (2003) A prospective, randomized, multicenter, controlled study of the safety of seprafilm adhesion barrier in abdominopelvic surgery of the intestine. Dis Colon Rectum 46:1310–1319
76. Menzies D, Ellis H (1990) Intestinal obstruction from adhesions–how big is the problem? Ann R Coll Surg Engl 72:60–63
77. Milingos S, Kallipolitis G, Loutradis D, Liapi A, Mavrommatis K, Drakakis P, Tourikis J, Creatsas G, Michalas S (2000) Adhesions: laparoscopic surgery versus laparotomy. Ann N Y Acad Sci 900:272–285
78. Vrijland WW, Jeekel J, van Geldorp HJ, Swank DJ, Bonjer HJ (2003) Abdominal adhesions: intestinal obstruction, pain, and infertility. Surg Endosc 17:1017–1022
79. Klingensmith ME, Soybel DI, Brooks DC (1996) Laparoscopy for chronic abdominal pain. Surg Endosc 10:1085–1087
80. Kresch AJ, Seifer DB, Sachs LB, Barrese I (1984) Laparoscopy in 100 women with chronic pelvic pain. Obstet Gynecol 64:672–674
81. Malik E, Berg C, Meyhofer-Malik A, Haider S, Rossmanith WG (2000) Subjective evaluation of the therapeutic value of laparoscopic adhesiolysis: a retrospective analysis. Surg Endosc 14:79–81
82. Rapkin AJ (1986) Adhesions and pelvic pain: a retrospective study. Obstet Gynecol 68:13–15
83. Coleman MG, McLain AD, Moran BJ (2000) Impact of previous surgery on time taken for incision and division of adhesions during laparotomy. Dis Colon Rectum 43:1297–1299
84. Van Der Krabben AA, Dijkstra FR, Nieuwenhuijzen M, Reijnen MM, Schaapveld M, van Goor H (2000) Morbidity and mortality of inadvertent enterotomy during adhesiotomy. Br J Surg 87:467–471
85. Parker MC, Ellis H, Moran BJ, Thompson JN, Wilson MS, Menzies D, McGuire A, Lower AM, Hawthorn RJ, O'Briena F et al (2001) Postoperative adhesions: ten-year follow-up of 12,584 patients undergoing lower abdominal surgery. Dis Colon Rectum 44:822–829
86. Luijendijk RW, de Lange DC, Wauters CC, Hop WC, Duron JJ, Pailler JL, Camprodon BR, Holmdahl L, van Geldorp HJ, Jeekel J et al (1996) Foreign material in postoperative adhesions. Ann Surg 223:242–248
87. Parker MC, Wilson MS, Menzies D, Sunderland G, Clark DN, Knight AD, Crowe AM (2005) The SCAR-3 study: 5 year adhesion-related readmission risk following lower abdominal surgical procedures. Colorectal Dis 7:551–558
88. Stewart RM, Page CP, Brender J, Schwesinger W, Eisenhut D (1987) The incidence and risk of early postoperative small bowel obstruction. A cohort study. Am J Surg 154:643–647
89. Tortella BJ, Lavery RF, Chandrakantan A, Medina D (1995) Incidence and risk factors for early small bowel obstruction after celiotomy for penetrating abdominal trauma. Am Surg 61:956–958
90. Gutt CN, Oniu T, Schemmer P, Mehrabi A, Buchler MW (2004) Fewer adhesions induced by laparoscopic surgery? Surg Endosc 18:898–906
91. Saed GM, Diamond MP (2003) Modulation of the expression of tissue plasminogen activator and its inhibitor by hypoxia in human peritoneal and adhesion fibroblasts. Fertil Steril 79:164–168

92. Zorio E, Gilabert-Estelles J, Espana F, Ramon LA, Cosin R, Estelles A (2008) Fibrinolysis: the key to new pathogenetic mechanisms. Curr Med Chem 15:923–929
93. Brokelman WJ, Holmdahl L, Janssen IM, Falk P, Bergstrom M, Klinkenbijl JH, Reijnen MM (2009) Decreased peritoneal tissue plasminogen activator during prolonged laparoscopic surgery. J Surg Res 151:89–93
94. Holmdahl L, Eriksson E, Eriksson BI, Risberg B (1998) Depression of peritoneal fibrinolysis during operation is a local response to trauma. Surgery 123:539–544
95. Hellebrekers BW, Trimbos-Kemper TC, Boesten L, Jansen FW, Kolkman W, Trimbos JB, Press RR, van Poelgeest MI, Emeis SJ, Kooistra T (2009) Preoperative predictors of postsurgical adhesion formation and the prevention of adhesions with plasminogen activator (PAPA-study): results of a clinical pilot study. Fertil Steril 91:1204–1214
96. Aarons CB, Cohen PA, Gower A, Reed KL, Leeman SE, Stucchi AF, Becker JM (2007) Statins (HMG-CoA reductase inhibitors) decrease postoperative adhesions by increasing peritoneal fibrinolytic activity. Ann Surg 245:176–184
97. Gago LA, Saed GM, Chauhan S, Elhammady EF, Diamond MP (2003) Seprafilm (modified hyaluronic acid and carboxymethylcellulose) acts as a physical barrier. Fertil Steril 80: 612–616
98. Matsuzaki S, Botchorishvili R, Jardon K, Maleysson E, Canis M, Mage G (2011) Impact of intraperitoneal pressure and duration of surgery on levels of tissue plasminogen activator and plasminogen activator inhibitor-1 mRNA in peritoneal tissues during laparoscopic surgery. Hum Reprod 26:1073–1081
99. Prushik SG, Stucchi AF, Matteotti R, Aarons CB, Reed KL, Gower AC, Becker JM (2010) Open adhesiolysis is more effective in reducing adhesion reformation than laparoscopic adhesiolysis in an experimental model. Br J Surg 97:420–427
100. Tarhan OR, Barut I, Sutcu R, Akdeniz Y, Akturk O (2006) Pentoxifylline, a methyl xanthine derivative, reduces peritoneal adhesions and increases peritoneal fibrinolysis in rats. Tohoku J Exp Med 209:249–255
101. Chegini N, Kotseos K, Bennett B, Diamond MP, Holmdahl L, Burns J (2001) Matrix metalloproteinase (MMP-1) and tissue inhibitor of MMP in peritoneal fluids and sera and correlation with peritoneal adhesions. Fertil Steril 76:1207–1211
102. Chegini N, Kotseos K, Zhao Y, Ma C, McLean F, Diamond MP, Holmdahl L, Burns J (2001) Expression of matrix metalloproteinase (MMP-1) and tissue inhibitor of MMP in serosal tissue of intraperitoneal organs and adhesions. Fertil Steril 76:1212–1219
103. Corona R, Verguts J, Binda MM, Molinas CR, Schonman R, Koninckx PR (2011) The impact of the learning curve on adhesion formation in a laparoscopic mouse model. Fertil Steril 96:193–197
104. Freeman ML, Saed GM, Elhammady EF, Diamond MP (2003) Expression of transforming growth factor beta isoform mRNA in injured peritoneum that healed with adhesions and without adhesions and in uninjured peritoneum. Fertil Steril 80(Suppl 2):708–713
105. Idell S, Zwieb C, Boggaram J, Holiday D, Johnson AR, Raghu G (1992) Mechanisms of fibrin formation and lysis by human lung fibroblasts: influence of TGF-beta and TNF-alpha. Am J Physiol 263:L487–L494
106. Amer MI, bd-El-Maeboud KH, Abdelfatah I, Salama FA, Abdallah AS (2010) Human amnion as a temporary biologic barrier after hysteroscopic lysis of severe intrauterine adhesions: pilot study. J Minim Invasive Gynecol 17:605–611
107. Petter-Puchner AH, Fortelny RH, Mika K, Hennerbichler S, Redl H, Gabriel C (2011) Human vital amniotic membrane reduces adhesions in experimental intraperitoneal onlay mesh repair. Surg Endosc 25:2125–2131
108. Alpay Z, Ozgonenel MS, Savasan S, Buck S, Saed GM, Diamond MP (2006) Possible role of natural immune response against altered fibroblasts in the development of post-operative adhesions. Am J Reprod Immunol 55:420–427
109. Alpay Z, Ozgonenel M, Savasan S, Buck S, Saed GM, Diamond MP (2007) Altered in vitro immune response to hypoxia-treated normal peritoneal fibroblasts. Fertil Steril 87:426–429

110. Saed GM, Zhang W, Diamond MP (2001) Molecular characterization of fibroblasts isolated from human peritoneum and adhesions. Fertil Steril 75:763–768
111. Saed GM, Diamond MP (2002) Apoptosis and proliferation of human peritoneal fibroblasts in response to hypoxia. Fertil Steril 78:137–143
112. Saed GM, Diamond MP (2004) Molecular characterization of postoperative adhesions: the adhesion phenotype. J Am Assoc Gynecol Laparosc 11:307–314
113. Alpay Z SSBSZJRYDMSG. (2008) Role of natural killer lymphocyte NKG2D receptor pathway in adhesion development. Reprod Sci 15(Suppl):150a
114. Kosaka H, Yoshimoto T, Yoshimoto T, Fujimoto J, Nakanishi K (2008) Interferon-gamma is a therapeutic target molecule for prevention of postoperative adhesion formation. Nat Med 14:437–441
115. de tes Souza AM, Rogers, Wang CC, Yuen PM, Ng PS (2003) Comparison of peritoneal oxidative stress during laparoscopy and laparotomy. J Am Assoc Gynecol Laparosc 10:65–74
116. Tsimoyiannis EC, Tsimoyiannis JC, Sarros CJ, Akalestos GC, Moutesidou KJ, Lekkas ET, Kotoulas OB (1989) The role of oxygen-derived free radicals in peritoneal adhesion formation induced by ileal ischaemia/reperfusion. Acta Chir Scand 155:171–174
117. Binda MM, Molinas CR, Koninckx PR (2003) Reactive oxygen species and adhesion formation: clinical implications in adhesion prevention. Hum Reprod 18:2503–2507
118. Hazebroek EJ, Schreve MA, Visser P, De Bruin RW, Marquet RL, Bonjer HJ (2002) Impact of temperature and humidity of carbon dioxide pneumoperitoneum on body temperature and peritoneal morphology. J Laparoendosc Adv Surg Tech A 12:355–364
119. Ott DE (2001) Laparoscopy and tribology: the effect of laparoscopic gas on peritoneal fluid. J Am Assoc Gynecol Laparosc 8:117–123
120. Taskin O, Buhur A, Birincioglu M, Burak F, Atmaca R, Yilmaz I, Wheeler JM (1998) The effects of duration of CO_2 insufflation and irrigation on peritoneal microcirculation assessed by free radical scavengers and total glutathion levels during operative laparoscopy. J Am Assoc Gynecol Laparosc 5:129–133
121. Taskin O, Sadik S, Onoglu A, Gokdeniz R, Yilmaz I, Burak F, Wheeler JM (1999) Adhesion formation after microlaparoscopic and laparoscopic ovarian coagulation for polycystic ovary disease. J Am Assoc Gynecol Laparosc 6:159–163
122. tes de Souza AM, Wang CC, Chu CY, Briton-Jones CM, Haines CJ, Rogers MS (2004) In vitro exposure to carbon dioxide induces oxidative stress in human peritoneal mesothelial cells. Hum Reprod 19:1281–1286
123. Jesch NK, Kuebler JF, Nguyen H, Nave H, Bottlaender M, Teichmann B, Braun A, Vieten G, Ure BM (2006) Laparoscopy vs minilaparotomy and full laparotomy preserves circulatory but not peritoneal and pulmonary immune responses. J Pediatr Surg 41:1085–1092
124. Molinas CR, Mynbaev O, Pauwels A, Novak P, Koninckx PR (2001) Peritoneal mesothelial hypoxia during pneumoperitoneum is a cofactor in adhesion formation in a laparoscopic mouse model. Fertil Steril 76:560–567
125. Molinas CR, Tjwa M, Vanacker B, Binda MM, Elkelani O, Koninckx PR (2004) Role of $CO(2)$ pneumoperitoneum-induced acidosis in $CO(2)$ pneumoperitoneum-enhanced adhesion formation in mice. Fertil Steril 81:708–711
126. Shukla A, Rasik AM, Shankar R (1999) Nitric oxide inhibits wounds collagen synthesis. Mol Cell Biochem 200:27–33
127. Dambisya YM, Lee TL (1996) A thromboelastography study on the in vitro effects of L-arginine and L-NG-nitro arginine methyl ester on human whole blood coagulation and fibrinolysis. Blood Coagul Fibrinolysis 7:678–683
128. Muriel P (1998) Nitric oxide protection of rat liver from lipid peroxidation, collagen accumulation, and liver damage induced by carbon tetrachloride. Biochem Pharmacol 56:773–779
129. Criado M, Flores O, Vazquez MJ, Esteller A (2000) Role of prostanoids and nitric oxide inhibition in rats with experimental hepatic fibrosis. J Physiol Biochem 56:181–188
130. Muscara MN, McKnight W, Asfaha S, Wallace JL (2000) Wound collagen deposition in rats: effects of an NO-NSAID and a selective COX-2 inhibitor. Br J Pharmacol 129: 681–686

131. Saed GM, bu-Soud HM, Diamond MP (2004) Role of nitric oxide in apoptosis of human peritoneal and adhesion fibroblasts after hypoxia. Fertil Steril 82(Suppl 3):1198–1205
132. Saed GM, Zhang W, Chegini N, Holmdahl L, Diamond MP (1999) Alteration of type I and III collagen expression in human peritoneal mesothelial cells in response to hypoxia and transforming growth factor-beta1. Wound Repair Regen 7:504–510
133. Saed GM, Zhang W, Diamond MP (2000) Effect of hypoxia on stimulatory effect of TGF-beta 1 on MMP-2 and MMP-9 activities in mouse fibroblasts. J Soc Gynecol Investig 7: 348–354
134. Witte MB, Barbul A (2002) Role of nitric oxide in wound repair. Am J Surg 183:406–412
135. Reed KL, Heydrick SJ, Aarons CB, Prushik S, Gower AC, Stucchi AF, Becker JM (2007) A neurokinin-1 receptor antagonist that reduces intra-abdominal adhesion formation decreases oxidative stress in the peritoneum. Am J Physiol Gastrointest Liver Physiol 293:G544–G551
136. Heydrick SJ, Reed KL, Cohen PA, Aarons CB, Gower AC, Becker JM, Stucchi AF (2007) Intraperitoneal administration of methylene blue attenuates oxidative stress, increases peritoneal fibrinolysis, and inhibits intraabdominal adhesion formation. J Surg Res 143:311–319
137. Ara C, Kirimlioglu H, Karabulut AB, Coban S, Hascalik S, Celik O, Yilmaz S, Kirimlioglu V (2005) Protective effect of melatonin against oxidative stress on adhesion formation in the rat cecum and uterine horn model. Life Sci 77:1341–1350
138. Hemadeh O, Chilukuri S, Bonet V, Hussein S, Chaudry IH (1993) Prevention of peritoneal adhesions by administration of sodium carboxymethyl cellulose and oral vitamin E. Surgery 114:907–910
139. Sanfilippo JS, Booth RJ, Burns CD (1995) Effect of vitamin E on adhesion formation. J Reprod Med 40:278–282
140. ten RS, van Den Tol MP, Sluiter W, Hofland LJ, van Eijck CH, Jeekel H (2006) The role of neutrophils and oxygen free radicals in post-operative adhesions. J Surg Res 136:45–52
141. Alpay Z, Saed GM, Diamond MP (2008) Postoperative adhesions: from formation to prevention. Semin Reprod Med 26:313–321

Chapter 17
Impact of Life Style Factors on Oxidative Stress

Peter T. Campbell

Abstract There are convincing epidemiologic data to support the role of increased physical activity and exercise with reduced risks of various chronic diseases, including cardiovascular disease, certain cancers, type 2 diabetes mellitus (T2DM), and osteoporosis. Reversing global population trends in physical inactivity is therefore a major public health priority. Better understandings of the biologic mechanisms that link sedentary lifestyle to poor health are needed to refine and support physical activity guidelines. Traditionally, the underlying mechanisms that have been proposed to link physical inactivity to various disease outcomes have been disease specific. For example, reduced circulating estrogens are often cited as one mechanism that links physical activity to reduced risk of breast cancer; in parallel, reduced insulin and insulin like growth factors are regularly offered to explain the link between physical activity and colon cancer. More recently, data are emerging to suggest that persistent physical activity may decrease resting levels of oxidative stress. At first, this connection may seem paradoxical since acute exercise transiently increases local and systemic oxidative stress levels. This paradox is resolved by the concept of hormesis: low-to-moderate level of reactant species that form during nonexhaustive exercise may lead to adaptations in the antioxidant defense system. Whether reduced resting levels of oxidative stress further translate to reduced risks of chronic disease is another emerging research question. While some studies support a connection between oxidative stress and chronic disease, prospective and well-powered studies are still relatively rare.

Keywords Physical activity · Oxidative stress · Exercise · Fitness · Risk of chronic disease

---

P. T. Campbell (✉)
Epidemiology Research Program, American Cancer Society,
250 Williams Street, NW Atlanta, GA 30303, USA
e-mail: peter.campbell@cancer.org

17.1 Introduction

Physical inactivity is the fourth leading cause of mortality worldwide, accounting for 6 % of all deaths [1]. Physical inactivity is associated with increased risks of: atherosclerotic cardiovascular diseases (CVD), including coronary artery disease, stroke, hypertension, angina, and heart attack; cancers of the breast, colon and endometrium; metabolic diseases, including type 2 diabetes mellitus (T2DM) and obesity; and poor bone health outcomes, such as osteoporosis and fracture. Increasing population levels of physical activity is therefore a major public health priority [2–7]. As part of the evidence base to help establish and support public health guidelines for physical activity, better understanding is needed for the mechanistic links between physical activity and reduced risks of mortality and morbidity. One potential link, in this regard, is oxidative stress: physical activity may lead to physiological adaptations that decrease resting levels of oxidative stress and subsequently decrease risk of illness, chronic disease, and death.

Oxidative stress occurs when the production of reactive species, derived largely from oxygen and nitrogen, exceeds degradation by the antioxidant defense and specialized repair systems. The ensuing damage to DNA, protein, and lipid has been implicated in a wide variety of diseases and conditions, including many of the same diseases and conditions associated with physical inactivity. Until recently, measurement of oxidative stress in humans was hindered by a lack of sensitive, specific, and reliable methods to assess oxidative stress in vivo: this limitation has been overcome in recent years by the discovery and refined measurement, largely via mass spectrometric-based technologies, of F_2-isoprostanes. F_2-isoprostanes are a family of isomeric F_2-prostaglandin–like compounds, derived from free radical–catalyzed peroxidation of arachidonic acid. This review will focus mostly on F_2-isoprostanes, a specific marker of lipid peroxidation and a general marker of oxidative stress, because of findings from a multi-institutional National Institutes of Health-sponsored validation study that reported that these compounds were the gold standard measurement of in vivo oxidative stress [8]. Where the literature on F_2-isoprostanes and certain outcomes is sparse or nonexistent, other markers that correlate with F_2-isoprostanes will be discussed, as needed.

Throughout this chapter, the focus on oxidative stress will take an epidemiologic, or public health, and *biomarker* perspective: A biomarker is defined as an objectively measured biologic characteristic that indicates a normal biological process, a pathologic process, or a response to a physiologic or pharmacologic intervention [9, 10]. In this respect, a biomarker of oxidative stress may be defined as a molecule whose structure can be modified by reactant species [10, 11]. In the context of physical activity, oxidative stress biomarkers can serve as indicators of acute exposure to exercise or they can serve as biomarkers of training adaptation to long-term exercise programs or to physically active lifestyles. Similarly, with respect to chronic disease, oxidative stress biomarkers may be often used as predictors of future disease risk, although data of this sort are still quite rare with

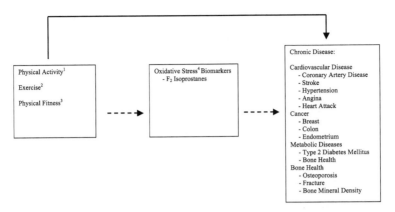

Fig. 17.1 Conceptual framework for the relationship of physical activity, oxidative stress, and chronic disease. *1* Physical activity: all leisure and nonleisure body movements resulting in an increased energy output from the resting condition [12]. *2* Exercise: structured and repetitive physical activity designed to maintain or improve physical fitness [12]. *3* Physical fitness: a physiological state of well-being that allows one to meet the demands of daily living or that provides the basis for sport performance, or both [12]. *4* Oxidative stress has several definitions including: (a) a disturbance in the prooxidant/antioxidant balance in favor of the former which may lead to tissue damage [13], (b) a serious imbalance between prooxidant reactive oxygen species (*ROS*), reactive nitrogen species (*RNS*), reactive chorine species (*RCS*) and antioxidant defenses [14], (c) an imbalance in pro-oxidants and antioxidants, which results in macromolecular damage and disruption of redox signaling and control [15]. *Dashed line* indicates emerging evidence. *Solid line* indicates convincing associations

respect to F_2-isoprostanes. Ultimately, two major goals of oxidative stress biomarker research as it relates to public health are to aid in the identification of persons at increased risk for morbidity/mortality and to identify safe and effective intervention methods to reduce oxidative stress levels, including effective lifestyle modification, perhaps through enhanced physical activity.

In this chapter, the interconnectedness of physical activity, oxidative stress and chronic disease will be reviewed, primarily as they relate to initially asymptomatic—or apparently healthy—individuals. The first part of this review gives an overview, largely from seminal papers, meta analyses and expert consensus panel reviews, concerning the associations of physical activity, exercise, and physical fitness with risk of CVD, certain cancers, T2DM, and bone health; in the second section, the effects of exercise on oxidative stress will be reviewed; and in the third section, the associations of oxidative stress with risk of CVD, certain cancers, T2DM, and bone health will be reviewed, where relevant studies are available. This chapter will conclude with a summary and highlights of some of the key issues surrounding physical activity, oxidative stress, and chronic disease. The underlying hypothesis to be pursued in this chapter is that oxidative stress, as measured by the gold standard of F_2-isoprostanes, is one of the links between physical inactivity and poor health outcomes (Fig. 17.1).

17.1.1 Physical Activity, Exercise, Fitness, and Risk of Cardiovascular Disease

The seminal work of Morris and colleagues in the early-1950s provided the first evidence that occupational physical activity was associated with reduced risk of coronary heart disease and all-cause mortality [16–18]. Morris and his colleagues began to track heart disease rates among male transport workers in London, England in 1949 after observing that bus, tram and trolley drivers sat for about 90% of their working hours while conductors climbed 500-to-750 stairs per shift. Their first data indicated that the sedentary drivers experienced higher incidence rates of fatal and nonfatal coronary heart disease [18]. These observations were replicated within the same publication by comparing active postal carriers to their less active supervisors and telephone operators [18].

The majority of studies on occupational physical activity, including seminal studies of longshoremen in the United States [19, 20], have agreed with these initial findings that men who engage in the most physically demanding occupations experience about one-half the risk of CVD and premature death than do their sedentary peers (for detailed review, see [21]).

Industrialization has rendered occupational physical activity increasingly rarer and more difficult to study. Additionally, there is a potential concern of self-selection, and the related *healthy worker effect*, with occupational physical activity studies. Workers who choose to engage in physically demanding occupations may be in better overall health at the onset of their jobs than their sedentary colleagues and the population from which they were drawn in general.

To counteract the inherent limitations in studies of occupational physical activity, leisure time physical activity is now most-often studied. Morris and colleagues in 1973 were again among the first to report on this topic [22]. In a prospective study of 16,882 male civil servants who would have been largely sedentary at work, the authors reported that men who engaged in regular vigorous leisure time activity ('swimming' and 'getting about quickly' were offered as examples) compared with men who were sedentary during leisure time had about 1/3 the risk of developing coronary artery disease [22]. At about the same time in the United States, Paffenbarger and colleagues began to publish results from the College Alumni Study. In a 1978 study, Paffenbarger and co-authors quantified physical activity as low (less than 2000 kcal/week) or high (more than 2000 kcal/week): men in the low activity group were at 64% higher risk of a first heart attack than were men in the more active group [23]. These findings were among the first to empirically support the health benefits of lifelong physical activity.

While early studies were remarkably consistent in their findings of increased physical activity and reduced risk of CVD, one limitation to this evidence base was a lack of studies on direct physical *fitness* measures. Physical activity and fitness are clearly correlated; however, one might expect aerobic fitness to be a more accurate indicator of disease risk since it is objectively measured. In perhaps the first study to address this gap, in an early study of railroad workers in the United States, the least-fit quartile of men relative to the most-fit quartile of men, evaluated by heart rate response to submaximal exercise, was about twice as likely to develop CVD during follow-up [24, 25].

Even more compelling results have been reported from the aerobics center longitudinal study (ACLS) where physical fitness was measured among men and women by a gold standard maximal treadmill exercise test [26]. The least-fit compared with the most-fit quartiles of men and women were at 3.4-fold and 4.7-fold, respectively, increased risks of death from any cause, mostly from CVD and cancer. More recent work from the ACLS has better defined the dose–response relationship between fitness and mortality: compared to the least-fit quintile, the second least-fit quintile had about $1/2$ the risk of death and the most-fit quintile had about $1/4$ the risk of death during follow-up [27]. Perhaps one of the largest advantages to a greater number of participants with direct fitness measures, in this context, is that it allows for examination of multiple rarer outcomes, such as specific types of CVD or cancer.

The relevance of musculoskeletal health was examined in a Canadian study of 8,116 adults with 13 years of follow-up [28]. In this study, musculoskeletal health was defined from sit ups, push ups, grip strength, and trunk flexibility. After adjustment for other relevant variables, only sit ups were predictive of future mortality in men and women. Relative to the immense amount of research on cardiovascular fitness and physical activity, the importance of musculoskeletal health on CVD and all-cause mortality is probably underappreciated.

In recent years, meta analyses of the links between physical activity and chronic disease have flourished as a means to systematically and quantitatively summarize research findings. As an example, a recent meta-analysis of walking and coronary artery disease identified 12 prospective studies with a total of 295,177 participants and 7,094 coronary artery disease events [29]. The results from this study indicated that 30 min of walking per day for 5 days per week was associated with a 19 % reduced risk of coronary artery disease, with evidence of a dose–response relationship. Similarly, another large meta-analysis of total leisure time physical activity, based on 26 studies, reported that moderate and high levels of leisure time physical activity were associated with 12 and 27 %, respectively, reduced risks of coronary artery disease [30].

Risk of stroke, too, has been assessed in meta analyses with physical activity [31, 32]. In a recent study, Wendel-Vos and colleagues [32] reported that moderate intensity occupational and leisure time physical activity, relative to being sedentary, reduced risks of total stroke by 36 and 15 %, respectively. Other meta analyses have summarized the effects of walking [33] and aerobic exercise [34] interventions on blood pressure from randomized, controlled trials. Both meta analyses [33, 34] reported beneficial effects of exercise intervention on resting blood pressure; a study by Whelton and colleagues reported a reduction in mean systolic and diastolic blood pressures of 3.84 and 2.58 mm Hg, respectively, for exercisers compared to nonexercisers [34]. In summary, there is incontrovertible evidence that physical activity, exercise, and aerobic fitness are associated with reduced risks of CVD, including coronary heart disease and heart attack, stroke, and hypertension.

17.1.2 Physical Activity, Fitness and Risk of Colorectal, Breast, and Endometrial Cancers

Physical activity is associated with reduced risk of cancer at several organ sites, particularly the colon, breast, and endometrium [35–38]. A 2007 expert report concluded that the evidence for a reduction in colon cancer risk with increasing physical activity was 'convincing' while the comparable level of evidence for cancers of the breast (post-menopausal) and endometrium were considered 'probable' [38].

Studies show a consistent reduction in risk of colon, but not rectal, cancer with increasing levels of physical activity [35, 39]. Even more compellingly, these associations have been observed across diverse populations and with many different measurements of physical activity. A 2005 meta-analysis of 19 cohort and 30 case-control studies summarized the associations of physical inactivity with risks of colon and rectal cancers [39]: among men, the summary estimates from cohort studies suggested 21% (occupational) and 22% (leisure time) reductions in colon cancer risk for high relative to low physical activity levels. Among women, the summary risk estimate from cohort studies suggested a 29% risk reduction for high versus low leisure time physical activity. There were no associations between physical activity and rectal cancer risk [39]. A more recent meta-analysis confirmed these findings and restricted analysis to leisure time physical activity only [35]: these authors reported summary relative risks from 14 articles, including data from 7,873 incident cases, of 0.80 for men and 0.86 for women when comparing highest versus lowest levels of leisure time physical activity and risk of colon cancer. Much like the previous meta-analysis, leisure time physical activity was not associated with risk of rectal cancer. Slattery and colleagues estimated that 14% of colon cancer may be caused by physical inactivity alone [40].

Studies of the association between physical activity and breast cancer risk have been fairly consistent and are suggestive of an overall inverse association, although the results are not as clear as the results for colon cancer. This ambiguity is partially clarified when results are stratified by menopausal status [38]. There is little empirical support that physical activity may protect against risk of premenopausal breast cancer, although many studies have been conducted; however, there is some evidence for a dose–response association between physical activity and reduced risk of postmenopausal breast cancer, albeit with a few inconsistencies [38]. A recent meta-analysis attempted to summarize data on this potential association, but because of considerable statistical and methodological differences among studies the authors did not summarize the results using traditional meta analytic techniques [41]. Instead, the authors took into consideration study quality and conducted a 'best-evidence synthesis' of 19 cohort and 29 case-control studies: they reported an inverse association between physical activity and postmenopausal breast cancer, ranging from 20 to 80 %, and they also concluded that there was limited evidence for an association with premenopausal breast cancer. A more recent report from the Nurses' Health Study, with 20 years of follow-up and 4,782 breast cancer cases, confirmed that increased physical activity, including brisk walking, was associated with reduced risk of postmenopausal breast cancer [42]. To date, the available evidence suggests that increased physical activity is associated with reduced risk of postmenopausal, and not premenopausal, breast cancer. There is little empirical evidence to explain this discordant association by menopausal status, although the relative contribution of adipose tissue to sex steroid synthesis in the postmenopausal period (high contribution) compared to the premenopausal period (low contribution) has been suggested as one potential explanation.

Endometrial cancer starts in the inner lining of the uterus; these malignancies are less common than colorectal or breast cancers [43] and they generally have a good prognosis. The 'probable' level of evidence identified by a 2007 expert report for the link between leisure time physical activity and endometrial cancer was largely based on case-control studies; in this expert report, high versus low physical activity was associated with an average 35% reduced risk of endometrial cancer [38]. Limitations to case-control studies include the potential for recall and survival biases; thus, prospective data are preferred when available. Indeed, more recent work from a meta-analysis of five prospective cohort studies, which included data from 2,663 endometrial cancer cases, found that more active relative to less active women experienced about a 30% lower risk of endometrial cancer [37].

Studies on physical fitness and primary cancer prevention are rare. While some studies have shown results between physical fitness and an 'all-cancer' outcome, these types of studies are less favorable because of the broad disease heterogeneity across various organ systems. Studies from the ACLS are an exception [44-47]. In a comparison of the highest 20th percentile of cardiorespiratory fitness to the lowest 20th percentile among men in the ACLS, risk of colon and rectal cancer mortality was reduced by 39% and 52%, respectively, although the latter estimate was not statistically significant [47]. Breast cancer mortality among women in the ACLS has been examined, too: the highest 40th percentile of cardiorespiratory fitness, relative to the lowest 20th percentile, experienced a 65% reduced risk of dying from breast cancer [46]. In both studies, these associations were independent of many potential confounders. There is a paucity of data on physical fitness and risk of the less common cancer sites, including the endometrium.

17.1.3 Physical Activity, Physical Fitness, and Risk of T2DM

Even when obesity is taken into consideration, there is ample evidence to support an independent association of physical activity and physical fitness with decreased risk of incident T2DM. Indeed, of the 20 prospective studies identified in a recent comprehensive review that examined the dose–response relationship of physical activity and fitness with risk of T2DM, all 20 identified an inverse association; on average, the risk reduction was 42 % for the most active relative to the least active study participants [21]. One of the earlier seminal studies to contribute to this knowledge-base used prospective data from 21,271 male physicians who were followed for an incident diagnosis of T2DM for 5 years [48]: men who exercised five or more times per week, relative to inactive men, had a 42% reduced risk of developing T2DM and the association was independent of body mass index. Similar results for women were published at about the same time [49]: during 8 years of follow-up, 1,303 T2DM cases were identified among a prospective study of 87,253 women. Women who were vigorously active at least one day per week had an age-adjusted relative risk for T2DM of 0.67 compared with women who did not exercise regularly. Adjustment for body-mass index attenuated the results somewhat (RR: 0.84) but they remained statistically significant, suggesting that the impact of physical activity on T2DM is partially, but not completely, dependent on its effects on reducing body fatness.

In a recent prospective study from Canada, Katzmarzyk and colleagues identified adiposity measures and physical fitness (musculoskeletal fitness and cardiorespiratory fitness) to be important predictors of T2DM over 15.5 years of follow-up time. Risks of T2DM were 70 and 61 % lower for each standard deviation increase in cardiorespiratory fitness and musculoskeletal fitness, respectively [50]. An earlier study from the ACLS similarly reported almost 2 and 4-fold increased risks of impaired fasting glucose and T2DM, respectively, among men in the least fit 20th percentile of cardiorespiratory fitness relative to men in the most fit 60th percentile. It is clear that physical activity and fitness are associated with reduced risk of incident T2DM.

17.1.4 Physical Activity, Exercise, Physical Fitness and Bone Health

There is little doubt that physical activity has positive effects on bone health, including increased bone mineral density and reduced risk of fracture. Studies of osteoporosis are rarer, however, but they are still in agreement. High impact and load bearing physical activities appear to confer the most benefits to bone health, likely because these types of dynamic activities are sufficient to induce osteogenesis [51, 52]. Meta analyses of randomized, controlled trials among pre- and post-menopausal women generally suggest that exercise interventions, including walking, step training, and resistance training, increase bone mineral density at the spine and femoral neck by a few percentage points [52–58]. These findings are largely in agreement with expert panel recommendations [59, 60] and a recent comprehensive review [18]. Although exercise interventions may decrease or reverse about 1 % of bone loss per year among older people [18], it has been debated as to whether this amount of bone preservation is clinically meaningful [61].

Risk of falling is reduced by physical activity intervention although it is unclear if the mechanism is through improved bone mineral density or through improved musculoskeletal and neuromuscular fitness: a 2005 meta-analysis of exercise interventions among the elderly indicated a 30% reduced risk of falling [62]. Similarly, a more recent meta-analysis of randomized, controlled trials suggested that exercise-alone interventions were about five-fold more effective at preventing falls than were multifactorial interventions, including those with drug therapy [63]. In line with these findings, moderate-to-vigorous physical activity from prospective cohort studies has been shown to reduce risk of hip fracture—by 45% in women and 38% in men [61]-and to prevent risk of incident osteoporosis [64, 65], although these data are more limited and restricted to cross-sectional [64] or case control studies [65] only.

17.1.5 Physical Activity, Physical Fitness, Exercise, and Oxidative Stress: Introduction

Physical activity and exercise clearly have a multitude of beneficial effects for physical and psychological health in young and older individuals. Exercise, however, especially when done strenuously, is associated with increased production of reactive oxygen and nitrogen species which are able to consume endogenous antioxidants and damage biological molecules and key cellular components [66, 67]. During exercise, whole body oxygen consumption can increase 10–15-fold and local oxygen consumption in working muscles can increase 100–200-fold, thus temporarily and dramatically increasing the rate of mitochondrial production of reactants species, most notably superoxide ions from complexes 1 and 3 of the mitochondrial electron transport chain [68]. Superoxide, in turn, can dismutate to create hydrogen peroxide. Hydrogen peroxide, in the presence of transition metals such as iron and copper, can create the hydroxyl radical. Hydroxyl radicals are generally assumed to be the most damaging and reactive species; they react with biomolecules in close proximity to where they were formed [68]. Reactant species produce a number of lesions in DNA, including base lesions, sugar lesions, DNA–protein cross links, single strand breaks, and double-strand breaks, as well as damage to lipids and protein [69–71]. While it is well-appreciated that excessive radical production is harmful, it is now becoming better appreciated that moderate levels of reactive species are essential for normal cellular signaling, co-ordination of gene expression, remodeling/adaptation, and modulation of force production during exercise [68, 72]. Indeed, the concept of *hormesis* has been often used to describe how exercise elicits an adaptive response of cells and organisms to a moderate, usually periodic, stressor [73]. In this section, the effect of acute and chronic exercise on F_2-isoprostane levels will be reviewed; where studies of other markers of oxidative stress offer particularly compelling findings, they too will be discussed. More detail on this topic, especially relating to the formation and measurement of F_2-isoprostanes, is offered elsewhere in a recent excellent review article [74].

17.1.6 Acute and Chronic Exercise and Oxidative Stress: Evidence From Laboratory Animals

Much of the evidence for a link between exercise and biomarkers of oxidative stress, including measures of F_2-isoprostanes, comes from studies in experimental animals. A three-day intensive 58 km run in sled dogs yielded significant increases in

F_2-isoprostanes on each of the 3 days of exercise relative to control dogs (rested) and a concomitant decrease in endogenous antioxidants [75], suggesting that acute and intense exercise increases lipid peroxidation in dogs. Acute and intense exercise (90 min, 32 m per minute, 8° grade—considered as near maximal effort) was also sufficient to dramatically increase plasma F_2-isoprostanes immediately postexercise (977 pg ml) compared with resting values (654 pg ml) and the values essentially returned to normal 24 h postexercise (753 pg ml); the increased oxidative stress occurred in concert with increased corticosterone, a general marker of the stress response and key glucocorticoid, and was observed to induce apoptosis of T lymphocytes in the large and small bowel [76]. In summary, acute and intense exercise can clearly increase F_2-isoprostanes in experimental animals.

Ten weeks of treadmill training in rats (training progressed to 60 min per day, 5 days per week, at 70 % VO_{2peak}) decreased resting values of F_2-isoprostanes among trained rats relative to sedentary control rats by nearly 3-fold at rest [77]. The improved oxidant status in the trained rats was attributed, in part, to increased nitric oxide release, which has antioxidant properties. Other training studies with rats have demonstrated increased antioxidant enzyme expression, including superoxide dismutase and glutathione peroxidase, postexercise [78]. A 10-week (15–60 min per day, 5 days per week) swim training program in rats demonstrated that while swimming did not have a significant effect on resting levels of oxidative stress (measured by 8-hydroxy-2′-deoxyguanosine–8-oxo-dG), it did decrease the oxidative stress response after exposure to a chemical carcinogen [79], suggesting that exercise can decrease risk of carcinogenesis. In a mouse model of T2DM, 8 weeks of moderate intensity treadmill exercise (1 h per day, 5 days per week, at 5.2 m per minute) decreased plasma F_2-isoprostanes among diabetic mice relative to diabetic control mice, but exercise had no impact on F_2-isoprostane levels among wild-type mice [80]. The decrease in F_2-isoprostane occurred in parallel with an increase in the antioxidant enzyme mitochondrial SOD in heart muscle [80]. Curiously, these adaptations were not correlated with an improvement in glucose control. Collectively, these data suggest that moderate exercise generally decreases resting levels of oxidative stress in animal models, perhaps through upregulation of endogenous antioxidant defenses.

17.1.7 Effects of Acute Exercise on Oxidative Stress Among Humans

There is evidence among humans (mostly athletes) that acute intense aerobic exercise is associated with an increase in urinary and sera F_2-isoprostane during exercise; levels have been reported to increase by as much as 88 % immediately after intense aerobic exercise [81]. Timing of blood or urine collection and the training status of the participants, in addition to intensity, are important parameters to consider in respect to acute postexercise changes in F_2-isoprostanes.

The impact of timing of blood or urine collection and exercise intensity has been demonstrated by a few studies [81–84]. Goto and colleagues conducted a 30 min bicycle ergometer study at 25, 50, and 75 % VO_{2max} [82]. The eight healthy male subjects had blood drawn immediately before exercise, 20 min into the exercise session, and 10 min after exercise. Participants were described as generally sedentary, although aerobic fitness data were not provided. The 25 and 50 % VO_{2max} intensities were insufficient to change F_2-isoprostane values during exercise or postexercise, compared to rest; however, the most intense exercise session, at 75 % maximal aerobic capacity, elicited a 1.66-fold increase in F_2-isoprostanes relative to rest, which reduced to a 1.33-fold increase relative to rest at the 10 min postexercise mark. The impact of timing of sample accrual, and F_2-isoprostane production and clearance by extension, was further demonstrated in a group of 11 highly trained extreme endurance athletes who participated in a 50 km ultramarathon [83]. In this study, plasma F_2-isoprostane increased at midrace, peaked immediately postrace, approached resting values at 1 h postrace, and returned essentially to normal resting values at 24 h post-race [83]. Similar pre- versus post-race results for F_2-isoprostane have been demonstrated in 90 and 160 km ultramarathons [81] and in a 2.5 h treadmill run at 75 % VO_{2max} [84]. Although intense aerobic exercise has been shown to almost double F_2-isoprostane values compared with resting values, normal resting values seem to be restored within a few hours of ceasing to exercise.

The impact of gender differences and resistance training with respect to exercise and oxidative stress is not well-appreciated. Because estradiol has been suggested to be a potent antioxidant [85], women compared with men may experience reduced oxidative stress levels at a similar workload. Kerksick and colleagues compared 8 men and 8 women before and after seven sets of 10 eccentric repetitions of the knee extensor muscles at 150 % of the participants' 1 repetition maximum (indicated to induce muscle soreness and damage) [86]. Blood samples were drawn before exercise, 6, 24, 48, and 72 h-postexercise. There was no indication of a gender by time interaction effect for F_2-isoprostanes. Indeed, F_2-isoprostanes did not change over the time period of study for men or for women. The authors did, however, observe a relatively marked difference at all time points when comparing men to women: that is, men had higher F_2-isoprostane levels at all timepoints [86]. While this study did not support the hypothesis that estradiol, or other estrogens, protected against oxidative stress during exercise, they do add some support that women generally experience lower resting levels of oxidative stress. Two hours of resistance exercise at moderate intensity (40–60 % 1 repetition maximum) also failed to increase F_2-isoprostane among 30 strength training acclimated subjects [87]. In contrast, a more recent study of eight healthy men with no history of regular exercise participation reported that a relatively intense resistance training session ($\sim$40 min duration at $\sim$70 % 1 repetition maximum) increased urinary F_2-isoprostane by about 40 % compared with resting values. It is difficult to draw firm conclusions from these studies; however, it appears that resistance training does not elevate oxidative stress to the same extent as aerobic exercise, except perhaps when performed by truly sedentary individuals. The role of gender and oxidative stress warrants further study.

17.1.8 Effects of Chronic Exercise on Oxidative Stress Among Humans in Non Controlled Studies

Long-term moderate physical activity or exercise training appears to exert an adaptive and favorable response with respect to lowered oxidative status; cross-sectional studies in women [88] and men [89] generally support this assertion. The cross-sectional inverse correlation between F_2-isoprostanes and VO_{2max} ($r = 0.37$), observed in a study of 173 postmenopausal women, add support this idea [90]. Further, regular habitual exercise relative to persistent inactivity reduces the acute exercise-induced oxidative stress in older men, apparently by preserving antioxidant function [91]. Prospective data on this topic are relatively rare, although some short-term combined diet and exercise interventions offered provocative results, as described below.

In a series of short-term combined diet and exercise intervention studies, Roberts and colleagues have demonstrated marked decreases in F_2-isoprostane in men [92–94]. In the first of these studies, 11 obese men (mean BMI at baseline $= 37.6$ kg/m^2) participated in an inpatient diet (diet: low fat, high fiber, no alcohol, no caffeine, no tobacco) and exercise (45–60 min per day at 70–85 % of maximal heart rate) intervention for 3 weeks. Over the course of the study, urinary F_2-isoprostane decreased by about 25 % [92]. In a similar study of 13 men with T2DM, F_2-isoprostane decreased by ~ 40 % compared to baseline after 3 weeks on the same intervention [93]. Among another 31 men (15 men diagnosed with the metabolic syndrome), the same intervention yielded a 35 % decrease in F_2-isoprostane [94]. While these studies yielded relatively dramatic results given their short durations, because they combined diet and exercise it is not possible to distinguish the independent effect of diet or exercise on oxidative stress.

To better understand the impact of exercise and dietary restriction on oxidative stress, Galassetti and colleagues conducted a 7 day training study where exercise was performed for 3 h per day at 75 % VO_{2max} in a group of 19 healthy young men [95]. Caloric intake was either 110 or 75 % of caloric expenditure. F_2-isoprostane values decreased similarly in the 110 % calorie (-23 %) and 75 % calorie (-31 %) groups, suggesting that a relatively brief and intense exercise regimen can decrease oxidative stress independent of dietary intake. Similar findings were offered in a recent non controlled study of 12 obese and 12 normal weight women [96]. All participants in this study underwent 12 weeks of aerobic training which progressed to three sessions per week, 60 min per session, at 65 % VO_{2peak} by the final week of the study. Both groups of women increased VO_{2peak} by the end of the trial, suggesting good adherence to the exercise protocol, and body weight remained essentially unchanged in both groups. Urinary F_2-isoprostane levels decreased postexercise by a similar amount in both groups. Likewise, Schmitz and colleagues examined the impact of a 15 week aerobic exercise intervention on F_2-isoprostanes and endogenous estrogens among 15 previously sedentary women aged 18–25 years [97]. Aerobic exercise resulted in a 10 % improvement in submaximal aerobic fitness and a dramatic 34 % decrease in F_2-isoprostanes over the trial period, despite having no impact on serum estrogens and relatively modest effects on body mass (−1.2 kg). Collectively, these studies suggest that relatively brief periods of aerobic training may have rather dramatic and beneficial effects on F_2-isoprostanes, and the effect is independent of fat loss or changes in diet.

While studies of generally moderate intensity exercise suggest a largely beneficial influence on oxidative stress, overtraining offers another perspective. Margonis and colleagues examined the impact of overtraining in a noncontrolled study 12 men who participated in a 12 week resistance training protocol with a progressively increased intensity and duration [98]. The most intense training period, the third of four periods, within this study consisted of six sessions per week for 3 weeks where training intensity was set at 85–100 % 1 repetition maximum. Overtraining was confirmed by a dramatic drop in performance in this third stage. Of the many oxidative stress biomarkers used in this study, F_2-isoprostane showed the most marked increase due to overtraining, with a 7-fold increase relative to baseline. Earlier stages of the progressive protocol elicited 2.4-fold and 4-fold increases in F_2-isoprostane, suggesting that F_2-isoprostane is a good marker for training intensity, and it may serve as useful marker in sport to monitor athlete overtraining. Further to this final point, F_2-isoprostanes were strongly correlated with the drop in performance. One commonality to all studies in this section is the lack of a control group. Randomized, controlled trials are generally considered the gold standard level of evidence in medicine; these types of studies are reviewed next.

17.1.9 Effects of Chronic Exercise on Oxidative Stress Among Humans in Controlled Studies

In what appears to be the first randomized, controlled study on this topic, Mori and colleagues studied patients aged 30–65 years who had T2DM [99]. In this trial, 55 patients with T2DM were randomized to one of four groups: control (participants performed 'light' exercise, defined as exercise at a heart rate below 100 bpm); exercise only (participants performed exercise for 3 days per week, 30 min per day, at 50–65 % VO_{2max}); low-fat diet only; or low-fat diet combined with exercise. After 8 weeks of intervention, the combined exercise and diet intervention group was the only arm of the trial to significantly decrease F_2-isoprostane values (−27.8 %) relative to controls (−4.9 %). Both the low fat and the exercise alone arms of the study resulted in nonsignificant decreases in F_2-isoprostanes at follow-up.

Similar findings were offered in a study of 19 overweight children who were randomized to a control group ($n = 9$) or to an aerobic exercise intervention [100]. The intervention group participated in cycle ergometer exercise for 4 days per week; by the 8th week of the trial, intensity progressed to 70–80 % VO_{2max} and duration increased to 50 min per session. Exercisers experienced a moderate gain in VO_{2max} (+12.8 %) while controls decreased in the same parameter (−12.6 %). Both groups gained a small amount of body weight (~1 kg). Although exercisers slightly decreased F_2-isoprostanes over the 8 week trial (−13.9 %), the effect was not statistically significant [100], perhaps owing to the relatively small sample size.

In a recent and relatively large randomized, controlled trial, Campbell and colleagues conducted a yearlong exercise intervention among 173 postmenopausal women in the Seattle, WA area [90]. Women were aged 50–75 years and they were overweight or obese and sedentary at the time of enrollment. The exercise intervention progressed to at least 45–60 min per day, 5 days per week, at 60–75 % maximal heart rate ($n = 87$); while the control group participated in once per week stretching and yoga classes. From baseline to the end of the study, F_2-isoprostane increased slightly among controls (+3.3 %) and decreased modestly in exercisers (−6.2 %), although the difference was not statistically significant. In planned subgroup analyses, however, F_2-isoprostane decreased linearly among exercisers with gain in VO_{2max}; exercisers who increased aerobic fitness >15 % decreased F_2-isoprostane 14.1 %, and the effect was statistically significant. These authors interpreted the inverse linear association between change in aerobic fitness and oxidative stress as a potential hormetic effect of exercise training [73]. Collectively, the sparse data from randomized, controlled trials do not generally support a strong, independent effect of exercise on F_2-isoprostane, although future studies in more diverse populations are clearly needed.

17.1.10 Oxidative Stress Biomarkers and Risk of Cardiovascular Disease

Oxidative stress appears to play an important role in the initiation and progression of atherosclerosis. Atherosclerosis, the underlying pathology of most CVD, is characterized by oxidation of lipid and protein in vascular walls [101, 102]. To date, the majority of the evidence linking oxidative stress and CVD are based on in vitro and laboratory animal studies, with some observational studies having been conducted among human participants (see [102] for detailed review). A gap in the current literature is that the majority of human studies are restricted to case patients where oxidative stress markers were measured in blood or urine *after* the onset of CVD. While these studies are helpful for generating hypotheses concerning prediction and etiology and they are valuable for studying secondary outcomes, such as disease progression and overall survival, it is not possible from these studies alone to determine if high oxidative stress is a cause or a consequence of CVD.

A recent nested case-control study, conducted within a prospective cohort of 10,529 men and women, identified 227 cases of fatal coronary heart disease and nonfatal heart attack outcomes and 420 matched controls after a median follow-up period of 5.6 years [103]. In this study, F_2-isoprostane was not associated with the mixed coronary heart disease outcome which included participants who had a prior history of heart attack; when these participants were excluded, however, the results were essentially unchanged, although the number of case patients necessarily grew smaller with this restriction. Lack of statistical power might have been one explanation for these results [103]. These findings were not in agreement with an earlier cross-sectional study of 2,850 healthy adults aged 18–30 years where coronary artery calcification, an indicator of coronary artery atherosclerosis, was the outcome [104]. After adjustment for sex, clinical site, age, and race, high versus low levels of F_2-isoprostane was associated with a 24 % increased risk of coronary artery calcification. Similarly, an earlier case control study, where urinary F_2-isoprostane was measured after diagnosis among 93 coronary heart disease patients and 93 age- and sex-matched controls, identified that high F_2-isoprostane was an independent marker for presence of coronary heart disease after controlling for a multitude of risk factors. Other cross-sectional studies have supported the hypothesis that oxidative stress, via measurement of F_2-isoprostanes, is associated with higher CVD risk and severity [105–107]. A recent nested case-control study further suggested that high urinary levels of F_2-isoprostanes were predictive of cardiovascular mortality in a cohort of postmenopausal women [108]. Relative to the more abundant literature in support of an association between oxidized low density lipoprotein (oxLDL) and CVD (reviewed in [102]), the literature with respect to F_2-isoprostanes is clearly sparse and more studies, preferably using prospective samples, are needed.

17.1.11 Oxidative Stress Markers and Risk of Cancer

The underlying mechanism for the effects of exercise on reducing risks of colon, breast, and endometrial cancers may include a free-radical scavenging mechanism [109]. To date, studies that have assessed the association among F_2-isoprostanes, using prospectively collected blood or urine samples, and risk of cancer at these organ sites are exceedingly rare. One exception is a nested case-control study drawn from the Shanghai Women's Health Study, a population-based prospective cohort study of 74,942 women aged 40–70 years [110]. In this study, prediagnostic urinary F_2-isoprostanes and its metabolite, 15-F (2t)-IsoP-M, were measured using gas chromatography mass spectrometry among 436 breast cancer cases and 852 matched controls. Overall, urinary isoprostanes were not different between cases and controls; however, body size appeared to be an effect modifier. Among women with a body mass index (BMI) in the obese range, the highest versus lowest tertile of 15-F(2t)-IsoP-M was associated with a 10-fold increased risk of breast cancer, while there was an inverse association between both isoprostane measures and breast cancer among nonoverweight women. Effect modification by adiposity status was not observed in an earlier case control study among women in the Long Island (New York, USA) Breast Cancer Project [111]. In this study, urinary levels of 15-F(2t)-isoprostanes were measured via ELISA among 400 cases and 401 controls: there was a statistically significant trend in increasing breast cancer risk with increasing quartiles of 15-F(2t)-IsoP levels, the 4th quartile relative to the 1st quartile of 15-F(2t)-IsoP experienced an 1.88-fold increased risk of breast cancer, and the authors reported no effect modification by a number of lifestyle and demographic factors, including BMI. Additionally, after excluding women who had initiated radiotherapy and/or chemotherapy, the results remained essentially unchanged, suggesting that treatment was not a major source of bias. In summary, increased levels of F_2-isoprostane appear to be associated with breast cancer; however, additional studies are clearly needed preferably with prospectively collected samples, especially to clarify whether risk varies by other lifestyle or anthropometric factors, such as body size.

Many of the major risk factors for colorectal cancer include a potential inflammation or oxidative stress-based mechanism, including ulcerative colitis and Crohn's disease. Indeed, both inflamed and malignant tissue in the colon exhibit increased reactant species, lipid peroxidation by-products, and 8-hydroxydeoxyguanosine (8-OHdG) [112, 113]. Thus, it is plausible that systemic markers of oxidative stress would be associated with increased risk of colorectal cancer, assuming that the oxidative insult that is observed in tissue is not limited to the affected tissue and it enters peripheral circulation. To date, no data are available on this topic with F_2-isoprostanes; however, a few studies have been conducted that assessed damage to DNA, including the markers 8-oxo-dG and 8-OHdG, and damage to oxLDL. In one of the earlier such studies, a nested case control study was conducted among participants in a Japanese cohort study [114]. Pre-diagnostic oxLDL levels were predictive of future colorectal cancer; men and women in the top quartile relative to the bottom quartile of oxLDL were at a 3.6-fold increased risk of developing colorectal cancer over the follow-up period [114]. Studies of the association between 8-OHdG and colorectal cancer have been more inconsistent, perhaps owing to their relatively small sample sizes. In a cross-sectional study of 36 colorectal cancer patients and 40 unaffected controls, serum 8-OHdG was slightly greater than 4-fold higher among case patients relative to the healthy controls [115]. These findings were somewhat corroborated by a recent case control study of patients with colorectal adenomas or cancer [116]. Multivariate analyses indicated that 8-OHdG was predictive of 'early' stage colorectal cancer (TNM stage 1), but not more advanced lesions. This study was based on a total of about 150 participants, divided among four strata; therefore, statistical power might have been limited. Clearly, more research, especially using large numbers of prospectively collected blood or urine samples, is needed to better define the connection between systemic oxidative stress and colorectal cancer.

Endometrial cancer shares some of the same risk factors as cancers of the breast and colon, including obesity, and physical activity. A recent meta-analysis of case-control studies suggested that dietary intakes of beta-carotene, vitamin C, and vitamin E were inversely associated with endometrial cancer risk [117], suggesting that oxidative stress might be relevant to the etiology of this disease. However, the only cohort study on this topic, as identified by the meta-analysis [117], reported no associations between endometrial cancer risk and these dietary antioxidants [118]. Endometrial cysts, premalignant and malignant tissues have been shown to have increased 8-OHdG, increased lipid peroxidation markers and decreased antioxidant enzyme expression relative to the normal endometrial tissue [119–121]. Thus, it again follows that systemic markers of oxidative stress may differ between case patients with endometrial cancer and controls. Unfortunately, prospective data and studies that evaluated F_2-isoprostanes relevant to endometrial cancer prediction do not yet appear to exist in the literature.

A hospital-based study on 103 women who attended a gynecologic clinic for symptomatic reasons or for routine screening reported that plasma lipid hydroperoxide levels, measured by a commercially available kit, were 41%, 52% and 57% higher among cases with simple hyperplasia, complex hyperplasia, and invasive adenocarcinoma, respectively, compared to healthy controls [122]. As described in-general elsewhere [123], it is not possible from cross-sectional studies alone to determine if higher oxidative stress levels are a cause of malignancy, and may therefore serve as appropriate markers for prediction, or if the high oxidative stress phenomenon among cancer patients is simply a product of tumors leaking reactant species, such as hydrogen peroxide [124], into the circulation. Prospective data are clearly needed to elucidate the role of oxidative stress biomarkers, including F2-isoprostanes, on the risk of incident endometrial cancer.

17.1.12 Oxidative Stress Biomarkers and Risk of T2DM

The relevance of oxidative stress in-general, and F_2-isoprostanes more specifically, to the etiology and progression of T2DM is relatively well appreciated, although prospective data on this topic, too, are rare [125]. An earlier clinic-based study identified a 3.3-fold difference in the plasma F_2-isoprostane levels of 39 patients with T2DM compared with 15 apparently healthy controls; curiously, there was no correlation between F_2-isoprostanes and HbA_{1c} or fasting glucose among patients with T2DM, suggesting that the increased plasma levels of F_2-isoprostanes among T2DM patients were not simply markers of hyperglycemia [126]. Subsequent work among 85 patients with diabetes (62 patients had T2DM and 23 had type 1 diabetes) and 85 age- and sex-matched controls confirmed the finding of increased F_2-isoprostane levels (albeit in urine) among patients with T2DM relative to controls (about a 2-fold difference) [127]. Unlike the earlier study [126], F_2-isoprostanes and blood glucose were strongly correlated, however, suggesting that lipid peroxidation and degree of glucose control may be related [127]. While these studies support the potential importance of oxidative stress on diabetes complications linked to CVD, since poor glucose control is a major predictor of macro- and microvascular complications, given their retrospective nature, these data do not contribute to the knowledge base on prediction of incident T2DM or impaired glucose intolerance. To that end, a recent nested case control study conducted among a cohort of Afro-Jamaicans over 3.9 years suggested that, among the 52 case patients who developed impaired glucose tolerance or T2DM, baseline urinary F2-isoprostanes were not predictive of future glucose intolerance and F2-isoprostanes did not correlate with insulin sensitivity or beta cell function [128]. These findings were essentially identical to a previous nested case control study of 26 T2DM patients and 26 controls, where baseline urine measures of F2-isoprostane were not predictive of a future diagnosis of T2DM [129].

Although the initial prospective studies on the association between F_2-isoprostane and T2DM have been null, future studies, which may require at least 10-fold larger sample sizes, will be required to adequately assess the connection between F_2-isoprostanes and risk of incident T2DM.

17.1.13 Oxidative Stress Biomarkers and Bone Health

There is some cross-sectional evidence that oxidative stress levels are relevant to bone health outcomes, including bone mineral density and osteoporosis; however, much like the CVD, T2DM and cancer outcomes discussed above, prospective data are especially rare. Oxidative stress is somewhat implicated as a plausible mechanism for osteoporosis since many of the risk factors for osteoporosis, including T2DM, smoking, and hypertension, are suspected to operate through increased oxidative stress. An earlier cross-sectional study of 101 men and women (mean age: 56 years) identified an inverse relationship of urinary F_2-isoprostanes with total body bone mineral density and bone mineral density at the distal forearm [130]. Similarly, a more recent cross-sectional study among 135 postmenopausal women identified moderate correlations between 8-OHdG and bone mineral densities at the lumbar spine ($r = -0.24$), total hip ($r = -0.26$), femoral neck ($r = -0.26$), and trochanter ($r = -0.25$) [131]. In a logistic regression model that adjusted for several other variables, a 1 standard deviation increase in 8-OHdG was associated with a 1.54-fold increased risk of osteoporosis, as defined by the World Health Organization (WHO) criteria [105]. These results are essentially corroborated by other cross-sectional studies [132–136]. In what appears to be the only prospective study on this topic, Ostman and colleagues recently reported that for each 1 standard deviation increase in baseline urinary F_2-isoprostanes levels at age 77 years, a 2–4 % decline in the mean adjusted bone mineral density was expected about 4 years later [136]. Curiously, these findings appeared to be slightly modified by serum alpha-tocopheral (i.e., vitamin E), an important antioxidant: men with low alpha-tocopherol levels and high F_2-isoprostane levels had a 7 % lower bone mineral density at the lumbar spine and a 5 % lower bone mineral density at the proximal femur [136]. While this initial prospective study offers support that oxidative stress is an indicator of decreased bone mineral density among older men, prospective studies of women, who are far more prone to osteoporosis, and fracture, are clearly needed.

17.2 Summary and Conclusions

- Physical activity, exercise and fitness are convincingly associated with reduced risks of CVD (e.g., coronary artery disease, stroke, hypertension, angina, and heart attack), certain cancers (breast, colon, and endometrium), T2DM, and poor bone health (osteoporosis, fracture, and low bone mineral density).

- Acute physical activity leads to an increase in F_2-isoprostane levels, especially when the activity is done at a high intensity. Aerobic activity appears to have a stronger effect on oxidative stress than resistance training types of activity.
- F_2-isoprostane levels return to normal values within a few hours of stopping exercise.
- Earlier noncontrolled studies of exercise intervention often suggested relatively marked drops in resting F_2-isoprostanes. These decreases in F_2-isoprostane levels occurred independent of diet or weight loss.
- Randomized, controlled trial data of the effect of exercise on F_2-isoprostane levels are sparse. Two studies suggest no effect. The largest study suggests that the decrease in F_2-isoprostane levels from exercise might occur only in the presence of large gains in aerobic fitness.
- Retrospective data suggest some associations between F_2-isoprostanes and chronic disease, perhaps especially for CVD. Prospective data on this topic are needed.

References

1. Global health risks (2009) Mortality and burden of disease attributable to selected major risks. World Health Organization, Geneva, p 62
2. Kesaniemi A et al (2010) Advancing the future of physical activity guidelines in Canada: an independent expert panel interpretation of the evidence. Int J Behav Nutr Phys Act 7:41
3. Physical activity guidelines advisory committee report (2008) To the secretary of health and human services. Part A: executive summary. Nutr Rev (2009) 67(2):114–120
4. Kushi LH et al (2006) American cancer society guidelines on nutrition and physical activity for cancer prevention: reducing the risk of cancer with healthy food choices and physical activity. CA Cancer J Clin 56(5):254–281 (quiz 313–314)
5. Global recommendations on physical activity for health (2010) World Health Organization, Geneva
6. Haskell WL et al (2007) Physical activity and public health: updated recommendation for adults from the American college of sports medicine and the American heart association. Med Sci Sports Exerc 39(8):1423–1434
7. O'Donovan G et al (2010) The ABC of physical activity for health: a consensus statement from the British association of sport and exercise sciences. J Sports Sci 28(6):573–591
8. Kadiiska MB et al (2005) Biomarkers of oxidative stress study II: are oxidation products of lipids, proteins, and DNA markers of CCl4 poisoning? Free Radic Biol Med 38(6):698–710
9. BAJ AJ, Definitions Working Group et al (2001) Biomarkers and surrogate endpoints: preferred definitions and conceptual framework. Clin Pharmacol Ther 69(3):89–95
10. Mateos R, Bravo l (2007) Chromatographic and electrophoretic methods for the analysis of biomarkers of oxidative damage to macromolecules (DNA, lipids, and proteins). J Sep Sci 30(2):175–191
11. Offord E, van Poppel G, Tyrrell R (2000) Markers of oxidative damage and antioxidant protection: current status and relevance to disease. Free Radic Res 33(Suppl):S5–S19
12. Warburton DE, Nicol CW, Bredin SS (2006) Health benefits of physical activity: the evidence. CMAJ 174(6):801–809
13. Sies H (ed) (1991) Oxidative stress: oxidants and antioxidants. Academic Press, New York

14. Halliwell B (2000) Oxidative stress markers in human disease: application to diabetes and to evolution of the effects of antioxidants. In: Packer L et al (eds) Antioxidants in diabetes management. Marcel Dekker, New York, pp 33–52
15. Sies H, Jones DP (2007) Oxidative stress. In: Encyclopedia of stress, F. G (ed) Elsevier pp 45–48
16. Morris JN, Heady JA (1953) Mortality in relation to the physical activity of work: a preliminary note on experience in middle age. Br J Ind Med 10(4):245–254
17. Morris JN et al (1953a) Coronary heart-disease and physical activity of work. Lancet 265(6796):1111–1120
18. Morris JN et al (1953b) Coronary heart-disease and physical activity of work. Lancet 265(6795):1053–1057
19. Paffenbarger RS Jr et al (1970) Work activity of longshoremen as related to death from coronary heart disease and stroke. N Engl J Med 282(20):1109–1114
20. Paffenbarger RS, Hale WE (1975) Work activity and coronary heart mortality. N Engl J Med 292(11):545–550
21. Warburton DE et al (2010) A systematic review of the evidence for canada's physical activity guidelines for adults. Int J Behav Nutr Phys Act 7:39
22. Morris JN et al (1973) Vigorous exercise in leisure-time and the incidence of coronary heart-disease. Lancet 1(7799):333–339
23. Paffenbarger RS, Wing AL Jr, Hyde RT (1978) Physical activity as an index of heart attack risk in college alumni. Am J Epidemiol 108(3):161–175
24. Taylor HL et al (1970) Coronary heart disease in seven countries IV. Five-year follow-up of employees of selected U.S. railroad companies. Circulation 41(Suppl 4):I20–I39
25. Slattery ML, Jacobs DR Jr (1988) Physical fitness and cardiovascular disease mortality. The US railroad study. Am J Epidemiol 127(3):571–580
26. Blair SN et al (1989) Physical fitness and all-cause mortality. A prospective study of healthy men and women. JAMA 262(17):2395–2401
27. Sui X et al (2007) Cardiorespiratory fitness and adiposity as mortality predictors in older adults. JAMA 298(21):2507–2516
28. Katzmarzyk PT, Craig CL (2002) Musculoskeletal fitness and risk of mortality. Med Sci Sports Exerc 34(5):740–744
29. Zheng H et al (2009) Quantifying the dose-response of walking in reducing coronary heart disease risk: meta-analysis. Eur J Epidemiol 24(4):181–192
30. Sofi F et al (2008) Physical activity during leisure time and primary prevention of coronary heart disease: an updated meta-analysis of cohort studies. Eur J Cardiovasc Prev Rehabil 15(3):247–257
31. Lee CD, Folsom AR, Blair SN (2003) Physical activity and stroke risk: a meta-analysis. Stroke 34(10):2475–2481
32. Wendel-Vos GC et al (2004) Physical activity and stroke. A meta-analysis of observational data. Int J Epidemiol 33(4):787–798
33. Murphy MH et al (2007) The effect of walking on fitness, fatness and resting blood pressure: a meta-analysis of randomised, controlled trials. Prev Med 44(5):377–385
34. Whelton SP et al (2002) Effect of aerobic exercise on blood pressure: a meta-analysis of randomized, controlled trials. Ann Intern Med 136(7):493–503
35. Harriss DJ et al (2009) Lifestyle factors and colorectal cancer risk (2): a systematic review and meta-analysis of associations with leisure-time physical activity. Colorectal Dis 11(7):689–701
36. Cummings SR et al (2009) Prevention of breast cancer in postmenopausal women: approaches to estimating and reducing risk. J Natl Cancer Inst 101(6):384–398
37. Moore SC et al (2010) Physical activity, sedentary behaviours, and the prevention of endometrial cancer. Br J Cancer 103(7):933–938
38. Cancer research fund/american institute for cancer research (2007) Food, nutrition, physical activity, and the prevention of cancer: a global perspective. AICR: Washington, DC
39. Samad AK et al (2005) A meta-analysis of the association of physical activity with reduced risk of colorectal cancer. Colorectal Dis 7(3):204–213

40. Slattery ML et al (1997) Physical activity and colon cancer: a public health perspective. Ann Epidemiol 7(2):137–145
41. Monninkhof EM et al (2007) Physical activity and breast cancer: a systematic review. Epidemiol 18(1):137–157
42. Eliassen AH et al (2010) Physical activity and risk of breast cancer among postmenopausal women. Arch Intern Med 170(19):1758–1764
43. Jemal A et al (2010) Cancer statistics 2010. CA Cancer J Clin 60(5):277–300
44. Byun W et al (2011) Cardiorespiratory fitness and risk of prostate cancer: findings from the aerobics center longitudinal study. Cancer Epidemiol 35(1):59–65
45. Sui X et al (2010) Influence of cardiorespiratory fitness on lung cancer mortality. Med Sci Sports Exerc 42(5):872–878
46. Peel JB et al (2009a) A prospective study of cardiorespiratory fitness and breast cancer mortality. Med Sci Sports Exerc 41(4):742–748
47. Peel JB et al (2009b) Cardiorespiratory fitness and digestive cancer mortality: findings from the aerobics center longitudinal study. Cancer Epidemiol Biomarkers Prev 18(4):1111–1117
48. Manson JE et al (1992) A prospective study of exercise and incidence of diabetes among US male physicians. JAMA 268(1):63–67
49. Manson JE et al (1991) Physical activity and incidence of non-insulin-dependent diabetes mellitus in women. Lancet 338(8770):774–778
50. Katzmarzyk PT, Craig CL, Gauvin L (2007) Adiposity, physical fitness and incident diabetes: the physical activity longitudinal study. Diabetologia 50(3):538–544
51. Lanyon LE, Rubin CT (1984) Static vs dynamic loads as an influence on bone remodelling. J Biomech 17(12):897–905
52. Martyn-St James M, Carroll S (2010) Effects of different impact exercise modalities on bone mineral density in premenopausal women: a meta-analysis. J Bone Miner Metab 28(3):251–267
53. Martyn-St James M, Carroll S (2006) Progressive high-intensity resistance training and bone mineral density changes among premenopausal women: evidence of discordant site-specific skeletal effects. Sports Med 36(8):683–704
54. Martyn-St James M, Carroll S (2008) Meta-analysis of walking for preservation of bone mineral density in postmenopausal women. Bone 43(3):521–531
55. Kelley GA, Kelley KS (2004) Efficacy of resistance exercise on lumbar spine and femoral neck bone mineral density in premenopausal women: a meta-analysis of individual patient data. J Womens Health (Larchmt) 13(3):293–300
56. Schmitt NM, Schmitt J, Doren M (2009) The role of physical activity in the prevention of osteoporosis in postmenopausal women-an update. Maturitas 63(1):34–38
57. Kelley GA (1998) Aerobic exercise and bone density at the hip in postmenopausal women: a meta-analysis. Prev Med 27(6):798–807
58. Kelley GA (1998) Exercise and regional bone mineral density in postmenopausal women: a meta-analytic review of randomized trials. Am J Phys Med Rehabil 77(1):76–87
59. American College of Sports medicine position stand (1995) Osteoporosis and exercise. Med Sci Sports Exerc 27(4):1–7
60. Brown JP, Josse RG (2002) 2002 clinical practice guidelines for the diagnosis and management of osteoporosis in Canada. CMAJ 167(10 Suppl):S1–S34
61. Moayyeri A (2008) The association between physical activity and osteoporotic fractures: a review of the evidence and implications for future research. Ann Epidemiol 18(11):827–835
62. Alessio HM (1993) Exercise-induced oxidative stress. Med Sci Sports Exerc 25(2):218–224
63. Petridou ET et al (2009) What works better for community-dwelling older people at risk to fall?: a meta-analysis of multifactorial versus physical exercise-alone interventions. J Aging Health 21(5):713–729
64. Robitaille J et al (2008) Prevalence, family history, and prevention of reported osteoporosis in US women. Am J Prev Med 35(1):47–54
65. Keramat A et al (2008) The assessment of osteoporosis risk factors in Iranian women compared with Indian women. BMC Musculoskelet Disord 9:28

66. Alessio HM (1993) Exercise-induced oxidative stress. Med Sci Sports Exerc 25(2):218–224
67. Ji LL (1995) Oxidative stress during exercise: implication of antioxidant nutrients. Free Radic Biol Med 18(6):1079–1086
68. Powers SK, Jackson MJ (2008) Exercise-induced oxidative stress: cellular mechanisms and impact on muscle force production. Physiol Rev 88(4):1243–1276
69. Ames BN, Shigenaga MK, Hagen TM (1993) Oxidants, antioxidants, and the degenerative diseases of aging. Proc Natl Acad Sci U S A 90(17):7915–7922
70. Dizdaroglu M (1992) Oxidative damage to DNA in mammalian chromatin. Mutat Res 275(3–6):331–342
71. Richter C, Park JW, Ames BN (1988) Normal oxidative damage to mitochondrial and nuclear DNA is extensive. Proc Natl Acad Sci U S A 85(17):6465–6467
72. Ji LL, Gomez-Cabrera MC, Vina J (2006) Exercise and hormesis: activation of cellular antioxidant signaling pathway. Ann N Y Acad Sci 1067:425–435
73. Mattson MP (2008) Hormesis defined. Ageing Res Rev 7(1):1–7
74. Nikolaidis MG, Kyparos A, Vrabas IS (2011) F-isoprostane formation, measurement and interpretation: the role of exercise. Prog Lipid Res 50(1):89–103
75. Hinchcliff KW et al (2000) Oxidant stress in sled dogs subjected to repetitive endurance exercise. Am J Vet Res 61(5):512–517
76. Hoffman-Goetz L, Quadrilatero J (2003) Treadmill exercise in mice increases intestinal lymphocyte loss via apoptosis. Acta Physiol Scand 179(3):289–297
77. Wang JS et al (2000) Role of chronic exercise in decreasing oxidized LDL-potentiated platelet activation by enhancing platelet-derived no release and bioactivity in rats. Life Sci 66(20):1937–1948
78. Oztasan N et al (2004) Endurance training attenuates exercise-induced oxidative stress in erythrocytes in rat. Eur J Appl Physiol 91(5–6):622–627
79. Nakatani K et al (2005) Habitual exercise induced resistance to oxidative stress. Free Radic Res 39(9):905–911
80. Moien-Afshari F et al (2008) Exercise restores coronary vascular function independent of myogenic tone or hyperglycemic status in db/db mice. Am J Physiol Heart Circ Physiol 295(4):H1470–H1480
81. Nieman DC et al (2003) Immune and oxidative changes during and following the western states endurance run. Int J Sports Med 24(7):541–547
82. Goto C et al (2007) Acute moderate-intensity exercise induces vasodilation through an increase in nitric oxide bioavailability in humans. Am J Hypertens 20(8):825–830
83. Mastaloudis A, Leonard SW, Traber MG (2001) Oxidative stress in athletes during extreme endurance exercise. Free Radic Biol Med 31(7):911–922
84. Steensberg A et al (2002) Prolonged exercise, lymphocyte apoptosis and F2-isoprostanes. Eur J Appl Physiol 87(1):38–42
85. Kendall B, Eston R (2002) Exercise-induced muscle damage and the potential protective role of estrogen. Sports Med 32(2):103–123
86. Kerksick C et al (2008) Gender-related differences in muscle injury, oxidative stress, and apoptosis. Med Sci Sports Exerc 40(10):1772–1780
87. McAnulty SR et al (2005) Effect of resistance exercise and carbohydrate ingestion on oxidative stress. Free Radic Res 39(11):1219–1224
88. Covas MI et al (2002) Relationship between physical activity and oxidative stress biomarkers in women. Med Sci Sports Exerc 34(5):814–819
89. Karolkiewicz J et al (2003) Oxidative stress and antioxidant defense system in healthy, elderly men: relationship to physical activity. Aging Male 6(2):100–105
90. Campbell PT et al (2010) Effect of exercise on oxidative stress: a 12-month randomized, controlled trial. Med Sci Sports Exerc 42(8):1448–1453
91. Meijer EP et al (2002) Exercise-induced oxidative stress in older adults as a function of habitual activity level. J Am Geriatr Soc 50(2):349–353
92. Roberts CK, Vaziri ND, Barnard RJ (2002) Effect of diet and exercise intervention on blood pressure, insulin, oxidative stress, and nitric oxide availability. Circulation 106(20):2530–2532

93. Roberts CK et al (2006) Effect of a diet and exercise intervention on oxidative stress, inflammation and monocyte adhesion in diabetic men. Diabetes Res Clin Pract 73(3):249–259
94. Roberts CK et al (2006) Effect of a short-term diet and exercise intervention on oxidative stress, inflammation, MMP-9, and monocyte chemotactic activity in men with metabolic syndrome factors. J Appl Physiol 100(5):1657–1665
95. Galassetti PR et al (2006) Exercise, caloric restriction, and systemic oxidative stress. J Investig Med 54(2):67–75
96. Devries MC et al (2008) Endurance training without weight loss lowers systemic, but not muscle, oxidative stress with no effect on inflammation in lean and obese women. Free Radic Biol Med 45(4):503–511
97. Schmitz KH et al (2008) Exercise effect on oxidative stress is independent of change in estrogen metabolism. Cancer Epidemiol Biomarkers Prev 17(1):220–223
98. Margonis K et al (2007) Oxidative stress biomarkers responses to physical overtraining: implications for diagnosis. Free Radic Biol Med 43(6):901–910
99. Mori TA et al (1999) Effect of dietary fish and exercise training on urinary F2-isoprostane excretion in non-insulin-dependent diabetic patients. Metabolism 48(11):1402–1408
100. Kelly AS et al (2007) In the absence of weight loss, exercise training does not improve adipokines or oxidative stress in overweight children. Metabolism 56(7):1005–1009
101. Flores-Mateo G et al (2009) Antioxidant enzyme activity and coronary heart disease: meta-analyses of observational studies. Am J Epidemiol 170(2):135–147
102. Strobel NA et al (2011) Oxidative stress biomarkers as predictors of cardiovascular disease. Int J Cardiol 147(2):191–201
130. Woodward M et al (2009) Association between both lipid and protein oxidation and the risk of fatal or non-fatal coronary heart disease in a human population. Clin Sci (Lond) 116(1):53–60
104. Gross M et al (2005) Plasma F2-isoprostanes and coronary artery calcification: the CARDIA study. Clin Chem 51(1):125–131
105. Vassalle C et al (2003) Evidence for enhanced 8-isoprostane plasma levels, as index of oxidative stress in vivo, in patients with coronary artery disease. Coron Artery Dis 14(3):213–218
106. Basarici I et al (2008) Urinary 8-isoprostane levels can indicate the presence, severity and extent of angiographic coronary artery disease. Acta Cardiol 63(4):415–422
107. Mueller T et al (2004) Serum total 8-iso-prostaglandin F2alpha: a new and independent predictor of peripheral arterial disease. J Vasc Surg 40(4):768–773
108. Roest M et al (2008) High levels of urinary F2-isoprostanes predict cardiovascular mortality in postmenopausal women. J Clin Lipidol 2(4):298–303
109. Rogers CJ et al (2008) Physical activity and cancer prevention : pathways and targets for intervention. Sports Med 38(4):271–296
110. Dai Q et al (2009) Oxidative stress, obesity, and breast cancer risk: results from the Shanghai women's health study. J Clin Oncol 27(15):2482–2488
111. Rossner P Jr et al (2006) Rossner P Jr et al (2006) Relationship between urinary 15-F2t-isoprostane and 8-oxodeoxyguanosine levels and breast cancer risk. Cancer Epidemiol Biomarkers Prev 15(4):639–644. Cancer Epidemiol Biomarkers Prev 15(4):639–644
112. Keshavarzian A et al (1992) Excessive production of reactive oxygen metabolites by inflamed colon: analysis by chemiluminescence probe. Gastroenterology 103(1):177–185
113. Keshavarzian A et al (1992) Excessive production of reactive oxygen metabolites by inflamed colon: analysis by chemiluminescence probe. Gastroenterology 103(1):177–185
114. Suzuki K et al (2004) Serum oxidized low-density lipoprotein levels and risk of colorectal cancer: a case-control study nested in the Japan Collaborative Cohort Study. Cancer Epidemiol Biomarkers Prev 13(11):1781–1787
115. Chang D et al (2008) Evaluation of oxidative stress in colorectal cancer patients. Biomed Environ Sci 21(4):286–289
116. Sato T et al (2010) Increased plasma levels of 8-hydroxydeoxyguanosine are associated with development of colorectal tumors. J Clin Biochem Nutr 47(1):59–63

117. Keshavarzian A et al (1992) High levels of reactive oxygen metabolites in colon cancer tissue: analysis by chemiluminescence probe. Nutr Cancer 17(3):243–249
117. Bandera EV et al (2009) Antioxidant vitamins and the risk of endometrial cancer: a dose-response meta-analysis. Cancer Causes Control 20(5):699–711
118. Jain MG et al (2000) A cohort study of nutritional factors and endometrial cancer. Eur J Epidemiol 16(10):899–905
119. Yamaguchi K et al (2008) Contents of endometriotic cysts, especially the high concentration of free iron, are a possible cause of carcinogenesis in the cysts through the iron-induced persistent oxidative stress. Clin Cancer Res 14(1):32–40
120. Punnonen R et al (1993) Activities of antioxidant enzymes and lipid peroxidation in endometrial cancer. Eur J Cancer 29A(2):266–269
121. Pejic S et al (2009) Antioxidant enzymes and lipid peroxidation in endometrium of patients with polyps, myoma, hyperplasia and adenocarcinoma. Reprod Biol Endocrinol 7:149
122. Pejic S et al (2006) Lipid peroxidation and antioxidant status in blood of patients with uterine myoma, endometrial polypus, hyperplastic and malignant endometrium. Biol Res 39(4):619–629
123. Tudek B et al (2010) Involvement of oxidatively damaged DNA and repair in cancer development and aging. Am J Transl Res 2(3):254–284
124. Szatrowski TP, Nathan CF (1991) Production of large amounts of hydrogen peroxide by human tumor cells. Cancer Res 51(3):794–798
125. Kaviarasan S et al (2009) F(2)-isoprostanes as novel biomarkers for type 2 diabetes: a review. J Clin Biochem Nutr 45(1):1–8
126. Gopaul NK et al (1995) Plasma 8-epi-PGF2 alpha levels are elevated in individuals with non-insulin dependent diabetes mellitus. FEBS Lett 368(2):225–229
127. Davi G et al (1999) In vivo formation of 8-iso-prostaglandin f2alpha and platelet activation in diabetes mellitus: effects of improved metabolic control and vitamin E supplementation. Circulation 99(2):224–229
128. Boyne MS et al (2007) Isoprostanes, a marker of lipid peroxidation, may not be involved in the development of glucose intolerance. Diabetes Res Clin Pract 76(1):149–151
129. Il'yasova D, Morrow JD, Wagenknecht LE (2005) Urinary F2-isoprostanes are not associated with increased risk of type 2 diabetes. Obes Res 13(9):1638–1644
130. Basu S et al (2001) Association between oxidative stress and bone mineral density. Biochem Biophys Res Commun 288(1):275–279
131. Baek KH et al (2010) Association of oxidative stress with postmenopausal osteoporosis and the effects of hydrogen peroxide on osteoclast formation in human bone marrow cell cultures. Calcif Tissue Int 87(3):226–235
132. Mangiafico RA et al (2007) Increased formation of 8-iso-prostaglandin F(2alpha) is associated with altered bone metabolism and lower bone mass in hypercholesterolaemic subjects. J Intern Med 261(6):587–596
133. Sendur OF et al (2009) Antioxidant status in patients with osteoporosis: a controlled study. Joint Bone Spine 76(5):514–518
134. Ozgocmen S et al (2007) Role of antioxidant systems, lipid peroxidation, and nitric oxide in postmenopausal osteoporosis. Mol Cell Biochem 295(1–2):45–52
135. Altindag O et al (2008) Total oxidative/anti-oxidative status and relation to bone mineral density in osteoporosis. Rheumatol Int 28(4):317–321
136. Ostman B et al (2009) Oxidative stress and bone mineral density in elderly men: antioxidant activity of alpha-tocopherol. Free Radic Biol Med 47(5):668–673

About the Editors

Ashok Agarwal is the Director of the Andrology Laboratory and Reproductive Tissue Bank and the Director of Research at the Center for Reproductive Medicine. He holds these positions at The Cleveland Clinic, where he is a Professor at the Lerner College of Medicine of Case Western Reserve University. Dr. Agarwal received advanced training in Male Infertility and Reproductive Endocrinology at the Brigham and Women's Hospital and served on the faculty of the Harvard Medical School. He has published over 500 scientific papers and reviews in peer reviewed scientific journals, authored over 50 book chapters, and presented more than 700 papers at scientific meetings. Dr. Agarwal is an editor of 10 medical text books/manuals related to male infertility, ART, fertility preservation, DNA damage, and antioxidants. He is the guest editor of four special journal issues and an Editorial Board member of over 20 peer-reviewed journals. Dr. Agarwal is active in basic and clinical research and his laboratory has trained more than 150 basic scientists and clinical researchers from the United States and abroad. In addition, over 150 medical, undergraduate, and high school students have worked in his laboratory. Dr. Agarwal is the Program Director of the highly successful Summer Internship Course in Reproductive Medicine. In the last 5 years, about 100 pre-med and medical students from across the United States and overseas have graduated from this highly competitive program. His current research interests have focused on molecular markers of oxidative stress, DNA integrity, and apoptosis in the pathophysiology of male and female reproduction, effect of radio frequency radiation on fertility and fertility preservation in patients with cancer.

Dr. Nabil Aziz is a Consultant in Gynecology and Reproductive Medicine and Lead Clinician at Liverpool Women's Hospital and the University of Liverpool, United Kingdom. The Hospital is the largest women's hospital in Europe and houses the largest assisted conception unit in the UK. Dr. Aziz received his medical degree from Ain Shams University School of Medicine in Cairo, Egypt. He became a member of the Royal College of Obstetricians in 1988 and was elevated to a Fellow in 2007. He was awarded the Doctorate degree (MD) in reproductive medicine from the University of Liverpool, UK in 1999. Dr. Aziz's primary research interest is in basic research of male infertility and the clinical aspect of assisted reproductive technology. His research work led to the routine use of the single-channelled needle in oocyte aspiration. He developed the concept of Sperm Deformity Index as a reliable predictor of male infertility. His methodology in assessing sperm morphology was adopted by the WHO Laboratory Manual for the Examination Human Semenin since 1999. Among his other interests are minimal invasive surgery and medical education. Dr. Aziz has published more than 80 scientific papers, abstracts, and book chapters. He has been the recipient of several clinical excellence awards for his endeavor in research and medical education.

Botros Rizk is Professor of Obstetrics and Gynecology and head of Reproductive Endocrinology and Infertility and Medical and Scientific Director of In Vitro Fertilization and Assisted Reproduction at the University of South Alabama. Mentored by Nobel Laureate Professor Robert Edwards and Professor Howard Jacobs, Dr. Rizk completed his prestigious fellowship in Endocrinology at Bourn Hall Clinic and joined Cambridge University as a clinical lecturer. Dr. Rizk is past chairman of the American Society for Reproductive Medicine international membership, Associate Editor of the Egyptian Fertility Society Journal, and president elect of the Middle East Fertility Society. Dr. Rizk authored and edited 12 medical textbooks on human reproduction and published over 400 peer reviewed articles, review articles, book chapters, and abstracts over the past 25 years. His research interest has focused on ovarian stimulation, endometriosis, and ovarian hyperstimulation syndrome.

Index

A
Angiogenesis, 71, 75, 95, 98
Antioxidant Defense
    Mechanisms, 62
antioxidant enzymes, 2, 6, 8, 9, 10, 41, 69, 76, 81, 82, 84, 89, 102, 103, 115, 146, 152, 155, 299
Antioxidant strategies, 237, 258
Antioxidant Treatment, 149, 157
Antioxidant vitamins, 13, 76, 181
Antioxidants, 34, 76, 82, 84, 95, 102, 132, 176, 206, 221, 238, 239, 257
Assisted reproductive
    technologies (ART), 205

C
Cardiovascular Disease, 286, 338
Carotenoids, 9, 115, 121
Chemiluminescence, 33, 34
Chorioamniotic membranes, 143, 144, 146
Clinical importance, 115, 116
Corpus luteum function
Curcuma longa, 181, 196
Curcumin, 149, 160

D
Decidualization, 62, 63
Deficient endogenous
    antioxidant, 181
Developmental defects, 2, 13
Dietary antioxidants, 2, 5, 9, 12, 13, 88, 350

E
ELISA, 33, 41
Endometriosis, 49, 68, 149, 150, 154, 156, 157, 162, 316, 318, 319
Endometrium, 48, 62
Environmental chemicals, 2, 3, 15, 17, 19
Enzymatic collagenolytic activity
    expressed, 143
Exercise, 87, 182, 192, 195, 198, 301, 335–339, 342–349, 352, 353

F
Female infertility, 315
Female Reproductive System, 33
Fertility of women, 169
Fetal membrane, 117, 133, 143, 144, 146
Fitness, 192, 335, 337–343, 345, 346, 348, 352, 353
Flow cytometry, 33, 39
Folliculogenesis, 51, 76
Forearm Ischemia/Reperfusion
Free radicals, 2, 3, 34, 87, 118, 119, 132, 149, 150, 152, 154, 155, 158, 174, 175, 185, 191, 193, 210, 223, 238, 257, 315, 316, 324, 325

G
Grape polyphenols, 181, 196

H
Herbal extracts, 181, 199
Hydrosapinx

I
Implications with chronic disease prevention
In vitro fertilization (IVF) procedures, 237
In vitro maturation (IVM), 81, 205, 206
Intrauterine Growth Restriction, 95
Ischemia/reperfusion, 324
Ischemic Preconditioning
IVF-embryo transfer, 83, 237, 248, 252

L
Loss of estrogen, 181, 184
Lycopene, 9, 42, 87, 120–122, 181, 196, 257, 257

M
Maternal antioxidant imbalance, 115, 118
Maternal deciduas, 72
Maternal foods, 2
Maternal nutrition, 3, 13, 19, 131, 136
Melatonin, 82, 149, 160, 198, 256
Menopause, 182–184, 186, 190, 192
Metabolic sequelae, 169, 171
Metabolic Syndrome, 169, 172, 173
Methods for Detection, 33
Miscarriage, 134
Mitochondria, 4, 6, 8, 11, 16, 53, 62, 81, 101, 117, 132, 136, 144, 174, 227, 239, 241, 285, 286, 289, 292, 293, 295, 299
Mitochondrial, 6
Myocardial Ischemia

N
NADPH oxidase, 5, 45, 101, 175, 191, 241, 285, 286, 288–290, 292–296, 300, 302, 325

O
Obesity, 135, 172
Older adults, 301
Oocyte cryo-preservation, 205
Oogenesis, 76

OS Induced Infertility, 149, 156
Oxidative Balance, 286
Oxidative insult
Oxidative Stress, 62–64, 68, 95, 100, 102, 103, 152, 154, 173, 183, 238, 291, 295, 316
Oxidative stress impact, 169
Oxygen, 34, 95, 152, 210, 227, 243

P
Pentoxifylline, 159, 255
Physical activity, 335, 338, 340–343, 352
Physiopathology of PE, 115, 116
Phytoestrogens, 181, 194, 195, 198
Placenta, 48, 95, 98
Placental antioxidant imbalance, 115, 117
Placentation abnormalities, 131, 133
Polycystic Ovary Syndrome, 169, 173
Postoperative adhesion, 315, 322, 324–326
Preeclampsia, 95
Pre-eclampsia, 51, 52, 115, 116, 133, 135, 248, 251
Pregnancy disorders and complications
Premature rupture of membranes (PROM), 143
Prenatal developmental outcomes, 2, 3, 11–20
Psychological problems, 169

R
Reactive oxygen species (ROS), 1, 4, 33, 34, 45, 49, 96, 101, 115, 116, 144, 152, 173, 205, 206, 227, 237, 238, 287
Redox in embryonic implantation, 131, 133
Reproduction failure, 315
Reproductive aging, 181, 182
Reproductive alterations, 169
Reproductive failure, 75, 76
Risk of cardiovascular disease, 184, 193–196, 335, 338, 348
ROS production, 1, 2, 4–6, 11, 13, 17, 40, 45, 52, 53, 79, 80, 86, 115, 117, 132, 133, 143, 153, 174, 175, 191, 210, 221, 222, 224, 239, 242–244, 253, 288–290, 292–296, 298, 300, 303

S
Secondary oocyte quality, 76
Sources of Oxidants, 286, 288

T
Tissue healing, 315, 323

U

Uncoupled NOS, 285, 286, 294, 295, 298–300, 302
Use of antioxidants, 115, 118

V

Vasculogenesis, 95, 98, 107
Vitamin A, 9, 42, 158, 181, 197, 198
Vitamin C, 9, 76, 82, 84, 85, 87, 115, 117–120, 137, 145, 152, 159, 181, 184, 186, 193, 194, 244, 249, 250, 251, 257, 350

Vitamin E, 9, 13, 42, 45, 49, 52, 66, 76, 80, 82, 83, 87, 119–121, 137, 145, 152, 154, 155, 158, 159, 161, 176, 181, 193, 194, 221, 249–252, 255–257, 325, 350, 352
Vitro fertilization (IVF) procedures

X

Xanthine oxidase, 5, 39, 45, 49, 78, 101, 154, 157, 210, 242, 285, 286, 321

Printed by Publishers' Graphics LLC